# Textbook of Radiographic Science

# Textbook of Radiographic Science

Edited by

## H. Brian Bentley
BA MEd MPhil FCR FRIPHH RGN
Principal, School of Radiography, The General Infirmary at
Leeds, Leeds, UK

## CHURCHILL LIVINGSTONE
EDINBURGH LONDON MELBOURNE AND NEW YORK 1986

CHURCHILL LIVINGSTONE
Medical Division of Longman Group UK Limited

Distributed in the United States of America by Churchill
Livingstone Inc., 1560 Broadway, New York, N.Y. 10036,
and by associated companies, branches and representatives
throughout the world.

First published 1986

ISBN 0-443-02550-9

British Library Cataloguing in Publication Data
Bentley, H. Brian
    Textbook of radiographic science.
    1. Diagnosis, Radioscopic
    I. Title
    616.07'57    RC78

Library of Congress Cataloging in Publication Data
Textbook of radiographic science.
    Includes index.
    1. Radiography, Medical.   2. Diagnosis, Radioscopic.
I. Bentley, H. Brian.   [DNLM: 1. Radiography.
WN 200 T3548]
RC78.T483  1986        616.07'57        85–19507

Produced by Longman Singapore
Publishers (Pte) Ltd.
Printed in Singapore.

# Preface

This textbook has been produced to cover some of the more specialised areas of radiography. The aim is to outline the scientific background to these special procedures so that it will be of use to the candidate studying for the Higher Examination of the College of Radiographers. It should also be of value to Diploma candidates and those studying radiographic technique for the Fellowship of the Royal College of Radiologists.

There are few radiography textbooks which provide details of these areas of radiography, and whilst this volume does not claim to be a major work in any one speciality, it aims to give a general scientific framework.

An attempt has been made to be as up-to-date as possible but this has been restricted by the rapid speed of change and the highly specialised nature of more recent imaging procedures. Indeed it has become clear that there remains much scope for future editions.

It is hoped that the chapter dealing with research should help those radiographers, and others, who are either contemplating research projects or are already engaged in research.

Leeds 1986                                        H. Brian Bentley

# Contributors

**H. Brian Bentley** BA MEd MPhil FCR FRIPHH RGN
Principal, School of Radiography, The General Infirmary at
Leeds, Leeds, UK

**Joan Broadley** BA DCR RT
Chief Technician, Montreal Neurological Hospital, Montreal,
Quebec, Canada

**Brenda Danby** HDCR
Radiographer, Southampton General Hospital, Southampton,
UK

**E. Higginbottom** FCR HDCR
Superintendent Radiographer, Robert Jones and Agnes Hunt
Orthopaedic Hospital, Oswestry, UK

**Susan Rouse** DCR
Superintendent Radiographer, Cardiovascular Suite, The
General Infirmary at Leeds, Leeds, UK

**B. C. Russell** FCR
Superintendent Radiographer, Department of Diagnostic
Radiology, The General Infirmary at Leeds, Leeds, UK

**Martin York** HDCR
Former Superintendent Radiographer, Newcastle Royal
Infirmary, Newcastle upon Tyne, UK

# Contents

1. Radiography of trauma in the cervical spine in patients with associated spinal cord injuries
   *E. Higginbottom*    1
2. Tomography of the petrous bone    29
   *B. C. Russell*
3. Paediatric radiography    50
   *Brenda Danby*
4. Accident and emergency radiography    74
   *Martin York*
5. Neuroradiography — new techniques    87
   *Joan Broadley*
6. Urodynamic studies in the patient with severe cord injury    101
   *E. Higginbottom*

7. Aspects of skeletal radiography    119
   *H. Brian Bentley*
8. Radiography of the heart and great vessels    141
   *Susan Rouse*
9. New concepts in angiography — including therapeutic radiology    180
   *Susan Rouse*
10. Research    198
    *H. Brian Bentley*
Index    211

# 1

# Radiography of trauma in the cervical spine in patients with associated spinal cord injuries

Over the past 30 years the number of patients admitted to spinal injury units throughout the world has gradually been increasing. The marked improvement in medical care in the immediate post-traumatic phase has meant a greater proportion of patients surviving to be admitted to these units. But more people are increasingly likely to suffer spinal injury as a result of road traffic accidents, sports injuries and other leisure-time pursuits involving a potential spinal risk factor. Rock climbing, gymnastics and rugby provide a regular quota of patients with neck injuries. Compulsory helmet-wearing for motor cyclists has meant a decrease in fatal head injuries but more of these patients survive to become tetraplegic patients although in this group the majority do present with lesions high in the dorsal region.

Historically, the earliest recorded case of tetraplegia with clinical details is mentioned in Edwin Smith's account of Papyrus after Herodotus. The building of the Great Pyramid involved a work force of some 100 000 labourers over a period of 30 years. Not surprisingly industrial injuries did occur and one of the labourers suffered severe cervical spine injuries; an account of the symptoms included

> Diagnosis: 'Thou should'st say concerning him: one having a dislocation in the vertebra of his neck while he is unconscious of his two legs and his two arms and his urine dribbles; an ailment not to be treated.'

In all the spinal injuries centres great emphasis is placed on rehabilitation of the patients so that he or she returns to the community. One of the most remarkable aspects has been the way that athletic competition, sometimes up to international level,

has been developed in a large number of these centres. This gradual change in policy and treatment which concentrates on the effective treatment, rehabilitation of patients and the development of sporting activities can be attributed to the drive, dedication and foresightedness of Sir Ludwig Guttman during his years as medical director of the National Spinal Injuries Unit at Stoke Mandeville Hospital. He foresaw the improvements that new knowledge and modern medicine would bring to give to the community a group of disabled, but otherwise fit, young people capable of achieving not only a high degree of rehabiliation but through participation in sport, a degree of competence and achievement that could be envied by able-bodied people. His efforts culminated in the establishment of the International Games at Stoke Mandeville and the Wheelchair Olympics.

Radiodiagnosis plays an essential role in providing the clinician with information on the extent of cervical spine injuries leading to cord damage with partial or total paralysis. The challenge faced by the radiographer is to provide the clinician with accurate, good quality diagnostic films achieved by plain radiography.

Anatomically the cervical spine is divided into the upper part consisting of two cervical vertebrae, atlas and axis, and the lower part which comprises the remaining five cervical vertebrae and one dorsal vertebra. From a functional point of view the cervical spine can be paired C1–C2, C3–C4, C4–C5, C5–C6, C6–C7 and C7–D1. Radiographically there are three divisions:

the upper — the first to the third cervical vertebrae;

the middle — the fourth to the sixth cervical vertebrae;

the lower — the sixth and seventh cervical vertebrae and the first dorsal vertebra.

## STABILITY AND ANATOMY OF THE CERVICAL SPINE

The architecture of the cervical spine has developed so as to allow a heavy ball (the skull) to rotate in either direction upwards and downwards at the same time as the cervical spine can be flexed and extended. The supporting elements are a series of ligaments interconnecting two or more vertebrae in their anterior, posterior and both lateral planes with the body and disc forming the links. The cervical spine is much more susceptible to disruptive force than that of the dorsal spine and lumbar spine; both the thoracic cage in the former and the strong lumbar muscles in the latter help to maintain stability in these areas. In the cervical spine the neck muscles play only a small part in support once the anterior or spinous complexes have been disrupted.

With the exception of the atlas and axis all the elements of the spine may be considered in pairs, each pair making a single motion segment with its neighbouring vertebra. The different anatomical and mechanical demands on the cervical, dorsal and lumbar spines show as structural differences related to the degree of weight-bearing on the one hand and the skeletal and muscular support on the other. There are sufficient structural similarities to have allowed early researchers to experiment initially with the lumbar spine motion segment and to extrapolate their findings to the whole of the spinal column with the exception of the atlas and axis. A later refinement was to build large models to experiment upon many times the size of the normal human vertebra.

More recently model construction of simulated vertebrae have been built using a large number of small bricks of various sizes rather than the earlier solid models, thus allowing pressure transducers and strain gauge instruments to be inserted into separate parts of the vertebra so that they can demonstrate forces and their relative directions within specific load bearing areas of each motional pair. This works very well with the dorsal and lumbar basic spinal pair, but in the more complex

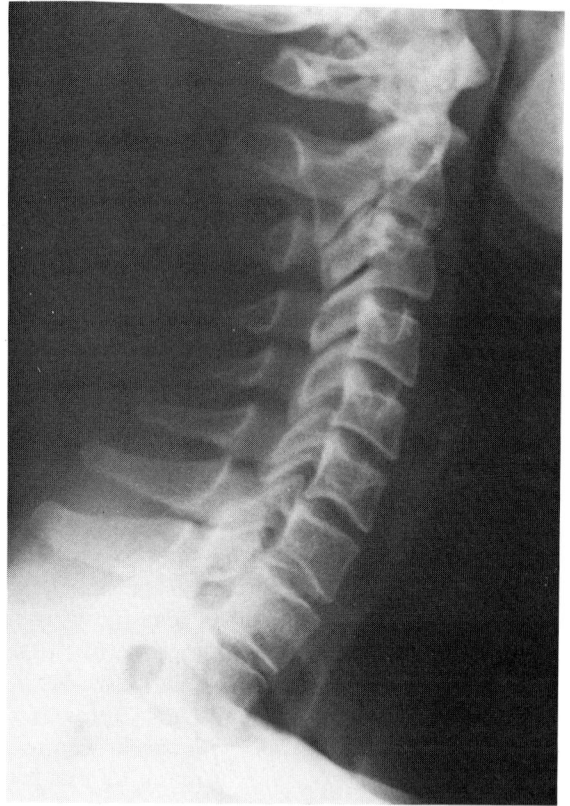

Fig. 1.1 The normal cervical spine.

cervical vertebral pair discrepancies do occur. To understand fully the effect of forces in different directions either unilaterally or combined, it is necessary to return to the cadaveric specimens as the experimental model in the cervical spine.

### The bony and ligmentous arrangements from the third cervical vertebra to the first dorsal vertebra

The cervical motion segment can be clearly divided into a posterior and anterior element with the dividing line just anterior to the articular facets. To permit the high degree of flexion with the cervical spine the posterior motion elements must have some elasticity of their ligaments and a small degree of laxity in the articular facet joints and their capsular ligaments. The smaller movements of each vertebra in extension will require very little take-up in the anterior motional segments. Consequently the anterior ligaments are

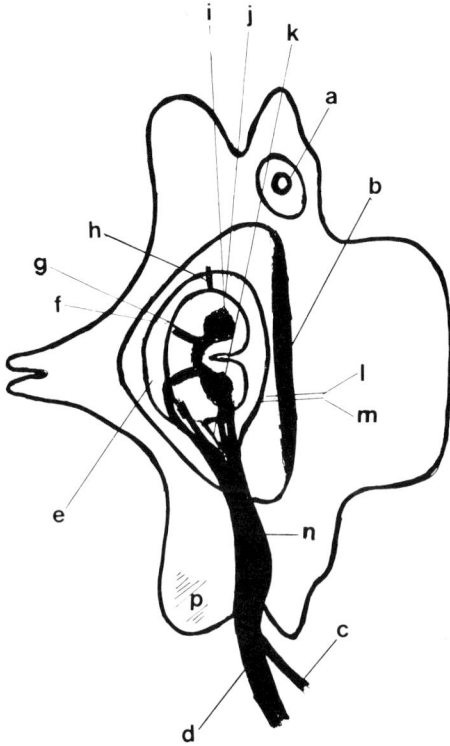

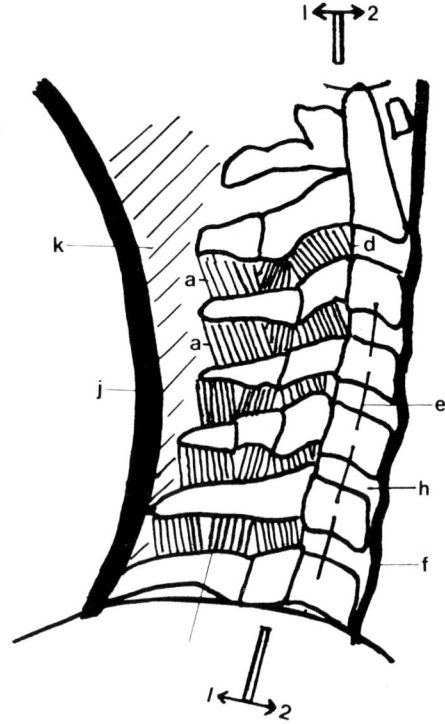

**Fig. 1.3A**

**Fig. 1.2** Oblique transverse section through a typical cervical vertebra. *Key*: (a) vertebral artery (b) posterior longitudinal ligament (c) dorsal nerve (d) ventral nerve (e) CSF (f) posterior horn, grey matter (g) white matter (h) dentine ligament (i) anterior horn, grey matter (j) piamater (k) dural septum (l) duramater (m) arachnoid mater (n) inferior aspects of the facet joint.

*Note.* The nerve root passes out of the canal from the cord just anterior to the superior facet, an important source of nerve root pain in facet dislocation.

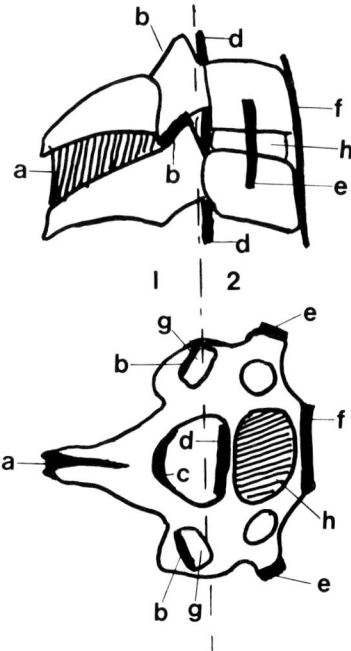

**Fig. 1.3** Ligaments of the cervical spine.
(A) The whole spine. The posterior spinous complex. *Key*:
(a) interspinous ligaments and the ligamenta flava
(b) capsular ligaments of the facet joints (c) yellow ligament
(j) supra-spinatous ligaments (k) interconnecting fibres.

*Note.* The ligamentum nuchae includes the supra-spinatous ligament, inter-connecting fibres and those forming the interspinous ligaments.

2. The anterior spinous complex. *Key*: (d) posterior longitudinal ligament (e) costo-transverse ligament (f) anterior longitudinal ligament (g) articular facet which intersects both the posterior and anterior spinous complexes (h) annulus fibrosus.

(B) A motional cervical spine pair from the lateral and the superior aspect. Key as Fig. 1.3A.

**Fig. 1.3B**

much more fibrous in nature and act principally as restraining bands from one vertebra to the next. Roaf (1960) has shown the importance of a normal nucleus pulposus in the lumbar spine in absorbing intervertebral stresses along the spine with the principal force in either the infero-superior or the supero-inferior planes.

By the very nature of the position of the spine it is the supero-inferior force which may severely disrupt the bony architecture of the spine; important elements here lie in the exact line of force through the cervical spine and the loss of the normal lordosis of the cervical spine at the time of impact. The combination of these two events leads to a vertical split in the vertebra. The normal nucleus pulposus can absorb quite heavy forces without changing shape or forcing a change in the shape of the annulus and often vertical end plates shear before the annulus is destroyed. In weakened and diseased discs posterior protrusion may be the precipitating cause of cord damage.

In experimental work to achieve physiologically accurate results, great care must be taken with the cadaveric specimen to maintain the normal texture of ligaments and soft tissue as far as possible. The specimen must be obtained as soon after death as possible, placed in a sealed polythene bag and must be frozen quickly down to a temperature in the range of $-18°$ to $-22°C$. This must then be thawed slowly and the experiments undertaken in a humid atmosphere to maintain, as far as possible, the natural moisture of the specimen. Experiments must preferably be undertaken in a chamber capable of this high humidity yet maintaining a cool temperature. The disadvantage of experimental work on human vertebrae lies in the measurements of very small movements of each motional segment involved and great care must be taken to firmly relate each vertebra to the measuring plates and the recording instruments.

One of the more surprising experimental results by White et al (1971) was that there was no recognizable pre-failure phase and all failures were sudden. What this means in practice is that if other injuries and bruising indicate either a mostly flexion injury or a hyperextension injury the manner in which the patients can be handled and radiographed must be adjusted to limit movements strictly in the direction of the applied force. Equal care is required in preventing any rotation of the neck as there are very few instances where the neck is subjected to a purely hyperextension injury without some degree of lateral force involved. Roaf (1960) showed that a very small rotational force may produce devastating effects on the spine compared with the greater forces involved in compression or extension. (Refer to a section on the handling of the patient.)

Piercey (1981) at the Nuffield Orthopaedic Centre in Oxford has used a bi-planar radiographic jig with X-ray tubes positioned in two planes in order to produce an antero-posterior and a lateral radiograph of the lumbar spine. He then draws related axes on the two projections, selects common anatomical parts on each and builds a three-dimensional model based on solid geometrical formulae. Another device he has used successfully is a flicker image technique where radiographs of the lumbar spine in neutral and flexed positions are each flashed quickly on to a projection screen and the eye perceives the two still films as a genuine production of movement. His work is aimed principally at research into the value of anterior grafts in the lumbar spine, but it has produced an interesting fact with possible reference to injuries of the cervical spine.

> A common feature of the motion segment above the graft was a much greater mobility of the posterior motion segment compared with that of the normal while the segment below the graft may give paradoxical movements. This could well be due to changes in the moment of the force involved posteriorly to the altered relationship of the spinal movements post-fusion. This gives some indication of why patients with congenitally fused cervical spines are susceptible to trauma and why these patients occasionally present with hyperflexion injuries producing flexion-type symptoms.

Recent research into small changes and slight abnormal movements of the spine has returned to ciné-radiography and video tape recording to produce dynamic studies. Each has its advantages and disadvantages. In ciné-radiography, processing and its limited other uses are outweighed by the facility of varying the projection speeds in either slowing down or speeding up the recorded movements. Video tape recording is very convenient and video display units are a common feature in most departments, but unfortunately on standard hospital models considerable degradation of the image occurs when there are changes in video head speeds and slow or fast motion studies.

There has recently been a renewed interest in stereo-radiography and parallax pairs because of the introduction of cardiographic techniques. The major problem here lies in the ability of the observer to maintain a stereo image in the mind whilst adjusting visually two moving light sources. Spatial awareness needs to be highly developed in the operator to achieve success. The radiography of severely disabled patients and the production of stereoscopic pairs of radiographs having the accuracy demanded by stereo-photometric methods trebles the degree of radiographic difficulty.

Another method now being used in lumbar spine motion pairs of appreciating bone changes similar to the flicker image technique on the lumbar spine in star pictures known by Piercey as the 'twinkle' effect. Further research into this needs to be undertaken to determine its value in such areas as the atlanto-axis pair, in pillar fractures or in small facet changes but accurate reproducibility is a major problem (Piercey 1981).

## NORMAL MOVEMENTS OF THE CERVICAL SPINE

It is relatively easy to measure the overall movements of the cervical spine in flexion, extension, rotation and lateral flexion, and it is not difficult to apportion the normal range of movement at the atlanto-occipital joints and the lower cervical spine for this is where the maximum movements occur. In extension the annulus fibrosus extends into the canal at the same time as the ligamentum flavum tends to bulge anteriorly into the posterior space of the canal. This ability of the canal to accept these encroachments without pressure on the cord is of paramount importance and any disease which reduces this gap between the cord and the inner circumference on the bony canal predisposes that individual to acute cervical cord injury often without any radiological evidence of bony injury.

The very complex series of ligaments which provide support at the atlanto-occipital joints and the atlanto-axial joints plays an important role in supporting the head and neck when fractures occur in the odontoid or there is dislocation at the atlanto-axial joint or there is a Jefferson (1920) fracture of the atlas. The transverse ligament has a major function in that it completes the fibro-osseous band around the odontoid in such a fashion that normally the odontoid always remains in contact with the internal arch of the axis. If rupture occurs there is forward displacement of the atlas giving luxation at the atlanto-axial joint compromising the cord at this level.

The normal rotation of the neck varies from individual to individual within the range of 80 ° to 90 ° about half of which occurs at the atlanto-axial joint. Simple rotation of the joint past 45 ° will narrow the spinal canal and going beyond this parameter puts the cord at risk. Lateral flexion must combine some degree of rotation of the spine with a small degree of movement at the atlanto-occipital junction, the greater part though occurring in the lower cervical spine. Each of these movements alters the shape of the canal reducing the width of the canal in cross-section and lateral flexion shortening the canal towards the flexion and lengthening it on the opposite side. Again it must be emphasised that anything which reduces the safety gap between the cord and the edge of the canal introduces a higher risk of cord involvement in any injury to the neck.

### The blood supply to the cord

There is still some disagreement on the true nature of the normal vascularisation of the cord. The main non-variant features are an anterior and a posterior supply — the former by a single anterior artery and the latter by two posterior arteries. The anterior artery at its beginning communicates with each vertebral artery, and the posterior inferior cerebellar artery receives contributions from some of the segmental arteries which in turn are fed by the vertebral artery, deep cervical trunk and costo-cervical trunk.

The blood supply to the cord in general can be divided into three areas: a superior, dorsal and inferior segment. The superior segment is that

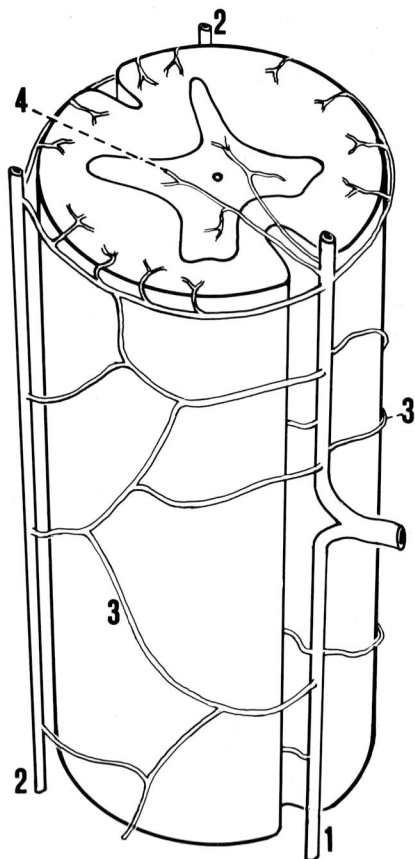

**Fig. 1.4** Blood supply to the cord. *Key*: (1) single anterior spinal artery arising from the anterior radicular artery. (2) bilateral posterior spinal arteries. (3) meningeal arteries within the coronal plane with pial perforators. (4) sulcal branches from the anterior spinal artery.

which provides a blood supply from the first to the eighth cervical segments. The major contributions come in at the third, sixth and eighth cervical roots to the anterior spinal artery stemming from the three arteries mentioned above, namely the vertebral, deep cervical and costo-cervical trunks.

In the sub-occipital area anastomoses between the vertebral, ascending cervical, deep cervical and occipital arteries form a complex of interlinked arteries. These are not very effective in rapid obstruction or trauma but develop a viable anastomosis best in slowly growing tumours. The arteries supplying the cervical segments of the cord are, however, well protected from trauma.

Blood supply to the cord itself is unevenly divided between the anterior and posterior spinal longitudinal arteries. The anterior spinal artery supplies the anterior two-thirds of the cord from its sulcal branch and also from its pial perforators arising from small meningeal branches that run in the coronal plane. These small meningeal arteries anastomose with similar ones leading from the posterior spinal longitudinal artery. The posterior third of the cord is supplied from the posterior longitudinal arteries. The nerve roots are supplied separately by branches from the segmental arteries.

**Venous drainage of the cord**

There are usually six longitudinal venous trunks connecting to extradural veins and draining through the intervertebral foramina to a plexus with a dual function since this also connects and drains from the vertebral bodies. Superiorly there are connections to the venous sinuses in the cranium; inferiorly drainage is into the azygos vein. The whole system is virtually valveless with only an occasional non-functioning valve. A series of small plexuses surround the cord and a generous collecting system normally precludes the venous collecting system from the effects of trauma.

## THE EFFECT OF TRAUMA ON THE SPINAL CORD

A considerable amount of research on the results of direct trauma to the spinal cord in primates, dogs and rabbits has been published. The cord is susceptible to impact from minor forces, the irreversible damage building up over a matter of hours, with grey matter showing greater areas of necrosis than the surrounding white matter. Four reasons have been suggested for this:

a. a force acting on the outer fairly tough casing (the pia) of the cord produces a significant centralised effect
b. capillary micro-vasculature in grey matter easily loses its conductability
c. micro-haemorrhages coalesce within 2 hours
d. neurones in grey matter are very much more sensitive to assault than those in white matter.

The speed at which changes develop and the extent of irreversible damage is related directly to the level of force inflicted. The areas of damage within the grey matter extend beyond that of the white which confirmed the original findings that grey matter was particularly sensitive to trauma. Ever present oedema at the site of the injury but not always isolated to any single segment of the cord contributes to sustained pressure within the confines of the pia. The significantly different anatomical microstructure of that between grey matter and white matter suggests that the grey matter is likely to be more susceptible to necrosis than white. Grey matter contains predominantly neurones with quite short axons and dendrites, but white matter on the other hand has a great proportion of closely packed long fibres providing an interrelated supportive system resistant to the effects of trauma.

Although researchers do disagree on the cause and effect of changes in the blood supply through traumatised areas there is no doubt that even minimal forces transmitted to the cord quickly disrupt blood circulation. This is in marked contrast to the protection from trauma of the external blood supply. The consequence of this is that pathological changes to the cord itself may increase over a period of 2 weeks even though there is evidence of clinical improvement in neurology. Further research also showed experimentally that some of the changes indicated were preventable. The main aim would be to reduce oedema by early dorsal myelotomy and intramedullary decompression and treating the injured cord by cooling it with saline perfusion giving local hypothermia as well as intramuscular dexamethosone.

Further experiments by Reed et al (1979) using Mannitol were at first encouraging and there was confirmatory evidence that there was a marked improvement of vascularisation of white matter in the cord within 4 hours of the injury, but this did not occur in grey matter, the more sensitive of the two to small traumatic forces, and at no time was there any improvement in perfusion in grey matter or in the posterior column. The lost time between a patient being injured, brought to hospital, diagnosed and treated with such a highly skilled medical regime precludes any hope of reversing a process already well established by the time that these interventions would be available.

## EARLY ASSESSMENT, RESUSCITATION AND THE TREATMENT OF PATIENTS

To arrive at hospital without damage to the cord with an unsuspected, unstable cervical spine requires a high degree of luck given to only a few; to achieve a correct diagnosis without further damage to the cord requires a high degree of expertise both in the casualty department and in the X-ray room. The initial assessment is based on sound clinical diagnosis and from this should stem instructions for the subsequent handling of the patient. A baseline of physiological responses must be mapped out so that progress or regression can be assessed from this early phase.

Early life-threatening conditions such as blood-loss complications, aspiration pneumonia and the prevention of further cord damage, especially if the lesion is high in the cervical spine, must be undertaken or dealt with. 'One-piece' lifting with the head supported and the neck in slight extension requires four handlers to ensure correct and safe movement of the patient. Pressure sores can develop very quickly especially if sharp objects from the pockets have not been removed. The very accurate and complete neurological assessment should be left until later to be undertaken in the ward situation.

In patients with high level lesions a careful check on airways is necessary since one possible complication in extension injuries is soft tissue swelling anterior to the cervical spine at the level of cervical vertebrae 2–4 for this may compromise the hypopharyngeal airway. General examination of the patient with particular reference to the chest, abdomen, head and limbs will indicate the need for additional radiographs of these areas. Nothing should be undertaken to show the extent of these injuries until the lateral radiograph of the cervical spine has been taken and examined.

In 1924 a Medical Research Council Report (No. 124) on spinal injuries showed clearly that survival was dependent on whether there was any sparing of neurological deficit, but in complete lesions there was a poor prognosis. The report's description of a patient with cord injury was as follows: 'The paraplegic patient may live for a few years in a state of more or less ill health.' Nissen (1942) in a paper to the Royal Society of Medicine was also very pessimistic about the value of extending the life of these patients and he quoted in this paper the following: 'Thou shall not kill; but needst not strive officiously to keep alive', as a maxim to be followed in the treatment of paraplegic patients.

It is only in the last 30 years that effective treatment regimes have been instituted so that the above statements are invalidated and that one can now say it is an ailment most definitely to be treated. Munro (1962) showed that in the 20 years between the two World Wars, 60% of all patients with cord injuries died in the first 3 years, mortality in cervical spine patients being that much worse. Mortality in patients with high cord lesions has been reduced to under 10% in this initial 3 year period, and life expectancy has also been extended markedly by better control of chest and urinary infections on the one hand and better skin care on the other.

## Radiographic and radiological pitfalls on the initial examination

Patients with severe injuries to the forehead and the face, and patients with forward buckling deformity fractures of the sternum should have lateral radiographs of the cervical spine taken before further radiography is undertaken (Park et al 1976). Park (1981) has suggested that incorrect diagnoses may stem from the following:

a. incomplete visualisation of the fracture site (radiographic)

b. failure to recognise the mechanism of the injury (clinical and radiographic)

c. incorrect interpretation of subtle signs (radiologic)

d. confused imaging caused by radiographic artefacts (radiographic)

e. misleading and unpredictable neurological deficit (clinical).

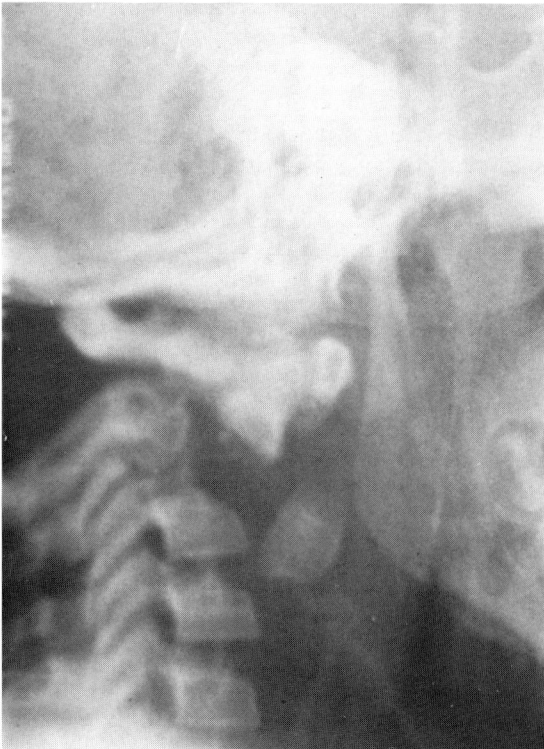
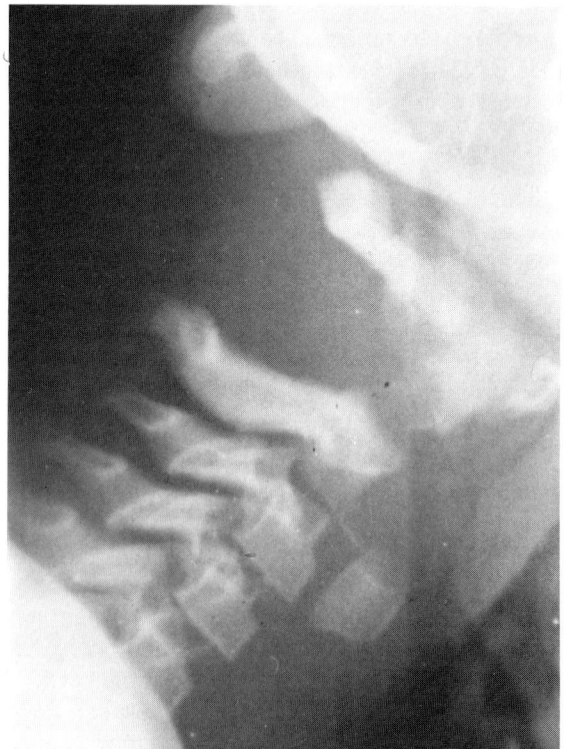

A                                                                B

**Fig. 1.5** Unsuspected fracture. RTA injury in a child having only skull radiography immediately after the accident, the cervical spine films taken some 2 days later. No neurology. (A) Pre-manipulation. (B) Post-manipulation.

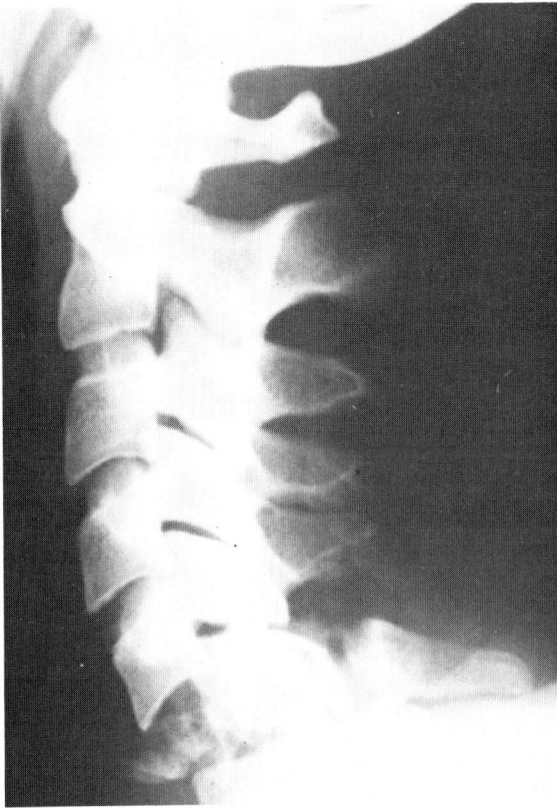

**Fig. 1.6** Burst of C7 missed on original lateral cervical spine film but shown on a subsequent film using a stationary grid technique with adequate pull on the shoulders.

He also goes on to list the more easily missed injuries as follows:

 a. odontoid and atlas fractures
 b. hyperextension fractures
 c. complete interspinous rupture
 d. articular mass fractures
 e. dislocation of cervical vertebra 6 and dorsal vertebra 1
 f. double fractures combined with sternal injuries.

The difficulty of producing high quality radiographic images of the upper and lower parts of the cervical spine in severely injured patients is obvious. An appreciation of the relationship between the direction of force and the mechanism of injury has not been discussed at all in radiographic papers or books on radiography. The need for adequate soft tissue detail anterior to the vertebral bodies has only recently been stressed sufficiently in radiological papers and books. (Hopcroft 1977, Penning 1981).

Penning (1981) in a survey of 30 patients with cervical spine trauma showed that 60% had evidence of pre-vertebral haematomas. There were two clearly defined groups:

 a. small changes due to relatively small haematomas associated with odontoid fractures and crush injuries of the vertebrae

 b. extensive pre-vertebral haematomas associated with extensive damage to the anterior ligament and probable rupture of larger blood vessels. Roaf (1960) has shown that small haematomas develop anterior to the vertebral bodies when compression takes place with blood being forced in between the superior and inferior plates in an anterior direction. It seems likely that the small haematomas in the crushed vertebra in the cervical spine could be due to extrusion of the blood anteriorly between the end plates and damaged vertebra.

Penning (1981) goes on to emphasise that the importance of the widening of the pre-vertebral space is as a 'pointing finger' to possible hidden fractures. He goes on to suggest that excessive widening of this space also points to the possibility of disruption of the anterior spinous ligament. If the lateral radiograph shows this without obvious bony injury, the handling of the patient must be very carefully undertaken and the radiographer should seek sufficient handlers and medical assistance to avoid further injury to the patient.

**Treatment of the patient on admission**

Neurological examination should elucidate sensory and motor changes, reflex activity and autonomic dysfunction. On the ward, monitoring of blood pressure, pulse, temperature, urinary output and abdominal distension is mandatory. Hypotension in spinal injuries above the level of dorsal vertebra 4 without associated injuries causes hypovolaemia due to autonomic hyper-reflexia. If the vital capacity is low and blood gases indicate a poor exchange, tracheostomy or endotracheal intubation may be necessary with breathing being assisted by the use of a ventilator. The state of the patient's lungs prior to the accident, loss of use in abdominal muscles, weakening of diaphragmatic movements, can reduce vital capacity to threshold levels. The patient will be poikilothermic and will assume the temperature of the surroundings. Cooling of the body may have to be undertaken as hyperthermia develops or more unusually warming if hypothermia occurs. A distended abdomen may indicate a paralytic ileus and a request for abdominal radiography in the supine and lateral decubitus position may subsequently be made. A paralytic ileus will also reduce vital capacity due to the build up of internal abdominal pressure.

Careful control of fluid intake is essential because the tetraplegic patient cannot accommodate a large volume intake. Pulmonary oedema can be induced and if this is suspected an increase in radiographic exposure factors will become necessary to overcome the increased opacification of the lung fields, when undertaking chest radiography. Segmental and lobar collapse can occur quite quickly if adequate physiotherapy is not undertaken. The collapse in this case is due to plugging of the bronchi by mucus. An overfilled bladder and bladder dyssynergia can also induce autonomic dysreflexia giving symptoms of severe headaches and leading to coma and eventual death. (Refer to Ch. 6 Urodynamics.) Intermittent catheterisation is the usual accepted procedure for controlling urinary bladder volumes.

Patients must be turned every 2 or 3 hours and it is helpful to ward staff if the timing of ward radiography coincides with the time when the patient is being nursed supine, and just before the patient is due to be turned. If the ward is fortunate in having turning beds then this is not so essential but it must be borne in mind that it does take time to set up the patient, who is usually in traction, for lateral cervical spine radiography in this supine position.

Various types of skull traction are available for use such as the Crutchfield skeletal traction equipment or Blackburn tongs. Varying weights may be added from 3–10 kilograms; and occasionally up to an unusual maximum of 20 kilograms. In children 1½–5 kilograms is the prescribed range. Weights in excess of 5 kilograms in children and 10 kilograms in adults together with very precise degrees of flexion and extension with careful

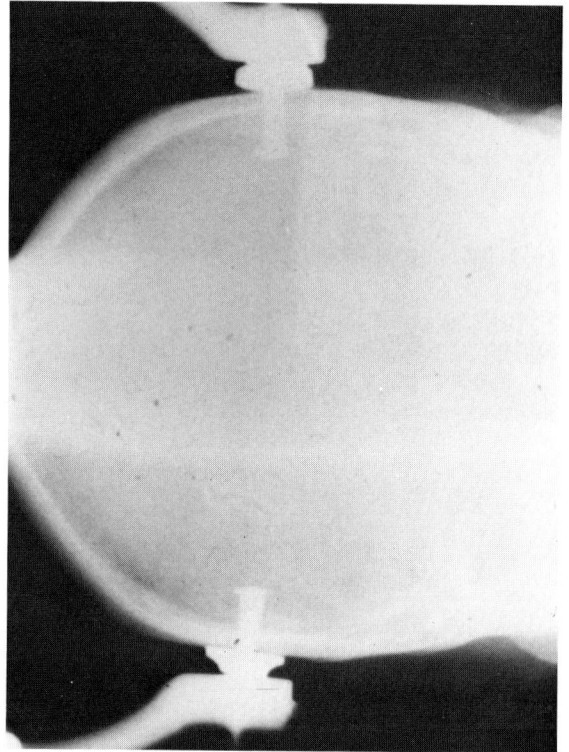

Fig. 1.7 CSF leaking from the burr holes of a patient on skull traction. Skull radiography in the ward situation. Tangential projection of each side of the skull, slight rotation of the skull away from the side radiographed. No stationary grid — a 24 × 30 cm cassette so that the film is placed under the skull once.

positioning of the pulley relative to the head are used to distract the apophyseal joints in order to reduce dislocation or to align fractures, especially in adults suffering from ankylosing spondylitis. Care must be taken especially in young children to avoid over-distraction, the lateral radiograph indicating when this has taken place for further neurological deficit may be induced in the patient if the spine is overstretched.

The lateral radiograph will also show misalignment of the spine in unstable fractures when the relationship between height and position of the pulley, the skull traction, the position of the skull and the position of supporting pillows at shoulder level counteract the optimal alignment of the spine above and below the fracture or fracture-dislocation. Traction tongs may pull free from the skull during normal nursing procedures and this usually necessitates an urgent visit by the radiographer to show if there has been any alteration in the position of the spine. Occasionally cerebrospinal fluid may be observed leaking through the burr holes and this is more likely to occur on the types of traction where the prongs are so positioned that the ends rest on the inferior surface of the cranium (Fig. 1.7).

When undertaking radiography no stationary grid is required as small radiographic fields can be used by making use of tangential projections. Separate exposures are made on each side of the skull on a single film, the head being rotated slightly towards the side to be radiographed, when the head is in the fronto-occipital position and a vertical X-ray beam is used. Other projections are not necessary as they provide no additional information, unless there is a suspicion that the burr hole has become infected, and in this case the appropriate lateral is taken.

### Classification of stable and unstable fractures

The forces applied to the spine at the time of the injury can be broadly divided directionally to those between flexion and extension although an element of rotational force is also invariably present. A number of patients will receive both flexion and hyperextension injuries as they are thrown about inside a car or if they fall downstairs. A very few will have a vertical force applied to the cranial vault at the time when the cervical spine has lost its natural lordosis. A number of definitions of 'stable' and 'unstable' abound in the literature, but the two which seem simplest to apply initially are as follows:

1. The *stable* spine is the one in which there is less likelihood of further severe cord damage and the patient usually has some sparing; it is the one in which nursing and diagnostic manoeuvres may be undertaken within a fairly aggressive nursing regime, without risk of aggravating the injury or giving further damage to the cord. An *unstable* fracture renders the cord liable to further damage by the application of very minor forces during clumsy handling, inexpert medical care, or to a too aggressive approach when undertaking diagnostic procedures.

2. Chester (1969), however, used a very simple approach based on conservative treatment and the

knowledge that some patients will present with *late instability* even though the injury may well be classified initially as stable and his definition of stability is as follows:

*Stability* is defined as the absence of any abnormal mobility between any pair of vertebrae with or without pain or other clinical manifestations, when lateral radiographs of the cervical spine are taken in extension and flexion at the conclusion (usually about 12 weeks) of conservative treatment of a fracture or fracture-dislocation.

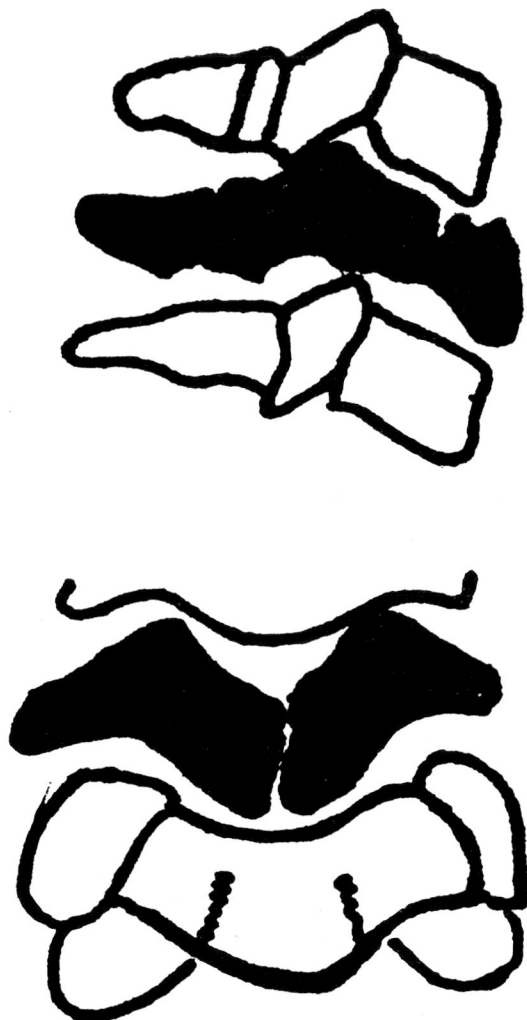

**Fig. 1.8** Wedge fracture of C5 with late and immediate post injury radiographs. Some 'settling' has taken place and there is bony bridging between C4 and C6.

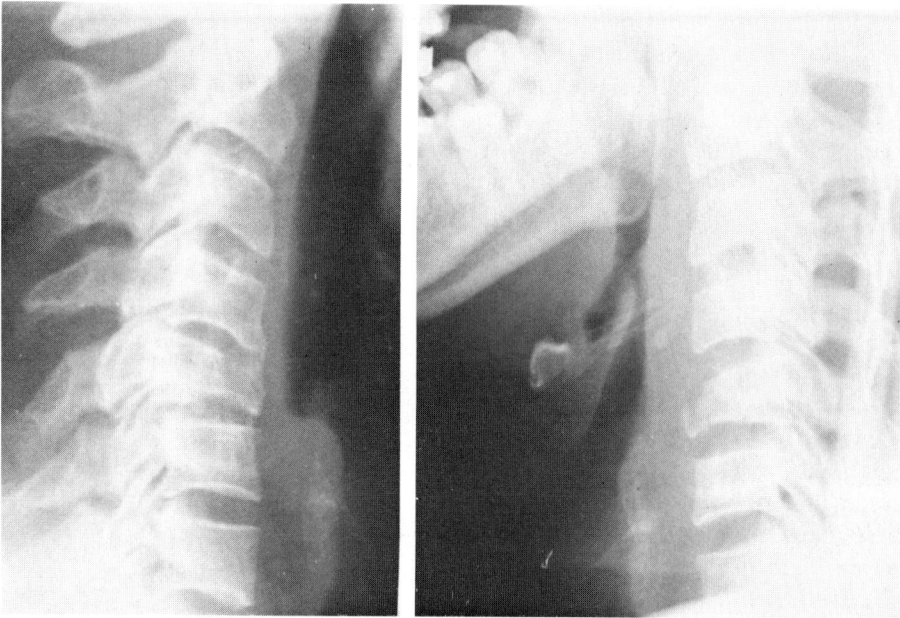

**Fig. 1.9** Flexion injury. Burst fracture of vertebral body, little change in vertebral height except in severely crushed vertebrae, changes in alignment of vertebral bodies. In the lower cervical spine in the antero-posterior position the vertical split in the vertebral body may be hidden by the air-filled larynx.

Patients with late instability syndrome show evidence of anterior subluxation or even dislocation of both facet joints some time after the initial event. Anterior subluxation encompasses both groups, partial subluxation due to laxity in the facet joints and some posterior ligamentous tearing but which does not dislocate on forced flexion. This manoeuvre must be undertaken with full medical supervision, preferably with the use of a mobile image intensifier and certainly with routine lateral radiography of the cervical spine.

### Group 1 Stable fractures

1. Wedge fractures without rupture of the anterior ligaments.
2. Posterior neural fractures of the atlas.
3. Articular mass (Pillar) fractures.
4. Clay shoveller's fracture.
5. Most burst fractures of the lower cervical spine.
6. Unilateral facet dislocation without fracture of the inferior facet.
7. Anterior subluxation with disruption of the posterior spinous complex with survival of the anterior part of the disc and an intact anterior ligament.

### Group 2 Unstable fractures

1. Anterior subluxation with disruption of the posterior spinous complex and failure of the inter-facetal ligaments.
2. Burst fractures where there has been a rotational element in addition to the vertical component of the forces involved.
3. Bilateral dislocation of the facet joints.
4. Jefferson fracture (multiple fractures of the atlas).
5. Hangman's fracture (fracture dislocation of cervical vertebrae 2 and 3).
6. Flexion tear drop fracture (torn posterior ligaments — disc damage and anterior ligaments).
7. Extension tear drop fracture (disruption of the anterior spinous complex).
8. Fracture dislocation after hyper-extension.

It is useful to tabulate all types of injuries associated with hyperextension and flexion injuries into:

a. *Flexion injuries — dominant force forwards*
   1. Anterior subluxation.
   2. Bilateral inter-facetal dislocation.
   3. Wedge fractures.

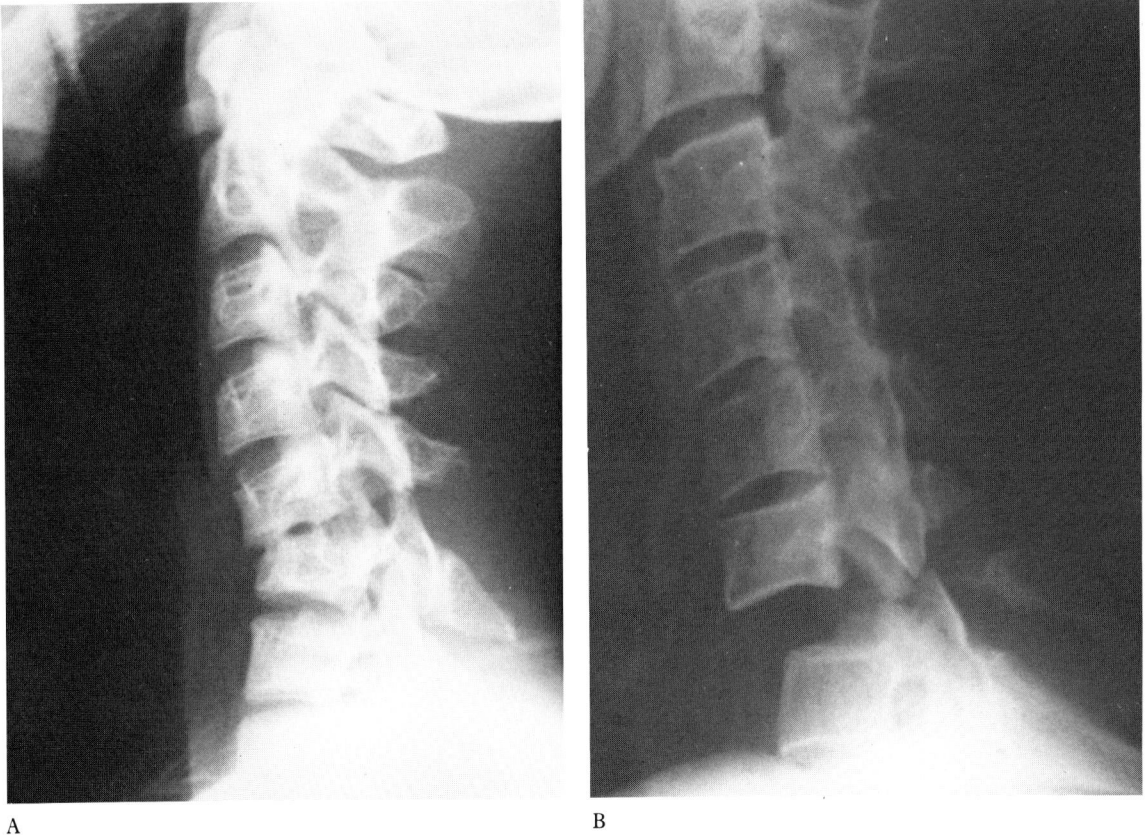

A                                              B

**Fig. 1.10** Facet dislocation. (A) Unilateral facet dislocation, change in angulation of vertebral bodies, slight rotation of vertebra above the dislocation. The 'step' between C5 and C6 is less than half a vertebral body. It is impossible to determine which facet is tip-to-tip in the lateral projection alone. (B) Bilateral facet dislocation. The 'step' is now at least half a vertebral diameter. The gap between the vertebrae indicates a bilateral tip-to-tip situation of the inferior and superior articular facets.

4. Flexion Tear Drop Fractures and in its most severe form:
5. Flexion Tear Drop Fractures with dislocation.

b. *Flexion or hyperextension injuries*
  1. Single dislocation of one facet joint (Flexion rotation).
  2. Extension rotation giving unilateral fracture of the pillar mass.
  3. Jefferson fracture of the atlas (vertical compression fracture).
  4. Burst (vertical split) fractures usually of a single vertebra
     (Cervical vertebrae 4, 5, and 6).

c. *Hyperextension injuries — dominant force backwards*
  1. Hyperextension with cord damage without visible bone involvement.

2. Hyperextension injuries with fracture-dislocation.
3. Posterior arch fractures of the atlas.
4. Extension tear drop fractures.
5. Hangman's fracture.

## RADIOGRAPHY

In their introduction Gerlock et al (1978) make the following statement:

There are few moments in medicine when the correct on-the-spot radiologic interpretation is more crucial to the management of a patient than the moment when you view this initial lateral radiograph of the cervical spine of an acutely injured patient.

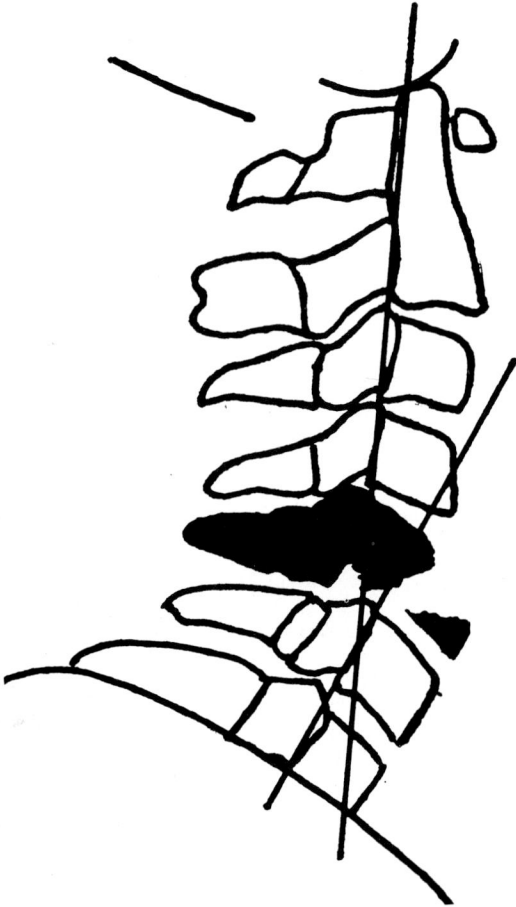

Fig. 1.11 Unstable hyperextension injury. Tear-drop fracture of C5 with posterior angulation of the spine above the fracture.

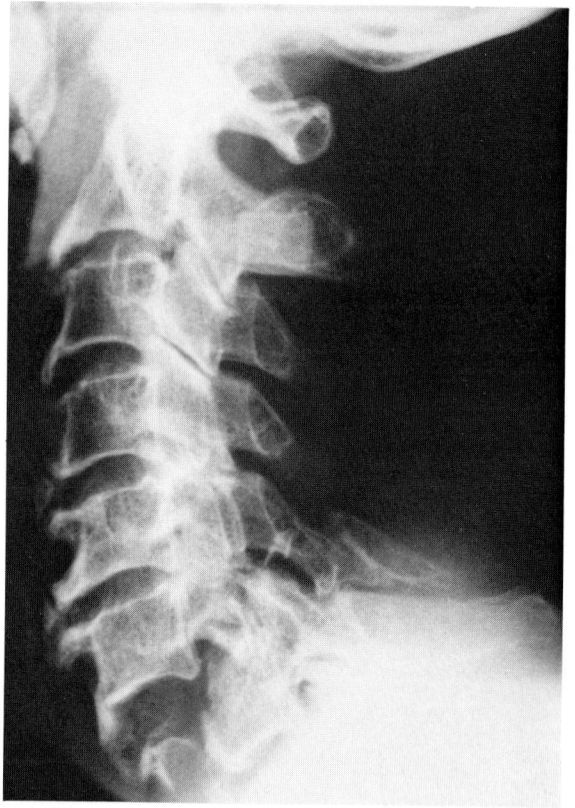

Fig. 1.12 Tear-drop fracture of C7 with anterior displacement of C6 on C7. The fractured superior vertebral plate is 'held' by the disc between C6 and C7. Anterior displacement of a whole vertebral diameter. Fractures through the small facet joints and also the posterior spinous process of C5.

The reader has only to substitute radiographic examination for radiological interpretation to make the importance of the statement applicable to radiography.

The basic projections of lateral, antero-posterior, 45° obliques in either the anterior oblique position or the reverse are well documented in Clark's *Positioning in Radiography* (1973). The importance of the true lateral view as the 'first sight' investigation cannot be over-emphasised. Unsuccessful projections which are not truly lateral should not be thrown into the waste film bin as these will sometimes show individual facet joints and the posterior arches of the axis. Radiographers are seldom fortunate enough to have patients with the correctly shaped anatomy, and

as co-operative as the patient whose cervical spine radiograph is shown in Figure 1.1.

One of the many difficulties is to outline clearly the whole of the cervical spine from cervical vertebrae 1 to 7 with sufficient of dorsal vertebra 1 to demonstrate the inter-facetal joints between cervical vertebra 7 and dorsal vertebra 1. Muscular spasm can force the shoulders upwards as far as cervical vertebra 5. It is an unfortunate fact of life that very muscular short-necked men arrive in the department with neurological signs suggesting injury at the level of cervical vertebra 7.

The Zimmer halter comprises two soft pads, one for under the occiput and one for under the lower mandible; each is joined at the edges by a cord which can be grasped or separated by a spreader and held in one hand. Two or three helpers are required to assist in this manoeuvre. One helper

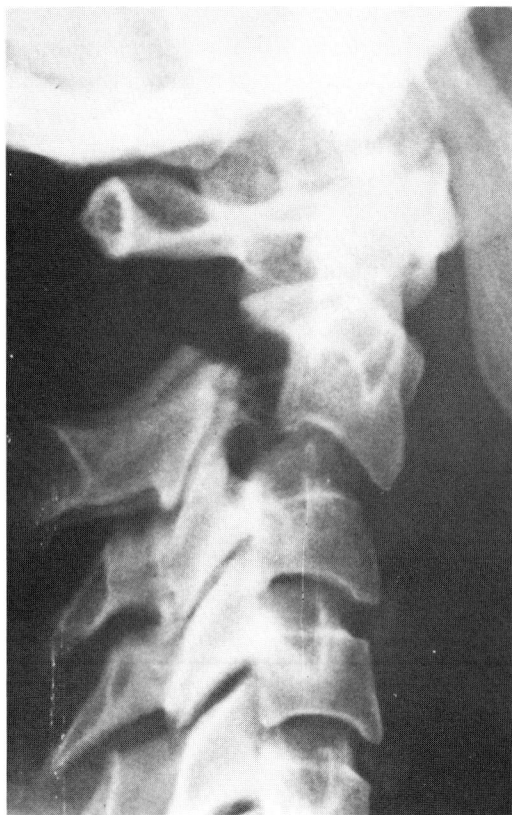

A

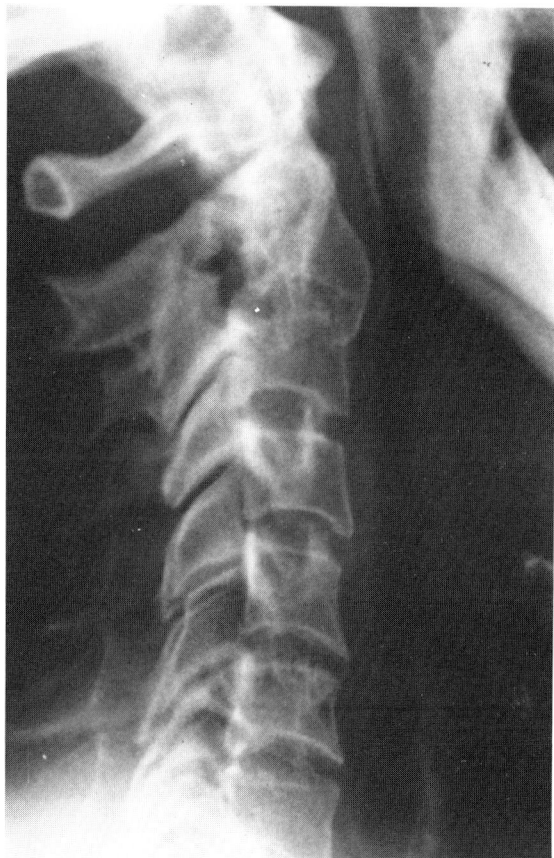

B

**Fig. 1.13** Hangman's fracture. (A) Immediate post-accident film. (B) 1 year later.
Hangman's fracture presents as an anterior displacement of the body of C2 on C3 with an intact odontoid and intact
occipito-atlantial joints. There are bilateral fractures through the pedicles anterior to the inferior facets.

stands at the head of the patient holding on to the
Zimmer halter with only sufficient force to prevent
the patient from sliding down the table or the
trolley or the bed, against the pull exerted by
either one or two helpers pulling on both arms. A
vital extra centimetre or two can be achieved by
alternately pulling and holding on each arm in
turn. It may be necessary to pull several times on
each arm gradually increasing the strength of the
pull while carefully noting the patient's reactions.

A difficult decision to make is when not to use
a stationary grid and to use a short exposure time,
or when to use a stationary grid and to use a long
exposure time risking patient movement or operator movement, both of which produce unsharpness in the radiograph. The arm pullers can impart

appreciable tremor to the patient when maximum
force acceptable to that patient is being exerted.
On heavily muscled, broad shouldered patients a
considerable amount of scattered radiation from
the shoulder girdle reaches the film at the level of
cervical vertebrae 6 and 7 and dorsal vertebra 1.
Blurred visible images are diagnostically more
acceptable than indistinguishable images lost in a
'sea of fogging' and therefore the use of a
stationary grid must be the technique of choice in
this type of patient.

If the lateral radiograph fails to show the appropriate vertebrae then an alternative projection is
the 'swimmers' view, with the arm nearest the X-
ray tube pulled caudally parallel to the side with
the arm nearest the film pulled forwards and flexed

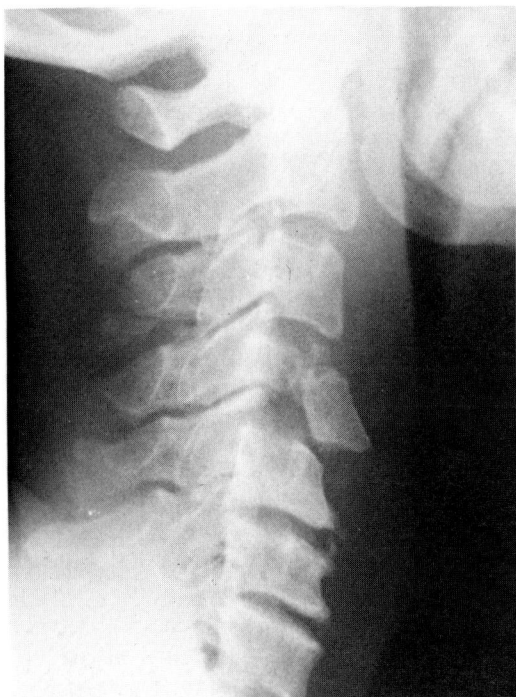

**Fig. 1.14** Tear-drop fracture of C4. Ruptured anterior ligaments with destruction of the discs superior and inferior to C4. Fracture of the posterior spinous process of C3.

around a pillow. It is not unusual for the patient to find this position intolerable. An added complication is that patients with injuries at the cervicodorsal junction may also have injuries to the shoulder girdle preventing movement of the arm. A more acceptable projection in this situation is an 'off-lateral' projection with only a little forward pull on the arm nearest the film and with the arm nearest the X-ray tube abducted gently posteriorly to the spine. In this projection each shoulder mass appears separated on the radiograph and not superimposed with the spine being projected in between. The flexed lateral projection with both arms brought forward and the neck fully flexed is a very dangerous manoeuvre for the patient to undertake and should rarely be used.

The standard 45° oblique projection will show the laminar line lateral to the intervertebral foramina (see Fig. 1.17). Any disruption in its smooth outline is indicative of trauma within the articular complex. This standard projection does mean that the patient must be rotated through 45°. If this is acceptable clinically, it is strongly recommended that a special trunk positioning pad be made 1 metre in length with two equal sides having a depth of not less than 18 centimetres, and the pad having angles of 90° and two times 45°. Some manufacturers of foam positioning pads now offer customers a choice of densities according to requirements, and here a firmer pad than normal is advantageous.

An alternative projection is the 45° supine position (McCall 1973) shown in Figure 1.15. An Additional 10° angulation cephalad will show the pillars in profile more certainly. Reversing the angulation to 10° caudad displays the facet joint space better from its superior aspect. The central X-ray beam is positioned with the centring point 5 cm above the sterno-clavicular joint. If a caudad angulation of 10° is used, then the centring point is 8 cm above the sterno-clavicular joint. If a cephalad angulation of 10° is used, the centring point is 2 cm above the sterno-clavicular joint. In each case the centring point should be 5 cm away from the mid-line towards the side under investigation.

The advantage of this projection lies in the fact that patients remain in the supine position and need not be lifted at all since the oblique ray projects the area of interest onto the centre of the displaced film. The body of the vertebra is shown as a distorted elongated shape, the distortion being proportional to the distance between a particular vertebra and the film. A single angulation of the X-ray tube will allow a stationary grid to be used and where double angulations are used an acceptable alternative is a 22 gauge tin plate so long as kilovoltage is kept below 100kV; above this a granular appearance is produced on the radiograph.

### The lateral oblique view

Orthopaedic surgeons differ in their approach to the treatment of patients with unilateral or bilateral facet dislocation. This ranges from 'leave well alone' on patients with unilateral facet dislocation, to possible open reduction on patients with bilateral dislocations. The type of fixation is determined by the surgical technique and the choice of either a posterior or an anterior approach to the site of the injury. Although single facet dislocation is usually stable, patients can suffer from severe root pain after the injury. The decision about reduction by manipulation either under general anaesthesia with muscle relaxants, or gentle manipulation and the appropriate use of skull traction with sufficient weight to distract the superior facet over the inferior facet is a matter of personal preference by the orthopaedic surgeon involved.

There is still a measure of disagreement between surgeons on the appropriate method to use. Before undertaking any of the above manoeuvres it is important to know the true state of the facet joints.

The straight lateral and the supine 45 ° films do not show the facet joint in profile. In these cases the projection of choice is the oblique lateral view of each of the facet joints in turn. Again this projection is possible with only minimal movement of the patient (not more than that in normal nursing procedures to avoid pressure sores of the skin). This projection is also particularly useful in the operating theatre when such manipulation under general anaesthetic might take place (Lodge et al 1966).

This projection is particularly useful for showing the facet joints of the cervical vertebrae 4·to 7 and dorsal vertebra 1. It shows clearly the relationship between the superior and inferior facets. The ideal projection requires either a 5 ° caudad or cephalad tilt to the X-ray tube but this is not essential to produce diagnostic radiographs and can certainly be dispensed with in theatre radiography where the information required relates only to the relative positions of the facet joints after manipulation (see Fig. 1.16). There is

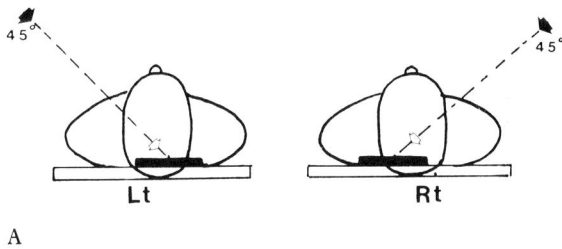

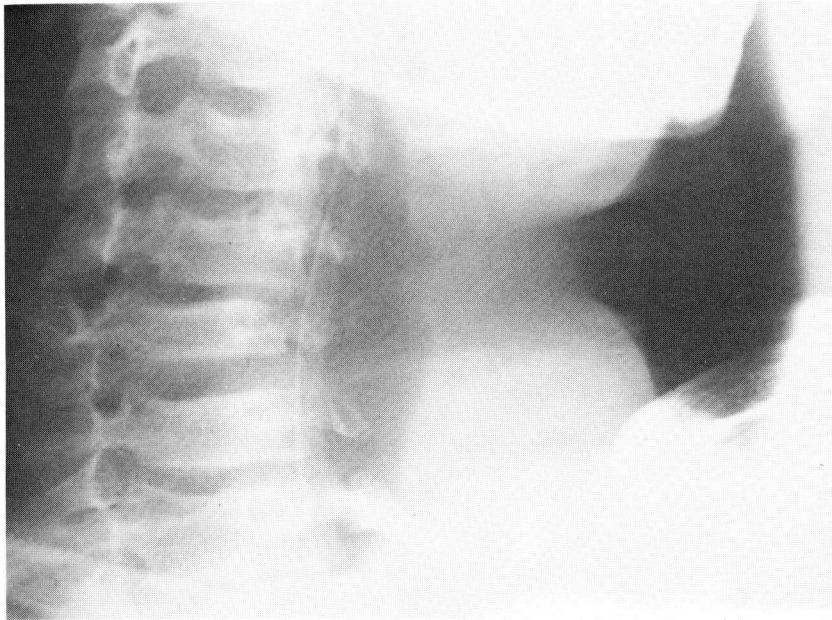

Fig. 1.15 The supine 45° obliques. (A) The supine 45° oblique projection of the cervical spine, showing the direction of the X-ray beam and position of the film. There is also a 10° cephalad tilt to the X-ray tube. *Note.* The degree of displacement of the film from the midline is away from the side under investigation. Two films are taken for comparison. The pillar masses are demonstrated on the side to which the X-ray tube has been centred. (B) Radiograph of the pillar masses on the right side.

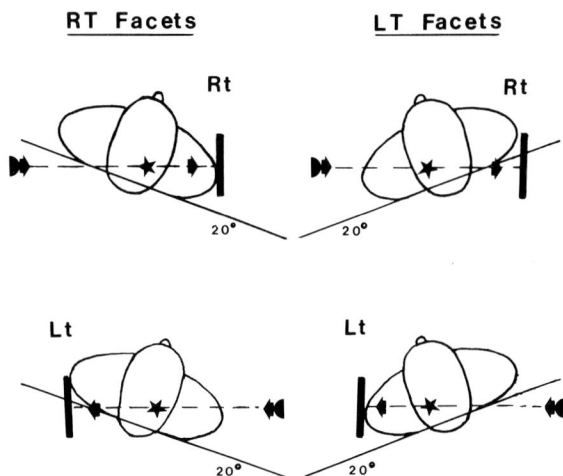

Fig. 1.16 The lateral oblique. The four projections in the 20°/5° lateral oblique for right and left facets with tube on the right and left of the patient. The facet shown is always on the side on which the shoulder is lowest. The marker must indicate the facet shown and not the particular lateral oblique. (See text for specific centering points.)

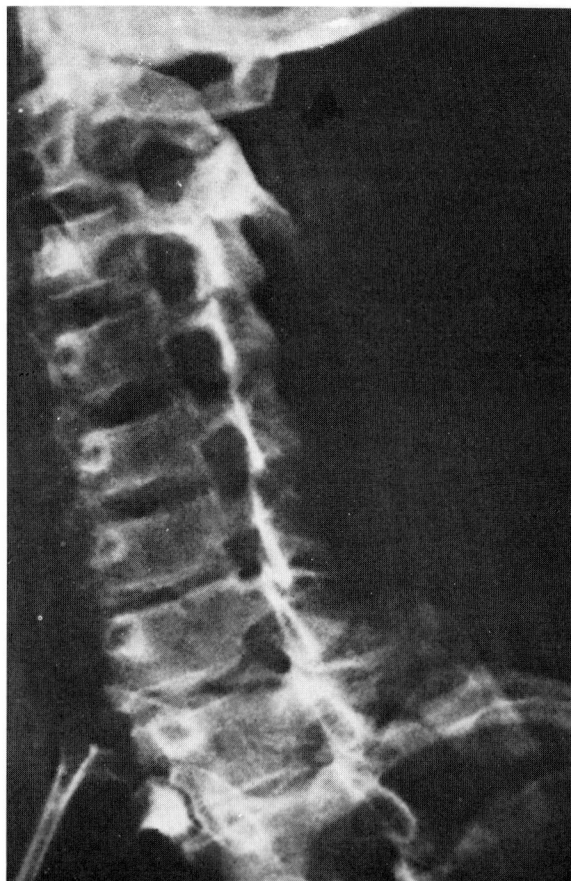

A

Fig. 1.17 Comparison of the routine 45° oblique in the antero-posterior position and the 20°/5° lateral oblique. (A) The 45° oblique showing the laminar line and the inter-vertebral foramina. (B) The 20°/5° lateral oblique showing the articular facets in profile.

a choice of techniques each being slightly different depending on whether radiography is undertaken from one side of a patient only or whether taken from one side and then the other with regard to the position of the X-ray tube. The radiographs are distinguished from each other by putting the right marker on that which is taken with the right shoulder down and the left marker on that which is taken with the left shoulder down. Both radiographs show the spine orientated in the same direction and are virtually indistinguishable (see Fig. 1.16).

If radiography is undertaken from one side only with the patient being rotated from one 20 ° lateral oblique projection to the other, the X-ray direction needs to be adjusted allowing for a change in inclination of the facet joint as the patient changes position. The X-ray tube angle is 5 ° cephalad when the side raised is nearest the film and 5 ° caudad when the side raised is nearest the X-ray tube. It is easier to achieve comparable results if the X-ray tube is moved from one side of the patient to the other so that each side is radiographed with the shoulder nearest the X-ray tube in the raised position. Centring in each case is to a point 5 cm above the sterno-clavicular junction and 2.5 cm from the posterior neck line with the patient rotated through 20°. Traction should be applied to the shoulders, especially when the X-ray tube is angled towards the feet. The introduction of this projection for use specifically for radiographs in theatre, by the author, was at the request of orthopaedic surgeons, especially Evans (1961) who developed closed manipulation of such dislocations to a very high level.

## Radiography of cervical vertebrae 1 and 2

The mechanisms affecting these two vertebrae in trauma are quite different from those in the rest of the cervical spine; structurally the two vertebrae have unique architecture with supporting ligaments arranged differently to all other vertebrae.

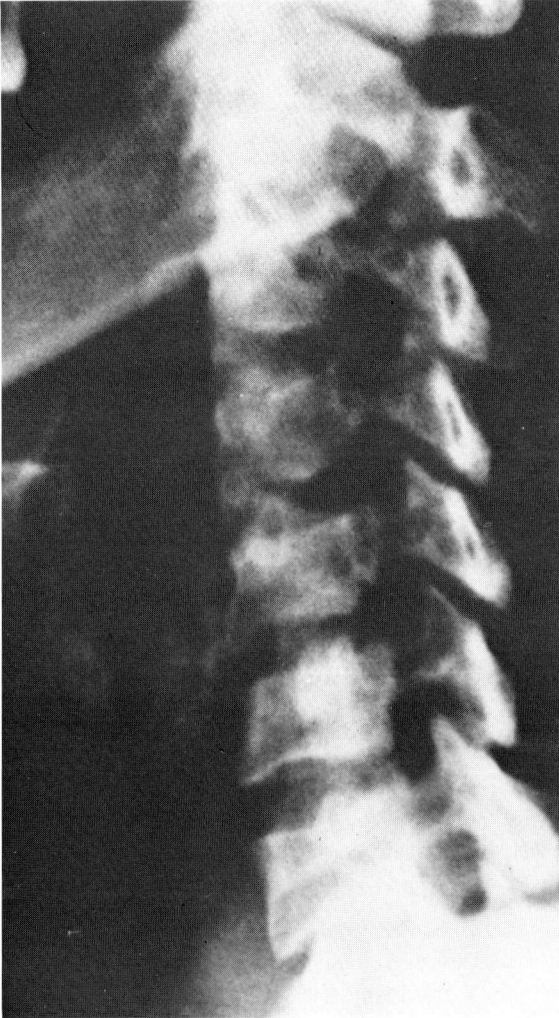

Fig. 1.17B

It is important to remember that the anterior liga-ment carries on and is attached to the anterior surface of the clivus; the posterior ligament terminates on the lower half of the posterior surface of the axis. There are a series of ligaments which are particular to the atlas and axis and main-tain their stability during varied movements. These are the alar ligament, transverse ligament of the atlas, cruciform ligament of the atlas and the accessory atlanto-axial ligament. The posterior group of interspinous ligaments are attached to the inferior surface of the spinous process of cervical vertebra 2 and to the superior surface of cervical vertebra 3.

The upper cervical spine can be subjected to the same forces as those which act on the rest of the cervical spine, i.e. hyperflexion, flexion, rotation or a combination of two or more. Injuries are more likely to occur during hyperextension because the normal range of movement is greater in flexion in the upper cervical spine than that in extension. Unlike injury to the lower cervical spine, there must be considerable sparing for life to continue and there can be neurological signs of either upper cord injury or brain stem injury to confuse the neurological picture.

Fractures of the odontoid peg are fairly common and usually occur across its base. Flexion injuries can be quite disruptive giving fracture of the odon-toid and the posterior arch and forward dislocation of the atlas. If the patient is to survive, it is necessary for the posterior arch to fracture allowing room for the body of the atlas to slide forward leaving behind a gap where the posterior arches are fractured and thus preventing damage to the cord. Hyperextension injuries in contrast give backward displacement of the odontoid and atlas with upward displacement of the fractured posterior arches. It must be borne in mind that there are fractures of the odontoid which are incapable of being shown radiographically, and even the best series of standard radiographic projections and multi-directional tomography fail to show these lesions.

More dramatic are the fracture dislocations of a very unstable nature known as the 'Hangman's fracture'. Vertical compression fractures of the atlas as in Jefferson's fracture occur when the lateral masses are driven apart. Whiplash injuries because of the alternating forces on both the anterior, posterior and the small ligaments of the atlas and axis may produce dislocations with apparent fractures and only by careful standard projections and the application of the appropriate radiological measurements can a diagnosis be given with any degree of assurance. Again the reader is referred to Clark's *Positioning in Radiography* (1973) for standard projections as further discussion will concentrate on the use of tomo-graphic techniques.

Linear tomography has limited use since this produces 'linear streaks' arising from the bones of the skull and the mandible and increasing the angle of tomographic movement exacerbates this

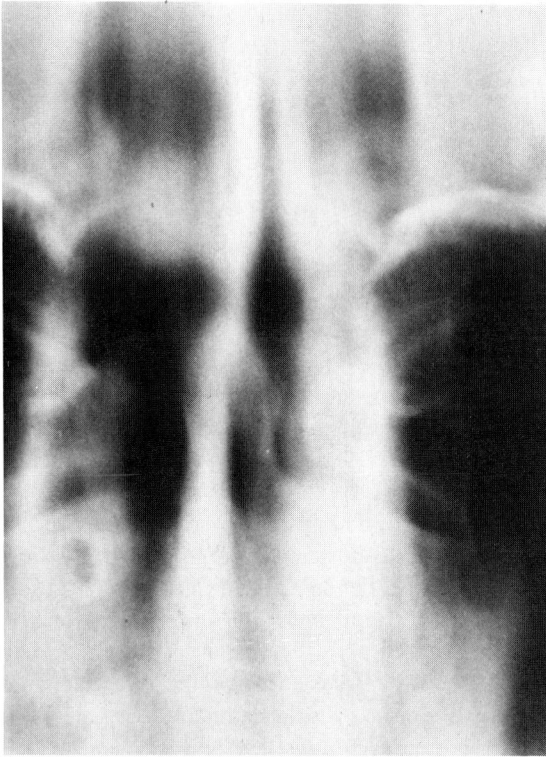
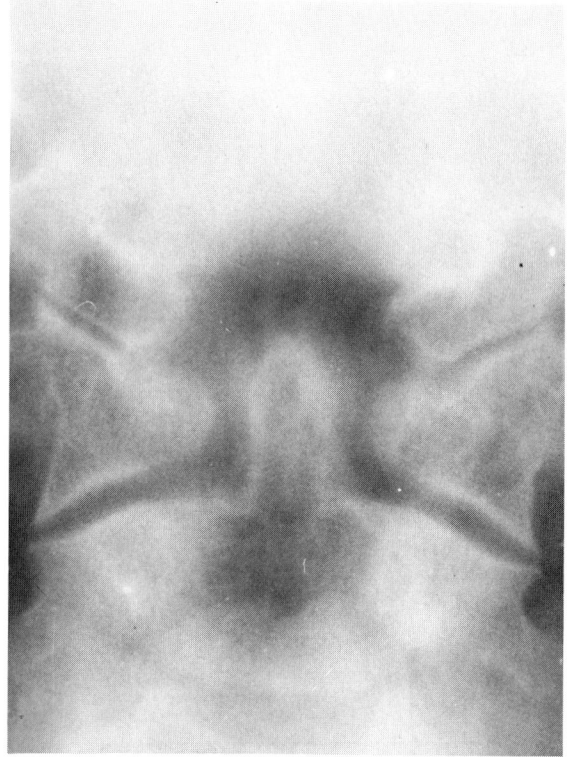

A

B

**Fig. 1.18** Tomography of the atlas and axis. (A) Linear tomography 40° angle on a tomographic attachment. (B) Circular tomography using a dedicated multi-directional tomographic unit. The superior blurring efficiency, lack of linear streaking and superior radiographic definition make for simple identification of the following: (1) the odontoid (2) atlanto-axial joints. (3) atlanto-occipital joints.

'linear streaking'. The usual alternative is to undertake multi-directional tomography, circular or spiral movements being enough to give a sufficient blurring factor to lose the bony structures above the atlas and axis. A 20° circular angle is required. If there is a diagnostic requirement to show the alignment of these two vertebrae and the atlanto-occipital joints, care must be taken to position the skull without any rotation whatsoever. If the interest is primarily of the odontoid itself, then small angle circular tomograms are useful so that a greater section of the odontoid is shown in the radiograph.

McInnes (1957, 1964, 1973) did suggest an alternative technique for those departments not having a multi-directional tomographic unit. He proposed that an asymmetrical linear tomographic movement be used. The full swing of the tomographic unit is not used with a symmetrical angle on either side of vertical. On some linear tomo-

graphic units the same angle can be used but adjusted asymmetrically around the vertical plane with its greater angle away from the head. Otherwise the patient must be placed on the table with the feet towards the start of the tomographic swing with careful use of exposure timer and not other devices which may determine the cessation of exposure by the tomographic unit. The time of exposure is selected so that the X-ray tube ceases to be energised at the vertical part of the swing. In this way difficult redundant images of the skull and mandible are not superimposed on the affected area and nearly all linear streaking is averted.

## Soft tissue radiography of the neck

The two occasions when soft tissue radiography is important are:

1. Initially at the time of injury when soft tissue

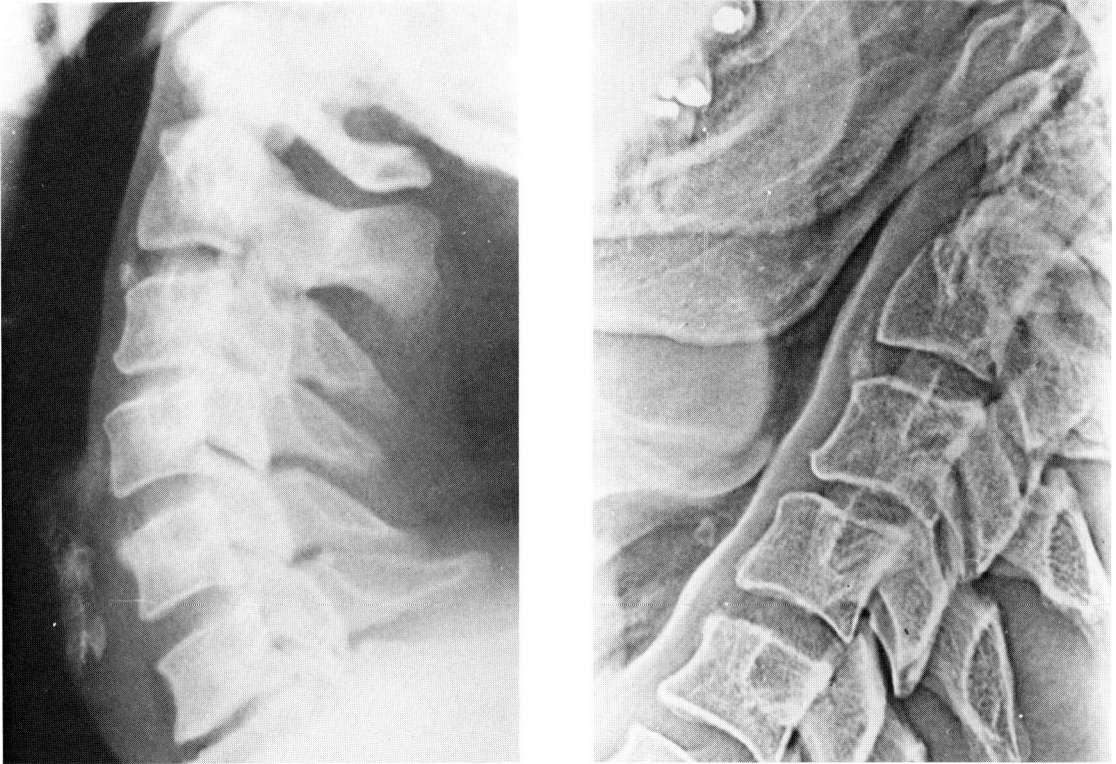

A                                                    B

**Fig. 1.19** Comparison of soft-tissue radiography and xero-radiography. (A) Soft tissue radiography showing new bone between C2 and C3. (B) Xero-radiogram. This gives edge-enhancement with greater clarity of soft tissue shadows.

swelling at cervical vertebra 2 to cervical vertebra 4 level may restrict the hypopharyngeal airway and is an indication of serious trauma to the anterior spinal ligament.

2. At a time varying from 6–9 weeks when a cloud of new bone forms anteriorly to the vertebral body which can be seen on the lateral radiograph. The importance of this is that it may be the first sign that the spine is becoming stable and that skull traction may be dispensed with in a short time and a more active nursing and rehabilitation regime may be commenced.

One of the image enhancement techniques to be employed successfully has been xero-radiography. A comparison in Figure 1.19 shows the differing edge contrasts firstly on a radiograph having a low kV penetration and low contrast and secondly a xero-radiogram (normally a blue tinted print). It must be borne in mind that xero-radiography is a high kV technique and if the X-ray unit cannot produce the appropriate kVs, the only choice is to undertake low kV soft tissue radiography. In xero-radiography the selenium plates once charged are very susceptible to slight pressures which produce artefacts on the resulting print and this alone detracts from its use in ward radiography.

A comparison of exposure factors using a 6 pulse generator is as follows:

1. Low kV soft tissue radiography
   60 kV 20 mAs 1 metre focus film distance
   High definition screens and fast film
2. Xero-radiography
   125 kV 32 mAs 1 metre focus film distance
   Charged selenium plate

### Radiography of the cervical spine in flexion and extension

Radiographs may be made on the ward with the

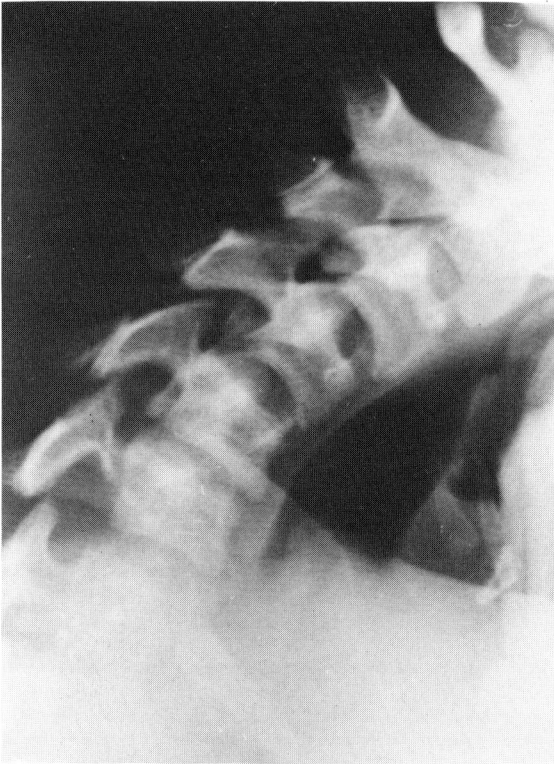

**Fig. 1.20** Spine in flexion. Flexion radiograph of the cervical spine in the lateral oblique position.

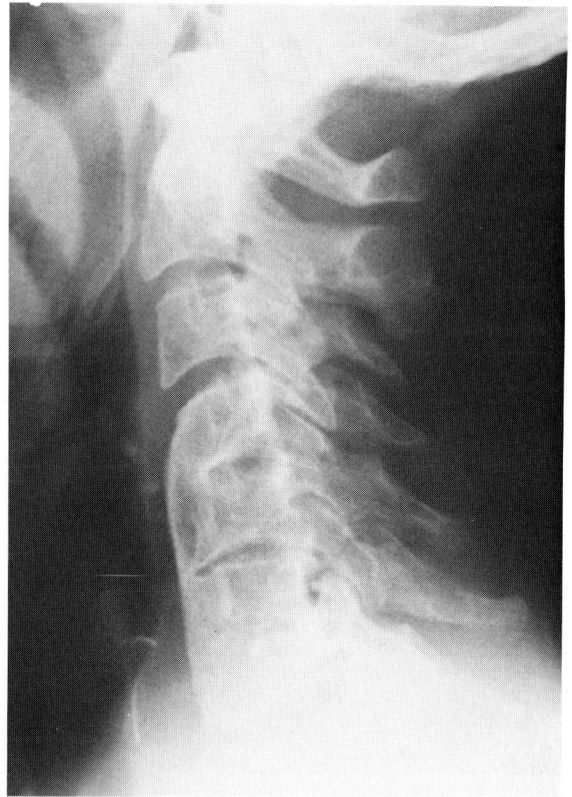

**Fig. 1.21** Patient with fracture C4 and C5 as well as congenital fusion of C6 and C7. Film taken 2 years after accident.

cervical spine first in flexion and then in extension. The procedure is as follows:

a. If the bed has a detachable headrest the patient is aligned in such a way that the shoulders are level with the top of the bed, the head being supported at all times, and especially so if the patient has been taken off traction for this examination. The neck is allowed to go into extension over the end of the bed whilst the head is lowered and supported by a clinician. Films are exposed in the appropriate lateral projection.

b. If the headrest on the bed cannot be removed, then the patient needs to be lifted in one piece to allow a firm lumbar pillow to be placed under the shoulders giving room for the neck to be extended and the head lowered as before, and films are exposed in the appropriate lateral projection.

c. The patient position for films in flexion are

the same, irrespective of the type of bed. A Zimmer halter is placed around the patient's head or a firm hold is taken of the traction bar and the patient's head is lifted so that the neck is flexed and some of the weight of the patient is taken up by the muscles of the neck. In this position it is more likely that the arms need to be pulled towards the feet in order to project the cervical spine clear of the shoulder masses. The appropriate lateral projections are taken.

*Patients with cervical spondylosis and ankylosing spondylitis*

Degenerative changes in the intervertebral discs lead to secondary changes in the vertebrae immediately above and below the affected part. These changes themselves can affect nerve roots and

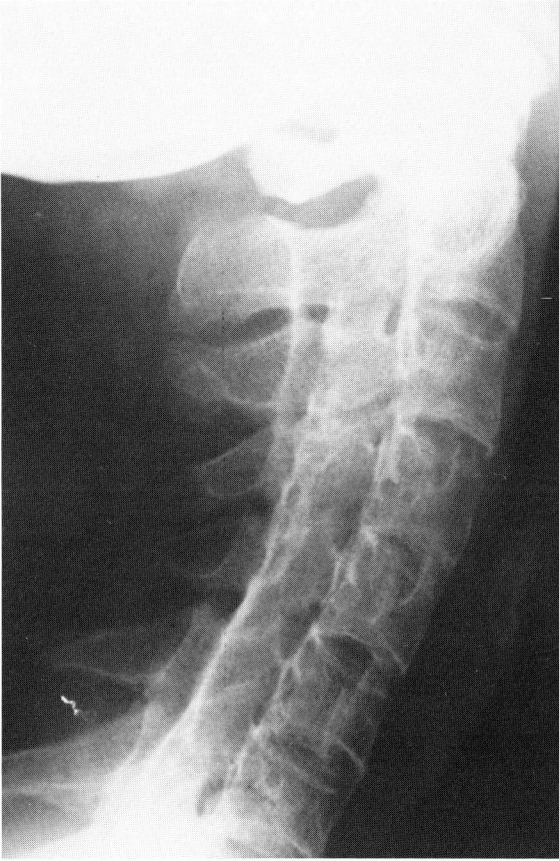

A

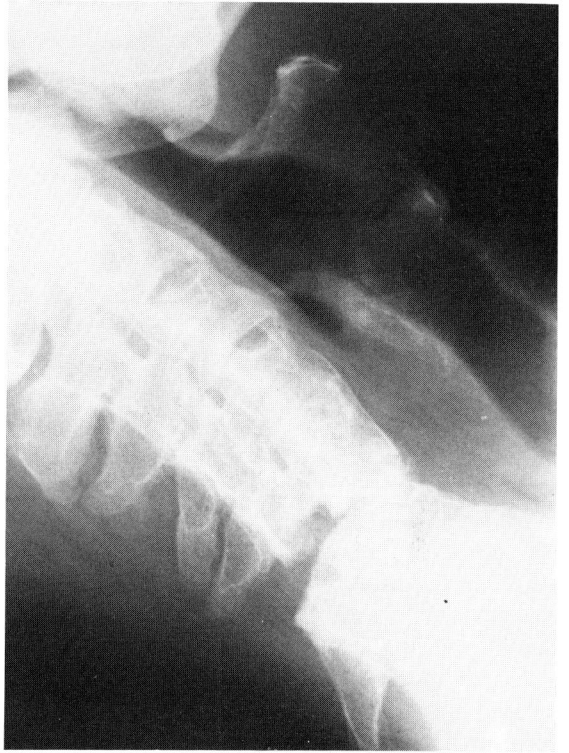

B

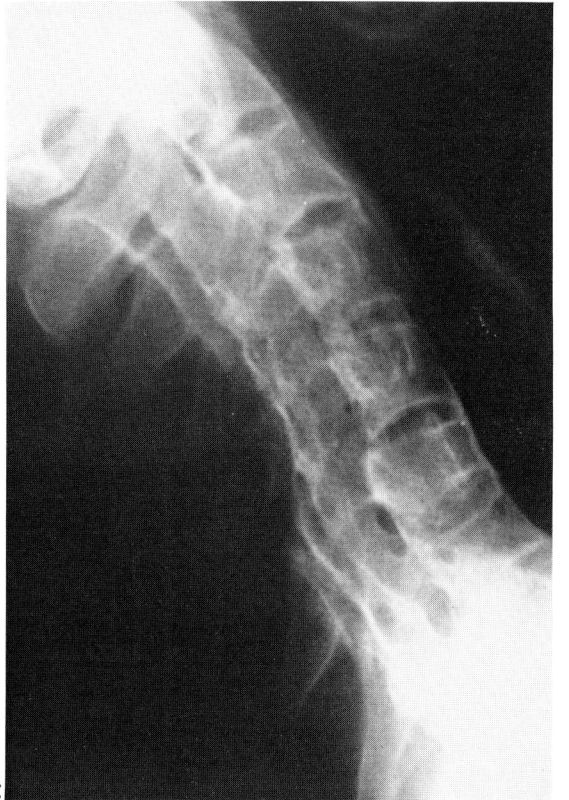

C

**Fig. 1.22** Fracture of the cervical spine in a patient suffering from ankylosing spondylitis. (A) Pre-injury film. (B) Immediate post-injury film. (C) Late film 1 year after the accident.

Note the way in which the spine has fractured completely across. Healing has taken place as if the spine were a long bone.

damage the cord. Changes in the sagittal diameter of the cord, changes in the dendate ligament reduction in adequate vascularisation of the cord all play a part in producing symptoms. Functionally the importance of a normal disc in supporting the skeletal column, in providing strength and absorbing transmitted force from trauma is immense.

In these patients a combination of osteophyte formation, root sleeve and cord adhesions, and the formation of posterior bars and bosses associated with affected intervertebral discs take away the 'slack' which normal young adults have in being able to absorb minor trauma to the neck. The same injury to a young person, which produces at most, a root lesion, in the case of the spondylitic patient inflicts severe cord damage and tetraplegia often with no radiological evidence of bony involvement. These patients are particularly susceptible to whiplash injuries and hyperextension injuries.

Braunstein et al (1980) showed that progressive degeneration and development of osteophyte formation continued some time after anterior fusion although these progressive radiographic findings did not correlate with clinical findings. These changes occurred above a lower cervical fusion and below an upper cervical fusion. He found that these changes were not apparent in patients with congenital fusion of the same vertebrae and it appeared that the patients' immediate adaption to stress starts off these progressive changes and are probably due to degenerative spondylosis. Braunstein et al go on to suggest that a long-term study of patients who have had anterior fusions for the treatment of unstable spines following trauma should be undertaken. No such changes have occurred in those patients who have had anterior fusions and long-term follow-up films, but insufficient numbers of these patients have been examined.

## RADIOGRAPHIC TECHIQUE

In these patients lateral radiography suffices and patients may be nursed with or without skull traction. Many of the patients are referred after falls ranging in degree from a fall downstairs to a fall off a chair, and in one case, simply walking into a door. Patients who are suffering from ankylosing spondylitis present the radiographer with very difficult problems in positioning for radiographic techniques. The fixed lordosis of the spine makes visualisation of the whole cervical spine difficult in the lateral, and impossible in the antero-posterior view. Gentle traction on the arms forwards will pull the shoulder masses away from cervical vertebra 5 to dorsal vertebra 1 area and the use of appropriate kilovoltages with a stationary grid is to be recommended.

Often no amount of X-ray tube angulation will allow the shadow of the mandible to be projected

clear of the cervical spine and one is left with the lateral projection only or very time-consuming multi-directional tomographic examinations for position and follow-up radiography. Because of the acute angle of the spine to the plane of the tomographic cut, anatomical interpretation can be made impossible. Great care in the selection of weights, heights of the pulley, back support of the patient is required to prevent too much distraction on the two parts of the spine. Fusion of all the vertebral bodies act like and heal like a long bone.

## Diagnostic procedure involving myelography and discography

Whatever care is taken in the handling of these patients when undergoing plain radiography this care needs to be tripled when patients with cord injuries undergo myelography in the prone position with the neck in hyperextension. A number of patients re-dislocate and others increase their cord lesions after myelography. If the investigation is undertaken too soon after injury, oedema of the cord prevents contrast media from reaching the site of interest and the contrast medium is held one segment below the site of injury (Fig. 1.23).

Discography, on the other hand, is a much safer investigation since the patient is radiographed in the supine position, and there has been some re-thinking of its usefulness especially in young patients with posterior disc herniations although it still remain an infrequent investigation in this group of patients. Chester (1969) found that this particular type of abnormality is a very infrequent cause of cord damage.

## Transverse axial tomography

Transverse axial tomography is a technique which has been available to departments for some years but has never been successfully introduced into the United Kingdom on any large scale mainly because of the difficulties in producing radiographs of sufficient contrast and the thin sections required in the spine. In the same way that conventional tomography has been refined in a number of centres, transverse axial tomography has been developed particularly in Japan. Here the Toshiba

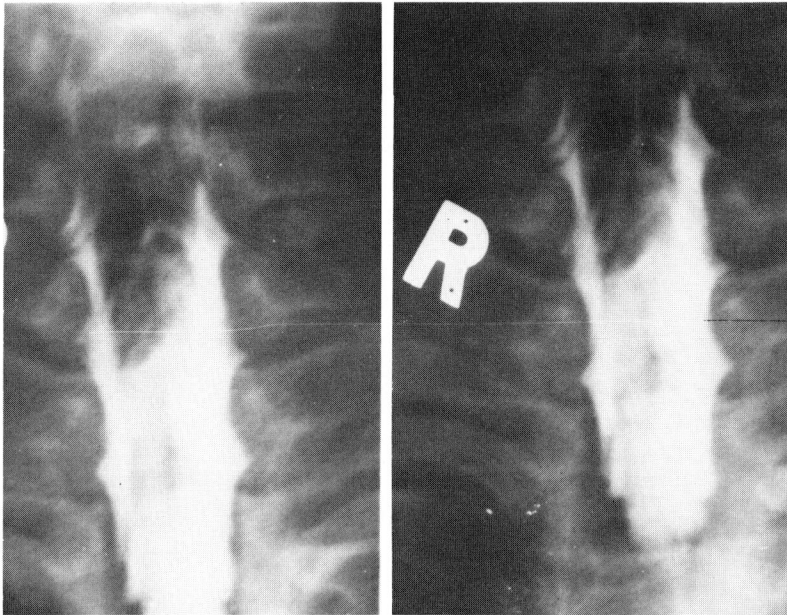

**Fig. 1.23** Cervical myelogram. 2 days post-injury. Oedematous swelling of the cord blocking egress of contrast medium and limiting visualisation to one full segment below the level of the bony injury.

system has produced transverse axial body sections of very high quality, especially under the direction of Matsadu (1976) and Takahashi (1969). Gorgano (1974) demonstrates its value in the cervical spine. Until image enhancement became available the very best transverse axial tomograms were superior to CAT scanning in terms of bony detail within the neuro-foramina and the delineation of trabeculae.

*The future application of computed tomography of the spine*

Even with the limited definition of the early whole body CAT scanners and the superior definition but limited use of the head CAT scanners it was apparent that computerised transverse axial tomography would play a major role in the radiology of the spine. The number of patients in the United States undergoing spinal investigation on these units has dramatically increased in the past 2 years so that now in a large number of centres over half of all referrals have spinal problems.

Computerised axial tomograms show excellent and clear information of the anterior arch, lateral masses, transverse processes and the posterior arch in their axial section (Donovan Post 1980). The major problems relate to:

1. The availability of CAT scanners in the hospital where the patient is referred following injury for the subsequent movement of the patient by road or helicopter is a major undertaking and puts the patient at risk. The provision of Mobile units as used in the United States is very dependent on the availability of roads capable of taking the very large trailers needed and relief from the very high insurance rates required for these vehicles.

2. The precise radiographic positioning required, the considerable handling and the moving of the patient needed to place him/her in the scanner before, during and after the investigation means that medical and nursing supervision is essential. The need to obliterate upper cervical lordosis by the use of firm pads under the neck in a fixed gantry is a considerable problem since this produces a flexion force on the cervical spine.

Lee et al (1978) outlined specific parameters for diagnostic uses in spinal radiography related to the

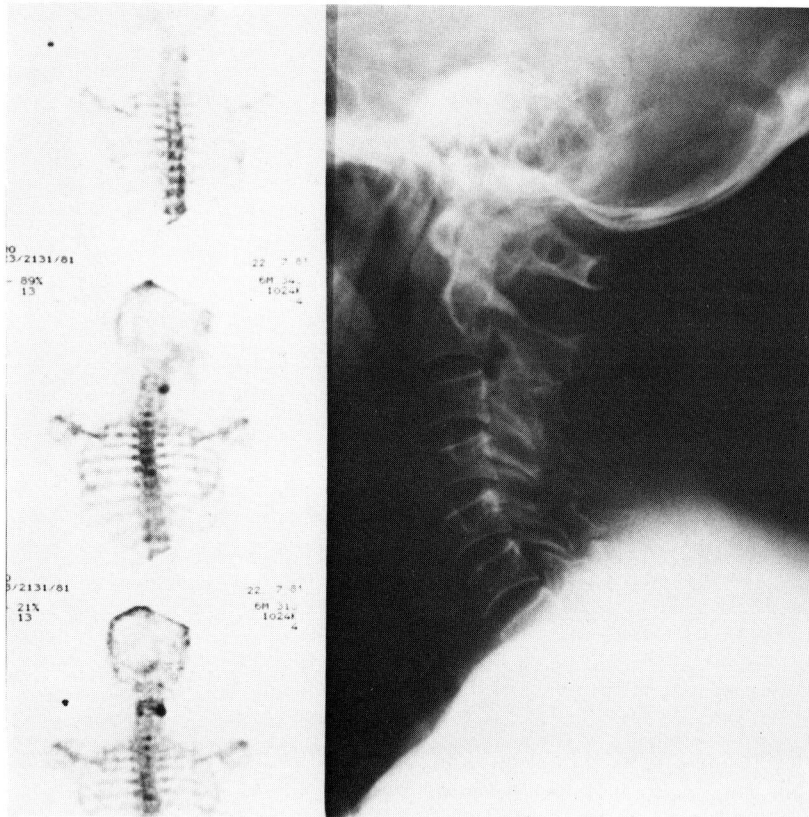

**Fig. 1.24** Value of radio-isotope imaging. A patient with a hidden pillar fracture at cervical vertebra 3 level which was clearly shown on the gamma camera scan using technetium 99$^m$ methyldiphosphonate.

scanners in use at that time. Donovan Post (1980) showed the impact of improved techniques and the upgrading of hardware and software technology.

The most useful 'state-of-the-art' improvements which Roub (1979) argued were necessary for further progress are now available and are as follows:

1. Gantry which has a ± 15° tilt as an aid to positioning.
2. Image enhancement and retrospective analysis for improved resolution.
3. 'Scout-view' capability to give greater reproducibility and selective angulations.
4. Reconstruction of transverse axial slices at a different angle to that scanned.
5. A dynamic scanning capability.
6. Digitiser software packages for subtraction techniques.

With improved image reconstruction the need for opaque media enhancement will be further diminished thus allowing a higher throughput of patients and the radiography of patients who would be at risk with metrizamide injections.

The improvement in recording systems and their ability to differentiate between cord and nerve root within the spinal canal by choosing the appropriate 'window' will add to the knowledge of the effects of trauma on the cord, especially in those patients suffering from spondylitis. The demonstration of uni-facet dislocation will also provide new information.

### The value of radio-isotope in undetected cervical spine injuries

Pillar fractures may be impossible to demonstrate on routine radiographic projections and at best the results may be equivocal on tomography of the suspect area. The injection of Technetium-99m methyl diphosphonates and subsequent radio-isotope bone scanning at 4 hours will indicate hot spots at small fracture sites in the cervical spine (Fig. 1.24).

Subsequent xero-radiograms of the area did reveal radiological evidence of fracture. This method of fracture imaging is very dependent on the improved quality achieved by gamma cameras dedicated to orthopaedic imaging. The other use of this technique could well be in the monitoring of the healing of fractures and especially in those patients who are difficult to demonstrate radiologically (Hughes 1980).

## CONCLUSION

In spite of much research, damage to the cord in severe trauma to the neck is still irreversible for it is thought that cord regeneration is still some 30 years away, and micro-surgery which may suggest ways of bypassing the injured area by micro-electronic circuitry is also a technique of the future. Modern attitudes to rehabilitation and more effective control of bladder dysfunction means that the future life of these unfortunate patients need not be as debilitating as in the past. Without this, there seems little point in extending life-spans after such accidents from the 3 years possible in the mid-1930s to the 20 or more years achievable today, if that life becomes nothing more than an intolerably long wait for death's relief.

The major emphasis has been upon achieving the maximum possible standard of life and in an attempt to stimulate research into ways and means of finding a cure for paraplegia, a charity foundation in the United States of America has put forward a 1 million dollar prize for the first research group to find an effective cure.

*Acknowledgements*

Acknowledgement is made to Dr P. Greck and Sir Thomas Lodge, formerly Consultant Radiologists to Lodge Moor Hospital, Sheffield, and to Dr I.W. McCall, Consultant Radiologist to the Robert Jones and Agnes Hunt Orthopaedic Hospital, Oswestry.

This chapter is dedicated to the memory of Dr W.M. Park, formerly Consultant Radiologist to the Robert Jones and Agnes Hunt Orthopaedic Hospital, Oswestry.

REFERENCES

Braunstein E M, Hunter L Y Bailey 1980 Long term radiographic changes following anterior fusion. Clinical Radiology 201–203

Chester D J E 1969 The stability of the cervical spine following conservative treatment of fractures and fracture-dislocations. Paraplegia 7 (3): 193–203

Clarke K C 1973 Positioning in radiography, vol IX Section 6 — Vertebral column. Heinemann, London

Donovan Post 1980 Radiographic evaluation of the spine. Current relevances with emphasis on computed tomography. Masson USA, New York

Evans D K 1961 Reductions of cervical dislocations. Journal of Bone and Joint Surgery 43B: 552–555

Gerlock A J, Kirchner S G, Hellar P M, Kay J J 1978 Advanced exercises in diagnostic radiology. II The cervical spine in trauma. W B Saunders, London

Hopcroft A J et al 1977 Potentially fatal asphyxia following a minor injury to the cervical spine. Journal of Bone and Joint Surgery 59B: 93–94

Hughes S 1980 Radionuclides in orthopaedic surgery. Journal of Bone and Joint Surgery 52B (2): 141–150

Jefferson G 1920 Fracture of the Atlas vertebra. British Journal Of Surgery. VII (27): 207–221

Lee B C P, Kazam E, Newman A D 1978 Computed tomography of the spine and spinal cord. Radiology 128: 95–102

Lodge T, Higginbottom E 1966 Fractures and dislocations of the cervical spine. X-ray Focus 7 (2): 2–9

Matsuda T 1968 Moving filter applied to axial transverse tomography. Tohoku Journal of Experimental Medicine 94: 163–167

McCall I W, Park W M, McSweeney T 1973 The radiological demonstration of acute lower cervical spine injury. Clinical Radiology 31: 49–53

McInnes J 1959, 1964 and 1973 Tavistock House, Ilford Ltd., Personal communication

Medical Research Council 1924 Report No. 124

Munro D 1962 In: French J P, Porter R W (eds) Basic research in paraplegia 210. Thomas, Springfield, Illinois

Park W M 1981 Personal communication

Penning L 1981 Prevertebral haematoma in cervical spine injury. American Journal of Radiology 136: 553–561

Piercey M J 1981 Motion studies. The painful lumbar spine symposium. Robert Jones and Agnes Hunt Orthopaedic Hospital, Oswestry

Reed J E, Allen W E, Dorman G J 1979 Effect of mannitol on the traumatised spinal cord. Spine 4 (5): 391–397. Harper & Row, Philadelphia

Roaf R 1960 A study of the mechanics of spinal injuries. Journal of Bone and Joint Surgery 44A: 810–823

Roub L W, Drayer B P 1979 Spinal computed tomography: Limitations and applications. American Journal of Roengenology August: 267–273

Takashashi S 1969 An atlas of transverse tomography and its clinical application. Springer-Verlag, New York

White A A, Johnson R M, Panjabi M M 1971 Biochemical analysis of clinical stability. Journal of Bone and Joint Surgery 58B: 322–327


BIBLIOGRAPHY

Albright J P, Moses J M, Feldick H G, Dolan K D, Burmeister L F 1976 Non-fatal cervical spine injuries in interscholastic football. Journal of the American Medical Association 236 (11): 1243–1245

Beatson T R 1963 Fractures and dislocation of the cervical spine. Journal of Bone and Joint Surgery 45B: 21–35

Binet E F, Moro J J, Marangola J P, Hodge C J 1977 Cervical spine tomography in trauma. Spine 2 (3): 163–171

Braakman R, Penning L 1971 Injuries to the cervical spine. Excerpta Medical Foundation

Braakman R, Penning L 1973 Mechanism of injury to the cervical cord. Paraplegia 10: 314–330

Brown R H, Burnstein A H, Nash C L, Stock C C 1976 Spinal analysis using three dimensional radiographic technique. Journal of Biomechanics 9: 355–365

Chesney M O 1952 Radiography of the cervical spine and upper thoracic vertebrae. Radiography XVIII (216): 243–251

Cook J B 1958 Radiography in spinal cord injury. Radiography XXIV (277): 12–14

Dohrmann G I, Wagner F C, Bucy P C 1971 The micro-vasculature in transitory traumatic paraplegia — An electron microscope study. Journal of Neurosurgery 35: 263–276

Ducker T B, Kinot G W, Kempe L G 1971 Pathological findings in acute spinal cord trauma. Journal of Neurosurgery 35: 700–707

Fairholme D J, Turnbull I M 1971 Micro-angiographic study of experimental cord injuries. Journal of Neurosurgery 35: 277–285

Forsyth J 1953 Radiography in traumatic paraplegia. Radiography XIX 21: 93–100

Guttman L 1966 Traumatic paraplegia and tetraplegia in ankylosing spondylitis Paraplegia 4:63

Guttman L 1978 Spinal cord injuries — Comprehension, management and research. W B Saunders, London

Hardy A J, Rossier A B 1975 Spinal cord injuries. Thieme, Stuttgart

Harris J H 1979 The radiology of acute cervical spine trauma. William and Wilkins, Baltimore

Herodotus. History Trans. Macauley cit. Hussein

Holdsworth F W 1963 Fractures, dislocations and fracture-dislocations of the spine. Journal of Bone and Joint Surgery 45B: 6–20

Jeffreys E 1980 Disorders of the cervical spine. Butterworths, London

Jones B F 1981 Personal communication

Kattam K R 1975 Trauma and non trauma of the cervical spine. Thomas, Springfield, Illinois

Lazorthes G, Gouvaze A, Zedeh J D, Santani J J, Lazorthes Y, Burdin P 1971 Arterial vascularisation of the spinal cord. Journal of Neuro-surgery 35: 253–261

Lodge T 1963 Radiology in the management of paraplegia. Clinical Radiology XIV (4): 365–380

Lodge T 1976 Radiodiagnosis in patients with spinal injuries. In: Vinken P J, Bruyn G W (eds) Handbook of clinical neurology, Vol 26. Injuries of the spine and the spinal cord. Part II Chapter 15. North Holland Publishing, Amsterdam

McKenzie G A 1973 A review of myelography. Radiography XXXIX: (460): 79–91

McNab I 1971 The whiplash syndrome. Symposium on disease of the cervical spine. Orthopaedic Clinics of North America 2 (2): 389–327

McSweeney T 1971 Early management of associated injuries. Paraplegia 9:137

Nissen K I 1942 Proceedings of the Royal Society of Medicine 35: 707

Park W M, McCall I W, McSweeney T, Jones B F 1976 Cervicodorsal injury presenting as a sternal fracture. Clinical Radiology 27: 335–340

Selvig G, Olsson T H, Willner S 1976 High accuracy analysis of movements of the spine with the aid of roentgen stereo-photogrammetric method. In: Komi P U (ed) Bio-mechanics U. B. 502–507. University Park Press, Baltimore, Maryland

Sher A T 1979 Anterior cervical subluxation; An unstable position. American Journal of Roentgenology August: 275–280

Von Torklas D, Gehle W 1972 The upper cervical spine. Grune and Stratton, New York

Wagner F C, Dohrmann G I, Bucy P C 1971 Histopathology of transitory traumatic paraplegia in the monkey. Journal of Neurosurgery 35: 263–276

Webb J K, Broughton R B K, McSweeney T, Park W M 1976 Hidden flexion injuries of the cervical spine. Clinical Orthopaedics and Related Research 109: 85–96

Wilkinson M 1971 Cervical spondylosis. W Heinemann, London

Yule A 1972 Two aspects of a road traffic accident. Radiography XXXVIII (450): 145–150

**2**

# Tomography of the petrous bone

## TOMOGRAPHIC TRENDS AND DEVELOPMENT

50 years ago tomography was first performed, using the simple movement of the jaw to demonstrate the cervical spine more easily. The lateral dorsal spine was shown when using a long exposure during which time the ribs were blurred out by the gentle breathing of the patient, and autotomography was used during an air-encephalogram examination to show the 4th ventricle.

With modern day specialized equipment, the tomographic examinations are more easily performed; nowadays the patient keeps still and the X-ray tube and film move instead.

Tomography has greatly assisted the radiologist in his diagnosis, especially when the specific area is in the middle ear. As the ossicles are so small in size, tomography removes the overlying shadows on the radiograph. Details of the anatomy are given in this chapter prior to the radiographic section to illustrate why the different positions of the patient are required to show more clearly the particular area under investigation.

## ANATOMY

### Development

The organs for the perception of balance and sound are situated within the cochlea, semicircular canals, saccule and utricle, which are formed within the membranous labyrinth, within the petrous bone. The vestibule is formed by the cochlea, saccule and utricle. There are eight ossific centres which form the cochlea, semicircular

*For key to numbers on Figures throughout chapter, see p. 49.*

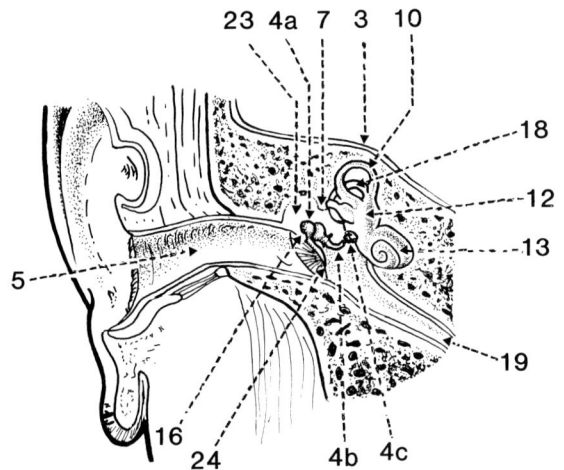

**Fig. 2.1** Petrous bone (not to scale).

canals and vestibule, and also the bony capsule but the ossicles, situated in the middle ear are developed from cartilage. A mucous membrane covers the ossicles. The ear drum, external auditory meatus, tympanic antrum and Eustachian tubes are eventually formed within a branchial arch which develops posteriorly from the pharynx.

This bony capsule, with its membranous labyrinth forming the inner ear, is fully developed at birth and is the only structure to be fully formed then. The structure of the internal part of the petrous bone is cancellous at birth, but becomes dense later on. The formation of the air cells in the mastoid section begins to form at birth.

Both petrous parts of the temporal bones are shaped like a pyramid and are situated in the base of the skull, one either side. Most of the petrous bone, where the middle and posterior fossae are situated, is within the base of the skull. The posterior part of the middle fossa is formed by one

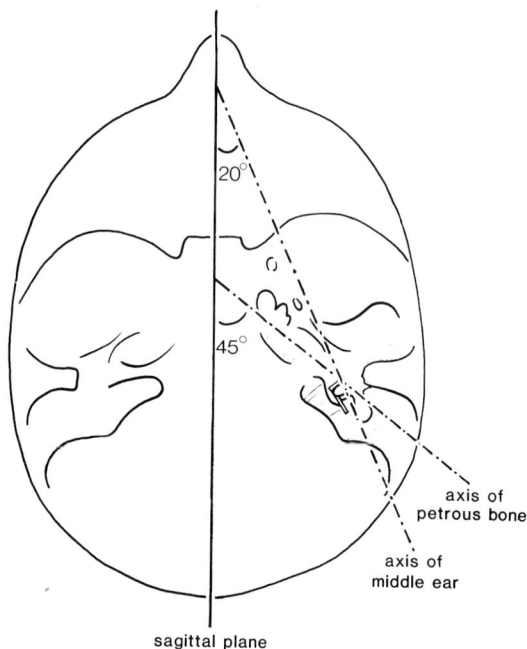

**Fig. 2.2** Relationships of the petrous bone and middle ear ossicles to the sagittal plane of the skull.

surface of the petrous bone; the anterior limit of the posterior fossa is formed by the second surface; the third petrous bone surface is projected on to the inferior aspect of the skull. The petrous part of the temporal bone houses important blood vessels and nerves, as well as the parts of the hearing organ all within a small area (see Fig. 2.2).

## Geometry

The axis of the petrous bone, to the sagittal plane of the skull, is approximately 45°, each bone being situated in about the same plane in the skull. The co-ordinated functions of the semicircular canals are influenced by these angles. The angle of the main axis of the middle ear is 20° to the sagittal plane of the skull. These two angles therefore influence the positioning for any planned radiographs.

The shape of the petrous bone resembles a three-sided pyramid, the base of which lies on the lateral aspect of the skull and the summit points anteriorly and medially within the skull. One surface of the petrous bone forms part of the lower aspect of the skull, the other two surfaces being

within the skull itself. Of the two internal surfaces, one faces posteriorly forming the anterior part of the posterior fossa, and the other lies anteriorly and superiorly within the posterior section of the middle fossa. The two petrous bones, one situated at each side of the base of the skull, are wedged between the greater wing of the sphenoid and the occipital bone. The body is adjacent to the occipital bone and the apex is beside the body of the sphenoid.

The length of the external auditory canal, from the opening of the external auditory meatus to the conical shaped eardrum which is membranous, is about 2.5 cm. There is a narrow cavity containing the three tiny ossicles — the malleus, incus and stapes — which are situated between the internal ear and the tympanic membrane. This is the tympanic cavity or middle ear, which is within the petrous portion of the temporal bone. The malleus and incus are in the epitympanic recess, which is above the level of the external auditory meatus. Fibres connect the shaft of the malleus to the upper part of the ear-drum. The head of the malleus and the head of the incus, which articulate with each other, are covered by mucous membrane. The bony process of the lower part of the incus joins the smallest ossicle, the stapes, which is situated inferiorly and posteriorly, to the chain of the ossicles. The stapes lies horizontally, having its base facing the oval window on the medial wall of the tympanic cavity, to which it is connected by fibres. There is also a membranous covering of the stapes. The sound vibrations are conveyed from the ear-drum to the cochlea via the ossicles.

The lateral wall of the tympanic cavity is mostly formed by the ear-drum and the spur, and the lateral part of the epitympanic recess also forms part of the lateral wall. The lateral semicircular canal bulges out of the medial wall. The tympanic bone and the descending process of the tegmen tympani combine to form the anterior wall of the tympanic cavity. A small opening joins the posterior part of the cavity with aditus, the tympanic antrum and the air cells. The Eustachian tube continues from the anterior section of the cavity. The superior limit is formed by the tegmen. A horizontal line, which is drawn from the spur tip, is the limit, inferiorly.

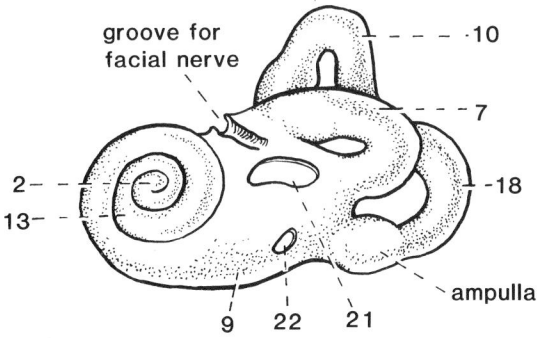

**Fig. 2.3** Cochlea, vestibule and semicircular canals.

The internal ear has the labyrinth, which contains the vestibule in the centre, communicating with the cochlea anteriorly and the three semicircular canals posteriorly. Medial to the middle ear is the cochlea in the shape of two and a half coils, like a snail's shell. A bulge, called the promontory, is formed by the first, largest coil, in the section where it joins the middle ear in the vestibule. This promontory can sometimes be seen on plain radiographs.

The basal diameter of the cochlea is 9 mm and it lies medially and posteriorly. The height of the cochlea is 5 mm, having the apex (or cupola) lying laterally and anteriorly. The axis of the cochlea is at 90° to the pyramid axis.

Situated between the tympany and the cochlea is the fenestra rotundum (or round window). The fenestra ovale (or oval window) has a long axis which is nearly horizontal, and it is situated only

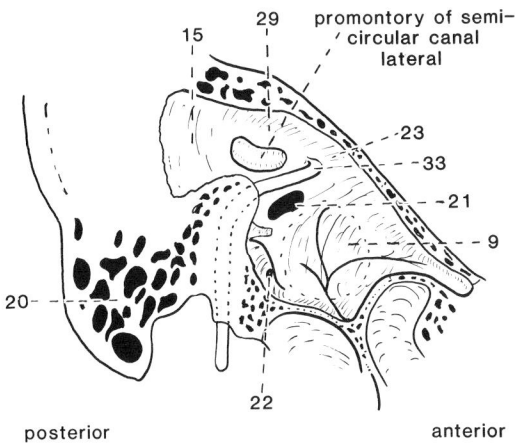

**Fig. 2.4** Medial part of the tympanic cavity showing facial nerve canal.

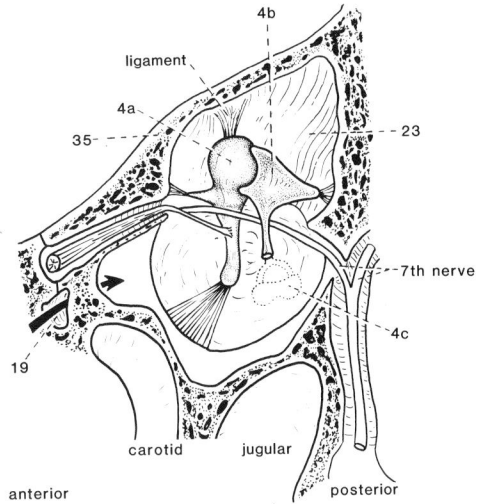

**Fig. 2.5(A)** Lateral half of the tympanic cavity.

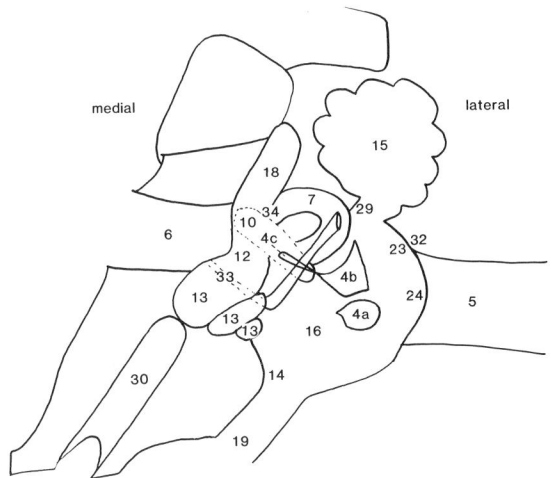

**Fig. 2.5(B)** Middle ear.

a few millimetres superior to the foramen rotundum — it supplies the communication between the tympanic cavity and the vestibule. The base of the stapes is attached by a ligament to the fenestra ovale. The vestibule is eliptical in shape and contains the interconnecting ducts of the labyrinth, the saccule and the membranous utricle. The vestibule lies immediately posteriorly to the cochlea. Radiographs demonstrate clearly the bony vestibule, which in its various axes, is 3–5 mm in diameter. There is a rectangular corner formed on the wall of the vestibule, by the semicircular canal entrances. The diameters of these canals are 4, 5 and 6 mm. The bulge formed by the lateral semicircular canal is called the prominenta which bulges out into the medial aspect of the tympanic (mastoid) antrum. The arcuate eminence is the name given to the bulge formed by the superior semicircular canal on the superior edge of the petrous bone. A figure of eight image, sometimes shown on a plain radiograph, is formed by the joint entrance into the utricle, of the posterior and superior semicircular canals.

The length of the internal auditory canal is approximately 1 cm and through it are transmitted the divisions of the eighth cranial nerve from the cochlea and vestibule. The internal auditory meatus also contains the seventh, or facial nerve, which after having passed through the meatus passes through the facial canal, which initially passes between the fenestra ovale and rotundum, lying superior to the tympanic cavity, prior to its sharp curve downwards posterior to the tympanic recess, and emerges via the stylomastoid foramen.

There is a meatus, which is almost parallel to the skull's coronal plane, in the posterior surface of the petrous bone, which is for the internal auditory canal. The cranial nerves conveyed through this canal are the seventh (facial) and the eighth (acoustic and auditory). The eighth nerve, at the lateral limit of the canal, passes to the cochlea, saccule, utricle and semicircular canals via the organ of Corti.

The crista transvera is a horizontal bony ridge which divides the superior from the inferior part of the fundus of the internal auditory meatus. This is sometimes visible on frontal tomograms.

## Nerves

Closely applied to the apex of the petrous bone are the third, fourth, fifth and sixth cranial nerves, and passing through into the petrous bone itself are the seventh and eighth cranial nerves. Lying side by side on the anterior wall of the jugular fossa, are the ninth, tenth and eleventh cranial nerves.

The canal for the facial nerve lies over the posterior and superior section of the cochlea, parallel with its axis; then at its second part it curves to lie parallel with the pyramid's axis; the third section is the vertical part forming the prom-

inentia canalis facialis in the tympanic cavity's medial wall, above the oval window and below the prominence formed by the lateral semicircular canal. Lying vertically, this third section is behind the tympanic cavity's posterior wall, and anterior to the mastoid air cells, opening posteriorly to the styloid process root, forming the stylomastoid foramen. It is here that the nerve enters the parotid canal.

## VESSELS

### Arteries

The entrance into the skull of the internal carotid artery is through the antero-medial section of the petrous bone via a canal which is parallel to the axis of this bone. It is the most important artery which supplies the cerebral hemispheres. At the apex of the petrous bone, the carotid artery enters the skull, not through the foramen lacerum, but over the top of it. The artery bends in an upward direction, close to the tympanic cavity. Immediately above the carotid canal is the upper coil of the cochlea.

### Veins

The sigmoid sinus is the most important vein of the many which are in this area. It curves over the posterior surface of the petrous bone, then on to the jugular foramen where it leaves the skull. The inferior petrosal sinus vacates the skull through the same foramen, lying on the medial side of the sigmoid sinus. The internal jugular vein, which is the major vein within the neck, is formed by the junction of these two venous channels, immediately after they have vacated the skull.

The superior petrosal sinus and the cavernous sinus are joined by the inferior petrosal sinus after it has passed downward over the posterior section of the apex portion of the petrous bone. The cavernous sinus forms the boundary between the posterior and medial fossae where it lies along the edge of the petrous bone. Within the petrous bone is a fossa for the jugular vein. The third section of the seventh nerve canal separates the mastoid air cells from the posterior wall of the tympanic cavity. The sigmoid sinus turns medially and within the jugular foramen, on the lowest section of the pars mastoidea on its medial surface, where it continues to enter the internal jugular vein. This vein is the route for the largest amount of the venous blood vacating the internal skull. (A modified sub-mento vertical view with the base line at 70° is required to demonstrate the jugular foramen on a radiograph. The foramen is shown on the dorso-lateral section of the fissure of the petrous bone, as an expanded area.)

The internal jugular vein has a superior bulbous area which lies within the jugular fossa. The floor of the tympanic cavity is formed by this fossa. The fossa then extends anteriorly to the foramen of the external carotid artery which is on the inferior section of the pars petrosa.

## Pathology and its effect on the petrous bone and skull

Tomography is required when the petrous bones have to be examined in detail to demonstrate if there is any disease, injury or malformation.

Infection from otitis media and mastoiditis, for example, can spread from the petrous bone so early diagnosis and treatment are essential. The ossicles can only be demonstrated by tomographs, which are required to see whether or not surgery can be performed and to demonstrate if there is an absence of these tiny bones, if they are malformed, or dislocated following injury. Cholesteatoma, or tumour of the middle ear, shows on a tomograph where the tumour is eroding the bony walls of the petrous bone. Acoustic neuroma, or tumour of the eighth cranial nerve, is demonstrated as an enlargement of the internal auditory canal and bilateral tomography is required to compare the size of the meatus of the petrous bones on both sides of the skull.

## Tomographic examinations

Because of the complex anatomy already described, the main structures can be demonstrated to greater advantage by means of tomography. The structures which obscure the area under examination can be blurred to a greater or lesser extent, and if there is co-ordinated movement between the X-ray tube and film, there is always one point which remains stationary, i.e. is the fulcrum. Only the areas which the X-ray beam strikes tangentially in tomography will be shown clearly on the radiograph.

## Points to be considered when using tomography

The quality of the X-ray beam is affected by the applied kVP, the type of rectification, the material of the X-ray tube target and the filtration of the X-ray beam as it passes from the target to the part under examination.

Scattered radiation is reflected from the delineating cone and there is secondary radiation from the object under investigation.

Consideration must be given to the actual contrast of both the object detail being examined, at the fulcrum level, and the structures outside this plane. The contrast may be minimal and there would be no diagnostic value in the tomograph. There is a loss of contrast on tomograms, which resembles an increase of film unsharpness. The permissible amount of unsharpness that the eye can tolerate is 0.6 mm (Medicamundi Vol II, 1966).

There are problems because of the photographic properties regarding the sensitivity, graininess and consequent unsharpness of the intensifying screens, and the film emulsion causing light diffusion, and also the gradation sensitivity and graininess of the film. The film shows a fogging with age, and the emulsion is also affected by the chemicals during the processing. There are many disadvantages to this means of examination which include irradiation of patient, increase in the cost of equipment, films and chemicals, extended length of examination time, possibility of movement of the patient between exposures. The film contrast is decreased and the equipment may have mechanical problems due to heat, X-ray tube and film judder and instability.

Such disadvantages in the X-ray tube are however outweighed by the advantages in that an increase in the diagnostic information can be obtained in some instances.

The slightest movement of the patient during the exposure can cause movement unsharpness on the radiograph. The smallest amount of movement, especially during an examination of the middle ear where the area is so small and the difference between each layer may be only 1 mm, can result in the patient having to be re-positioned and the series of tomograms re-taken.

*Magnification of the object*

Magnification on the film can be calculated by the following formula:

Magnification ratio = $\dfrac{\text{distance from X-ray tube focus to film}}{\text{distance from X-ray tube focus to pivot point (fulcrum)}}$

Using the Philips Polytome Universal Unit:

| | |
|---|---|
| the distance from focus to film | 146.5 cm |
| the distance from focus to pivot (fulcrum) | 110.0 cm |
| the distance from pivot (fulcrum) to film | 36.5 cm (37 cm) |

Therefore, the magnification using the above figures is:

$$\frac{4}{3} = \frac{1.33}{1}$$

*Layer thickness*

This can be calculated from the overall angle of X-ray tube arm swing (angle of exposure) of the Polytome Unit, using the following formula:

$$E = \frac{0.6}{\tan \frac{1}{2}\alpha}$$

where E = layer thickness in mm
alpha $\alpha$ = angle of X-ray tube arm swing

When using the hypocycloidal movement, the actual length of travel of the X-ray tube is 451 cm. This movement gives the best overall blurring of all the structures outside the selected plane, and the best prospects of obtaining a clear cross-section, because the shadows of masking shadows are then spread over a wide area.

*Protection for the eyes because of the radiation dose*

To estimate the radiation dose received by the eyes, tests were carried out on patients. Thermo-luminescent dosemeters (lithium fluoride) were positioned on the patients' closed eyelids. Above these dosemeters were placed a pair of plastic goggles, having a 2 mm lead protection fitted over each eye section.

As the hypocycloidal movement necessitates a six second exposure, and as tomography requires a number of films to be taken of the same area, the dose to the eyes, which are very sensitive to X-radiation, mounts up and can be quite high. For a six exposure series, the average radiation dose to the eye was 1.44 rads beneath the lead (or lead equivalent).

The radiographer must keep the amount of the exposures as low as possible; when using a 2 mm lead eye shield, the exposure factor has to be raised by approximately 8 kV, but the dose received by the eye is reduced to approximately one-sixth of the amount received when no lead protection is used.

**Tomography of the petrous bones**

Thin section tomography is required, on the whole, to demonstrate the ossicles in the middle ear. A Polytome Universal Unit can be used with hypocycloidal movement which is a 48° angle of swing of the X-ray tube, giving a very thin in focus layer of 0.9 mm, and a constant magnification of 1.3:1. The same cassette is used throughout the examination to counter differences in cassettes due to the mounting of the screens, the difference in fluorescence of the screens which occurs with age, and also to give the same 'mottling' effect, caused by the screen crystal on each film. The radiographs (Figs 2.6D, 2.7E, 2.9E, 2.10D, 2.11D, 2.11E, 2.12C, 2.12D) were taken using 3 M Trimax cassettes fitted with alpha 4 intensifying screens and 3 M XUD rapid processing films.

The middle ear is an air-filled cavity, with bone inside it. The tomograph demonstrated when using this equipment and movement, shows as a very thin layer only in focus. The contrast therefore tends to be very low but this is improved by covering a very small field.

The ossicles are normally demonstrated in the negative contrast of the middle ear, but pus in the same cavity, caused by otitis media, restricts this demonstration on a tomograph.

The actual sizes of the areas of the middle/inner ear shown on the tomograph are very small:

| | |
|---|---|
| crura stapes | 0.3–0.4 mm |
| manubrium mallei | 0.7 mm |
| corpus stapes | 0.9 mm |
| semicircular canals | 1.0 mm |
| saccules, utricles and a single convolution of the cochlea | 1.0–2.0 mm |

The tegmen is very thin and is not usually demonstrated on the tomograph.

To obtain a satisfactory series in the tomographic examination it is necessary to have the co-operation of the patient, which is achieved by giving an explanation of the procedure, and placing them in as comfortable a position as is possible. Sedation or general anaesthetic may have to be used, especially when examining very young children.

A 5 cm foam mattress is used and foam head pads with Velcro straps, Spiers bolus bags (40% rice flour and 60% bicarbonate) to mould around the head, and lead rubber strips to reduce the scattered radiation as much as is possible. These strips are placed on top of the bolus bags at the approximate height to the table top of the centre film of the tomographic series. Lead goggles are also used (made by the Radiotherapy Department at Cookridge Hospital, Leeds).

## RADIOGRAPHIC TECHNIQUE

Basic tomographic projections are obtained in the antero-posterior (AP) and lateral positions.

1. The AP projection shows the cochlea and vestibular planes.
2. The lateral view demonstrates the external auditory canal, the attic and ossicles, internal auditory canal and the vertical portion of the facial canal, as well as the jugular canal.

Three important cuts in this series are the areas which show the internal auditory canal, the round window and the ossicles.

Further views are:

1. Guillen's which demonstrates well the middle ear ossicles, wall of the attic, the promontory of the cochlea and the jugular canal.
2. Stenver's position is used to show the round window, posterior canal and the carotid canal.
3. The sub-mento vertical position (SMV) is awkward for many patients to achieve. This series of films should therefore be taken as quickly as is possible. It is an excellent view for demonstrating congenital malformations, fractures and tumours — both sides being shown simultaneously.
4. Pöschl's projection is at 90° to the Stenver's position. It is the ideal projection to examine the inner ear, especially the cochlea. The superior semicircular canal is parallel to the plane of these films.
5. Chausse III is one of the best projections to demonstrate the mastoid antrum, middle ear ossicles and the lateral attic wall.

*Single antero-posterior view (each side is demonstrated separately)*

The patient is placed with the skull in the true antero-posterior position. The central X-ray beam is located between the mid-sagittal plane of the skull and the external auditory meatus. The first control film is taken at the level of the crease of the tragus, which usually shows the cochlea. The initial series of six radiographs is taken, going posteriorly from the cochlea, with a 2 mm separation between each cut. This is followed by a further series of the intervening 1 mm cuts of the area required for that particular examination.

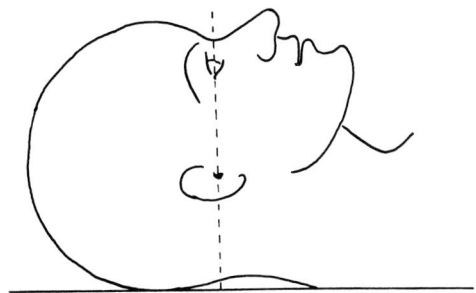

A

Fig. 2.6 (A) Position for anterior posterior. (B) Anterior posterior projection. (C) Anterior posterior tomographic section — dotted line indicates plane shown. (D) Radiographic views.

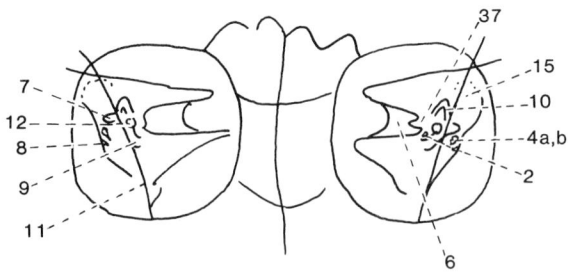

B

C

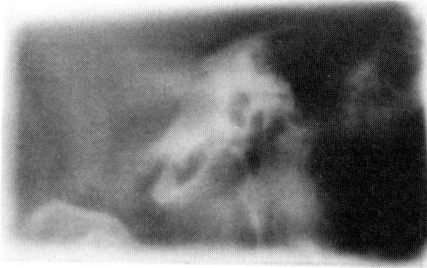

11.6

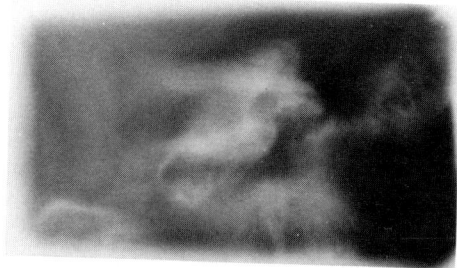

11.8

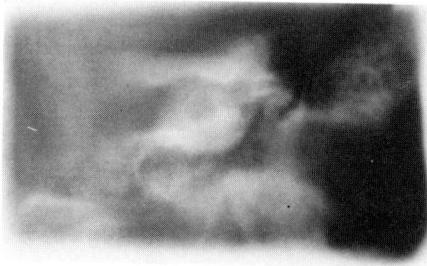

12.0

12.2

D          12.4

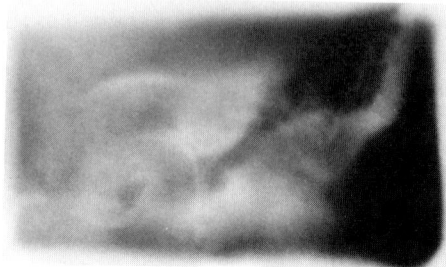

12.6

**Fig. 2.6** (contd)

*1st cut.* The areas well demonstrated on the anterior cut of the series are the spur, labyrinth, tympanic cavity with the ossicles and external auditory meatus being visualized side by side, clearly separated from each other (see Fig. 2.6D sections 11.8, 12.2, 12.4). The lateral semicircular canals are also shown. The internal auditory canal is seen in its long axis. The cochlea lies obliquely in this view, with the facial nerve canal lying superiorly and laterally to the cochlea. The vertical portion of the carotid canal lies inferiorly and anteriorly to the cochlea. The tympanic membrane is not demonstrated, but it is attached to the point of the spur and the upper part of it with the handle of the malleolus.

*2nd cut.* This shows the internal auditory canal, spur, facial nerve canal, crista transversa and mastoid antrum.

*3rd cut.* Demonstrates the superior and lateral semicircular canals just beginning to show; the oval window is shown inferiorly to the lateral semicircular canal. Immediately lateral to the oval window is the facial nerve canal, which is barely visible on tomographs, as the bone is very thin. The epitympanic recess is shown between the middle ear and the mastoid antrum.

*4th cut.* Shows the vestibule, superior and lateral semicircular canals, jugular fossa, internal auditory canal. The niche of the oval window (which is better demonstrated in a 20° oblique positon, when it is perpendicular to the film), the round window which faces infero-laterally is not very well shown in this view.

*5th cut.* Demonstrates the segments of the lateral and posterior semicircular canals; the superior semicircular canal is seen end-on as a small lucency with a rim of bone. The jugular fossa is near to the middle ear, sometimes separated from it only by a very thin piece of bone. Various parts of the descending section of the facial nerve canal, in a nearly vertical course, can be seen, opening out to end in the stylomastoid foramen.

*Bilateral antero-posterior view*

The radiographic base line and the sagittal plane of the skull are perpendicular to the film, and the interpupillary line is parallel with the table top. The central X-ray beam is centred over the nasion, and is localized to the orbits. It must be a true antero-posterior projection, with the vestibules equidistant from the innominate lines. The fulcrum is set at the level of the crease of the tragus of the ear, for the control film. The bilateral antero-posterior radiographs show a good comparison of both sides, of the internal auditory canals, the external auditory canals and the petrous tips simultaneously.

*Transorbital or Guillen's projection*

This view demonstrates the middle ear very well. The patient is positioned as for the antero-posterior projection, but with the central X-ray beam passing through the middle and outer third

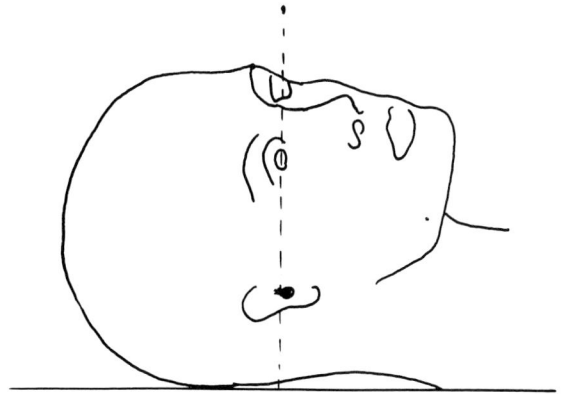

Fig. 2.7 (A) Position for Guillen's projection.

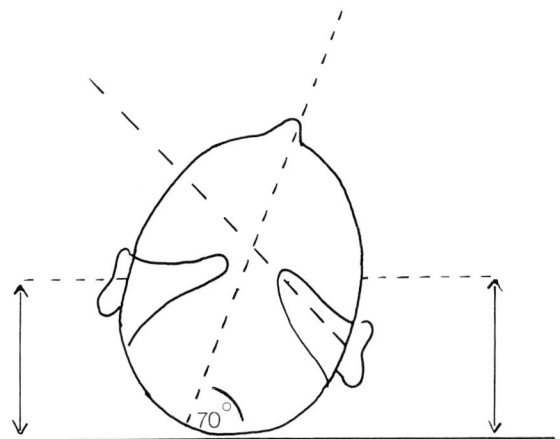

Fig. 2.7(B) Rotation of the head.

**Fig. 2.7** (C) Guillen's tomographic section — dotted line indicates plane shown.

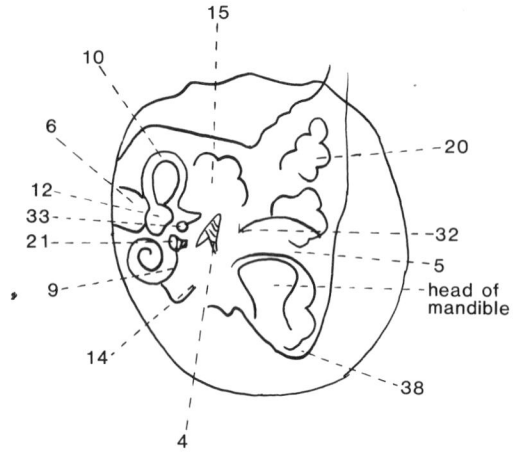

**Fig. 2.7** (D) Section guide.

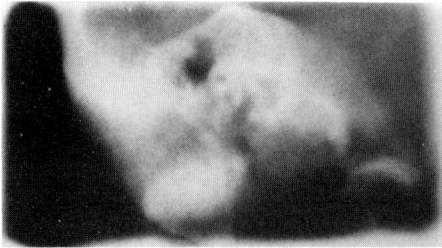

10.5

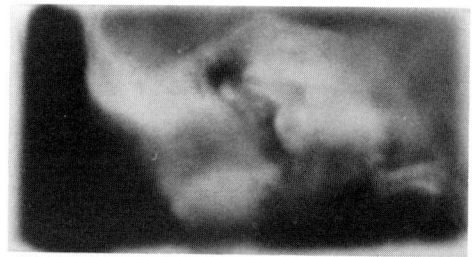

10.7

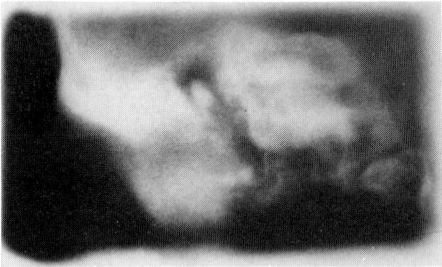

10.9

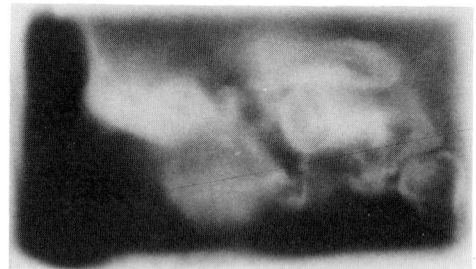

11.1

11.3

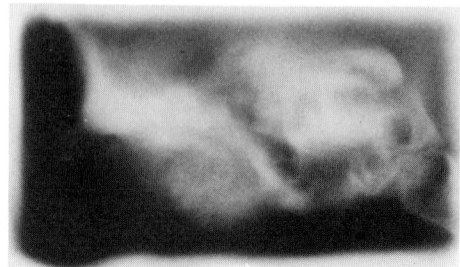

11.5

**Fig. 2.7** (E) Tomograms of petrous bones in Guillen's position.

of the orbit of the side under examination. The head is rotated approximately 20° towards the side being examined (so that the central X-ray beam is perpendicular to the middle ear), the central X-ray beam now passing through the supero-medial quadrant of that eye. The antero-superior junction of that tempero-mandibular joint and the tragus of the opposite side are equidistant from the film.

A trial radiograph is taken 5 mm anterior to the tip of the tragus of the side under examination. This demonstrates, in general, the oval window and the vestibule. If the vestibule is shown overlying the inner wall of the orbit, the rotation of the head has been too great.

The Guillen's projection demonstrates the internal walls of the middle ear and it projects the image of the ossicles between the lateral semicircular canal and the internal wall of the tympanic antrum (see Fig. 2.7E section 10.7). The ossicles are shown very clearly as rounded shadows and are cleared from the other structures. It also demonstrates the promontory of the cochlea, vestibule, superior and lateral semicircular canals, the lateral semicircular canal being foreshortened and part of the internal auditory canal. The cochlea is demonstrated on cuts taken more anteriorly. (see Fig. 2.7E sections 10.7, 10.9, 11.1 and 11.3).

*Chausse III*

The patient lies supine with the head rotated 10–15° away from the side under examination; the line from the external auditory meatus and the inferior border of that orbit forms an angle of 25–30°, with the chin being depressed. The

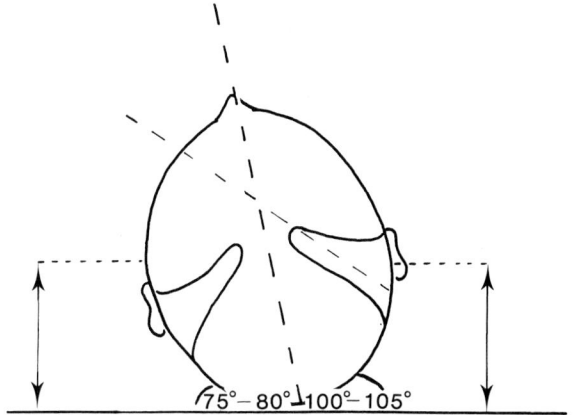

Fig. 2.8 (B) Rotation of the head.

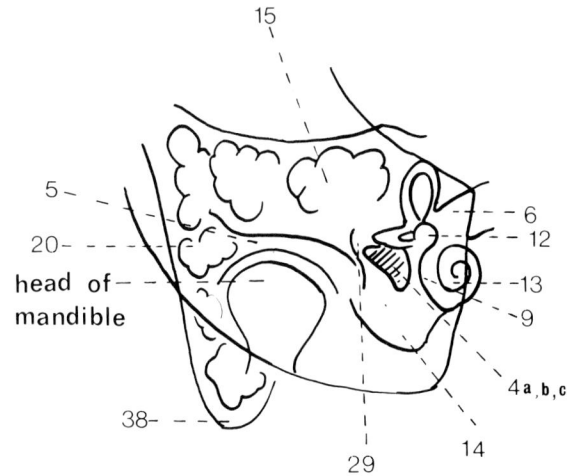

Fig. 2.8 (C) Diagram to illustrate the radiographic anatomy of the middle ear.

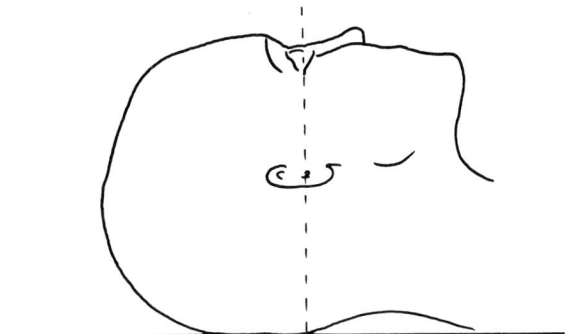

Fig. 2.8 (A) Position for Chausse III.

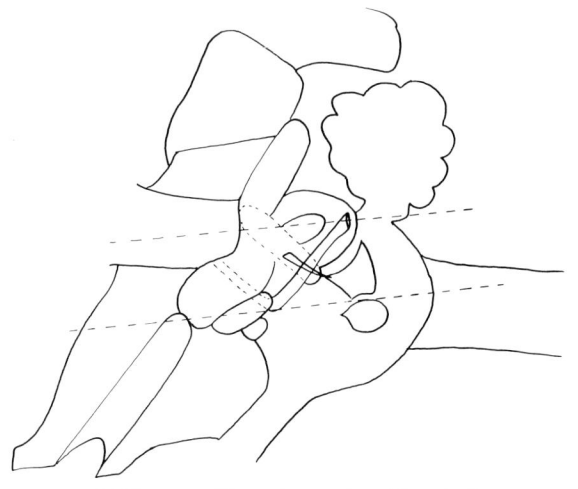

Fig. 2.8 (D) Tomographic section — dotted lines indicate plane shown.

central X-ray beam passes through a point just lateral and superior to the lateral edge of that orbit. This now demonstrates, in fine detail, the middle ear and especially the external half of the attic, seen as a triangular spur. (Partial or total destruction of this spur is often the first radiological sign of a cholesteatoma.).

### Stenvers and Chausse IV

(Chausse IV is a slight variation of the Stenver's view, when the head has to be rotated slightly more away from the side being examined, if the petrous bone is overshadowed by the outer border of the orbit; it also slightly uncoils the superior semicircular canal.)

The patient is supine, with the head rotated 45° away from the side under examination, so that the long axis of the petrous bone is parallel with the table-top/film. The mastoid process of this side and the outer border of the opposite orbit are equidistant from the film. The central X-ray beam is perpendicular to the long axis of the petrous bone and is centred over the tempero-mandibular joint of the side under examination. The trial cuts are taken; one at the level of the tragus and the second one 15 mm posteriorly to the tragus. The posterior cut shows the tympanic antrum and the mastoid process, and the anterior cut demonstrates the cochlea and internal auditory canal.

This position shows the jugular fossa, labyrinth, vestibule, the superior and posterior semicircular canals (the whole of the posterior semicircular canal can be shown on one cut) the course of the facial nerve canal and the relationship of the jugular fossa to the middle ear. Also demonstrated are the tip of the petrous bone, the malleous and incus which are seen to be side by side, the cochlea showing its three coils and the round window

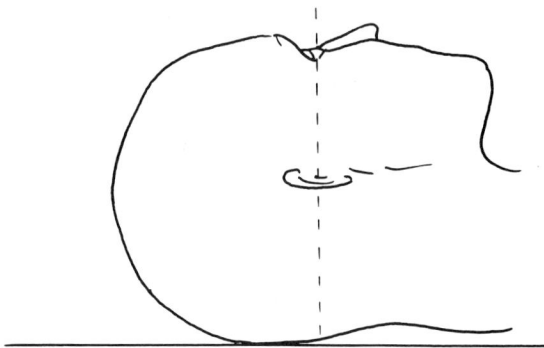

Fig. 2.9 (C) Diagram to illustrate radiographic anatomy of the petrous bone in the Stenver's position.

Fig. 2.9 (A) Position for Stenver's (or Chausse IV).

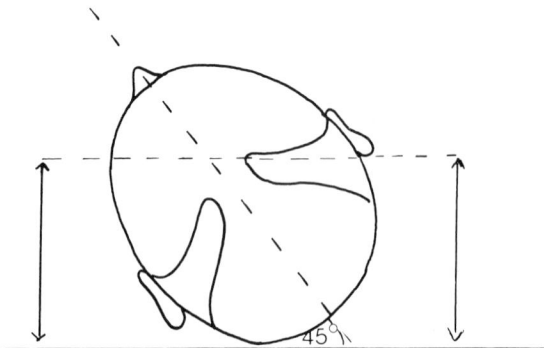

Fig. 2.9 (B) Rotation of the head.

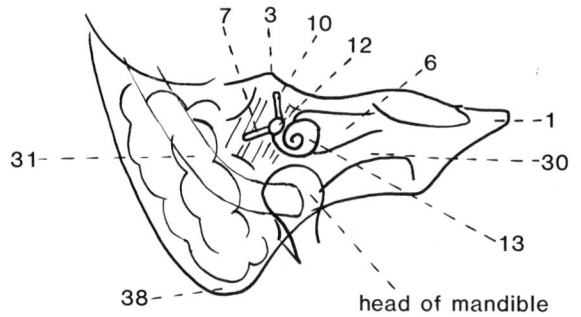

Fig. 2.9 (D) Tomographic section — dotted lines indicate plane shown.

12.4

12.7

12.8

12.9

13.0

13.1

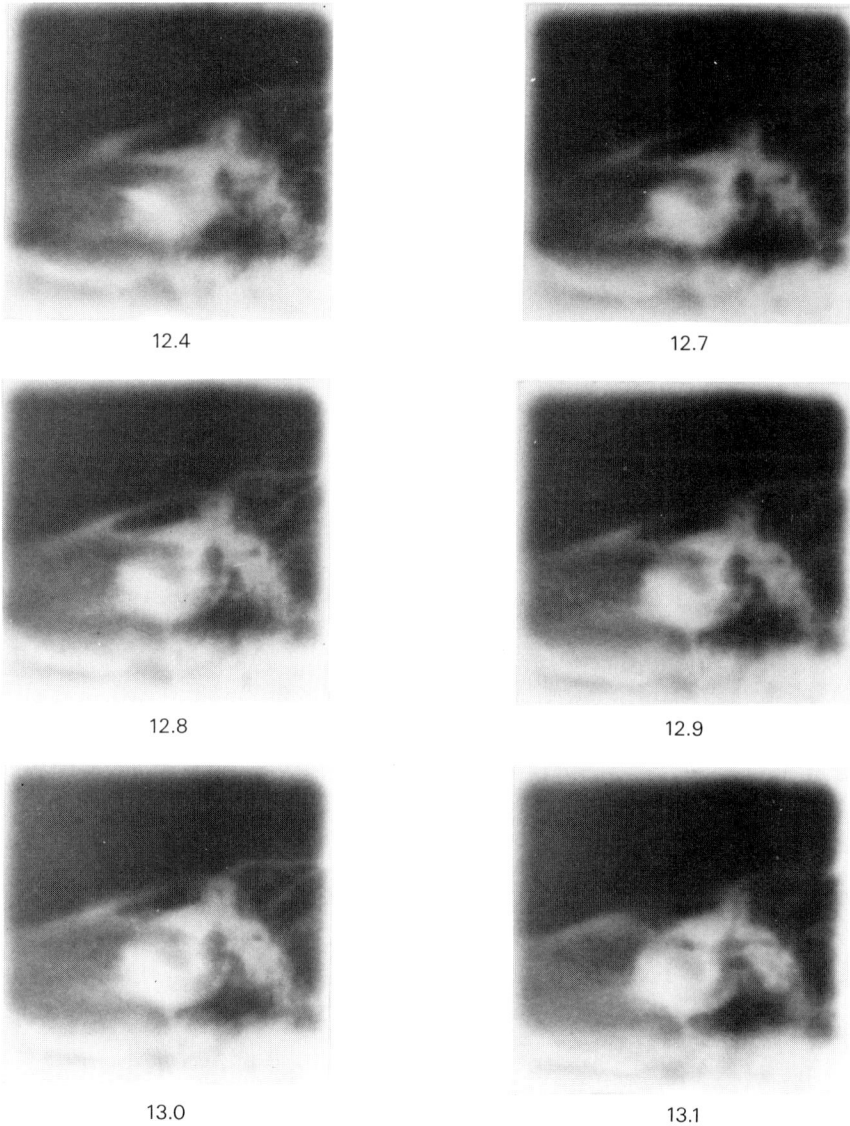

**Fig. 2.9** (E) Tomograms taken with patient in Stenver's position.

which appears funnel shaped 1–2 mm in size and connecting the base of the cochlea with the tympanic cavity. The oval window is seen projecting onto the lower half of the vestibule. The malleous is demonstrated in its entire length.

*1st cut.* Shows part of the external auditory canal and the middle ear, with part of the mastoid process and some mastoid air cells.

*2nd cut.* Going posteriorly should show the relation between the middle ear, the superior and lateral semicircular canals, mastoid antrum and

facial nerve canal (this is inferior to the lateral semicircular canal and crosses the superior portion of the middle ear). This cut also demonstrates the arcuate eminence and styloid process. The axis of the cochlea is cut perpendicularly by the central ray, so appears as a circle or spiral on the film (see Fig. 2.9E section 13.1).

*3rd cut.* This should demonstrate further the course of the facial nerve canal as it bends posteriorly and inferiorly. The niche of the oval window to the vestibule is clearly seen. The carotid canal is in sharp focus and its postero-lateral wall is in close proximity to the hypotympanum, separated from it by only a thin lamella of bone.

*4th cut.* Shows the descending portion of the facial nerve canal which terminates at the stylomastoid foramen. The lucency seen at the inferior aspect of the cochlea is most likely to be the niche of the round window.

*5th cut.* Shows part of the semicircular canal (this area is best demonstrated on these projections). The internal auditory canal is seen obliquely to its long axis, and the petrous apex is also visible. The jugular fossa is well demonstrated and part of the crista transversa has just come into focus.

*6th cut.* The internal auditory canal is shown more medially. The posterior and superior semicircular canals are seen where they unite to form the crus commune (common limb).

*Axial pyramidal view — Pöschl's* (pairs with Stenver's view)

Stenver's view shows the areas along the axis of the petrous bone, and Pöschl's shows the areas down its axis, through the image of the opposite orbit. It separates the attic from the labyrinth.

The patient is placed in the antero-posterior position and the head is then rotated 45° towards the side being examined. The central X-ray beam is centred over the outer canthus of the opposite orbit. The middle of the mastoid of the side being examined and the middle of the outer canthus of the opposite eye are superimposed.

This projection demonstrates the external auditory canal, internal auditory canal, middle ear, ossicles and cochlea, and the tip of the petrous bone (see Fig. 2.10D sections 10.4, 10.6, 10.8 11.0).

A trial 'cut' is taken 25 mm medial to the tragus which shows the ossicles and a 'cut' at 35 mm medial to the tragus shows the cochlea as a clover leaf. The whole of the superior semicircular canal can be seen in 1 'cut'.

*Lateral view* (pairs with the antero-posterior view)

The patient is placed prone with the head in the true lateral positon. The central X-ray beam is

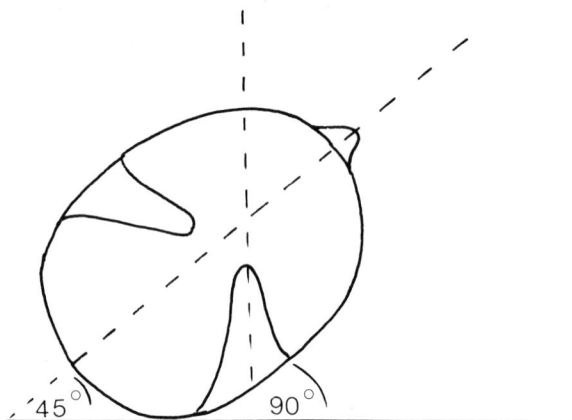

Fig. 2.10 (A) Position for Pöschl's.

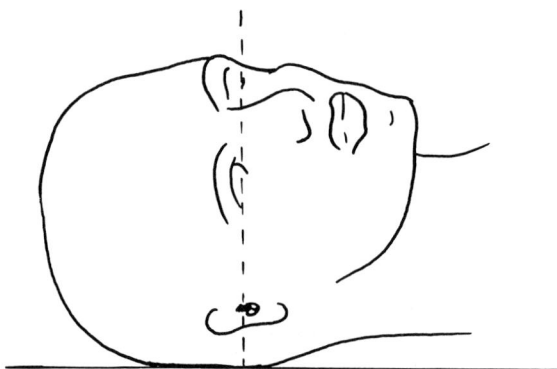

Fig. 2.10 (B) Rotation of the head.

Fig. 2.10 (C) Tomographic section — dotted line indicates plane shown.

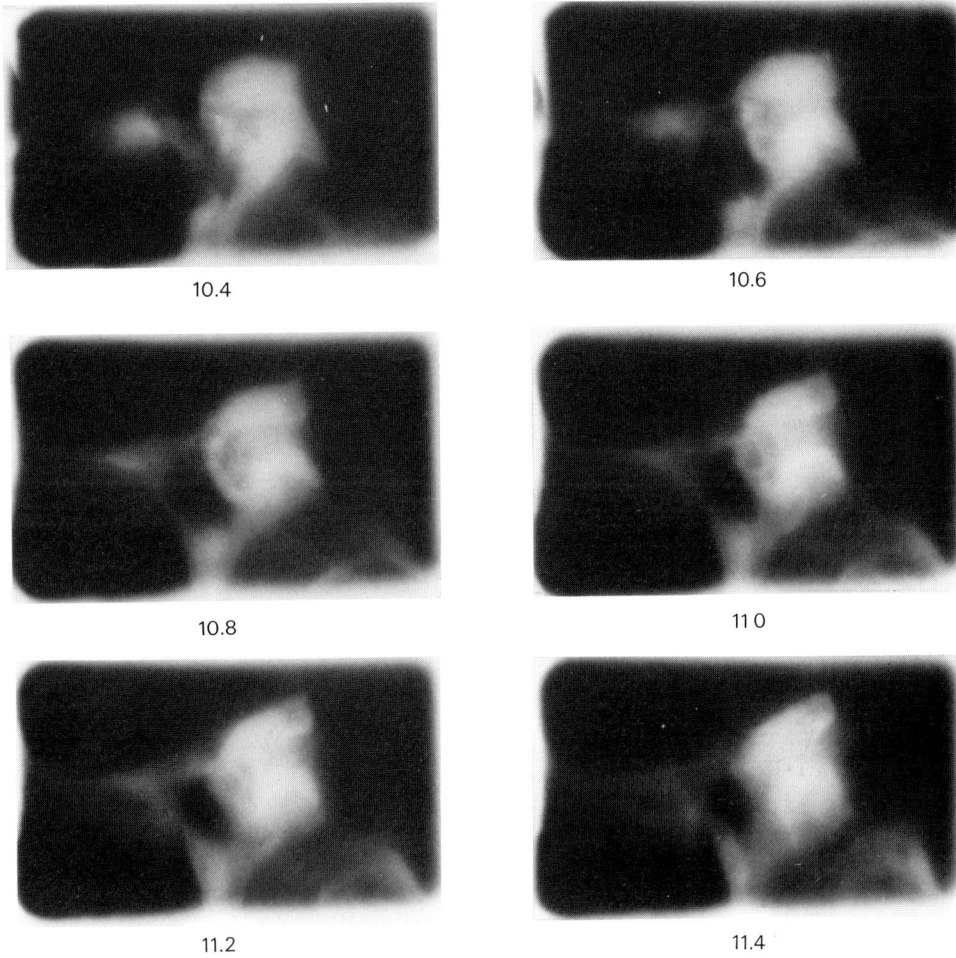

10.4

10.6

10.8

11 0

11.2

11.4

**Fig. 2.10** (D) Tomograms taken with patient in Pöschl's position.

centred slightly above the external auditory meatus. The 'cuts' are usually 1.5 cm–3 cm medial from the skin surface near to the external auditory canal. The lateral view demonstrates the relationship of the attic to the mastoid antrum and also demonstrates the mastoid segment of the facial nerve canal. A good separation of the incus and malleus is achieved so that this is a good projection when evaluating ossicular dislocations. The external auditory canal is demonstrated as a large bony ring and the internal auditory canal as a small one. The malleus and incus are visualized on the

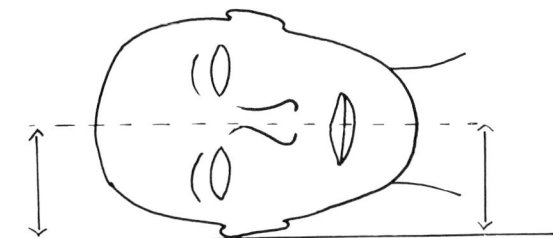

**Fig. 2.11** (A) Position for lateral view.

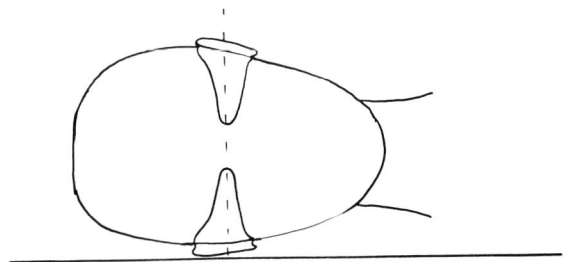

**Fig. 2.11** (B) Position of patient with head in true lateral position.

**Fig. 2.11** (C) Tomographic section — dotted line indicates plane shown.

upper border of the middle ear as a molar tooth appearance.

*1st level.* 2.0–2.5 cm below the surface of the skin. The external auditory canal is demonstrated as a bony ring. Just above this canal, the epitympanic recess can be seen as it extends slightly laterally and posteriorly from the middle ear. The mastoid antrum is visible and the aditus ad antrum (aperture between the mastoid antrum and the epitympanic recess). The mastoid segment of the facial nerve is well shown. The tegmen is very thin (Fig. 2.11D sections 12.9, 13.0, 13.2).

*2nd level.* Taken 1 mm medially from the first level, at a point just within the termination of the external auditory canal. This level shows the middle ear, the aditus ad antrum, lateral semicircular canal, facial nerve canal (where it makes its infero-posterior bend). The malleus and incus show as the classical molar tooth appearance, with the handle of the malleus forming the anterior and the long process of the incus forming the posterior part of the tooth (see Fig. 2.11D section 12.9).

*3rd level.* 1 mm medially to the previous 'cut' shows the two limbs of the lateral semicircular canal and part of the posterior semicircular canal and the tympanic segment of the facial nerve canal (see Fig. 2.11E sections 13.0, 13.2).

*4th level.* Demonstrates the plane of the vestibule, the promontory of the middle ear (formed by the lateral wall of the basal turn of the cochlea) the oval window, round window and the posterior semicircular canal. The jugular fossa lies inferiorly to the hypotympanic region.

*5th level.* This shows parts of the cochlea, the relationship between the carotid canal anteriorly and the jugular fossa directly behind and slightly inferior to this. There is a bony spur between these two structures (see Fig. 2.11E section 12.8).

*6th level.* This level, if taken 3 mm medially to the previous 'cut', demonstrates the internal auditory canal. The carotid canal, being anterior to the internal auditory canal, is cut obliquely in this projection. A slight bulge on the anterior wall of the internal auditory canal is the most medial part of the crista transversa (see Fig. 2.11E section 12.4).

*Sub-mento vertical view* (Fig. 2.12)

This is a very difficult position in which the patient must lie. A solid foam mattress may be used to a depth of about 23 cm with the vertex of the skull on a thin foam pad and secured with a Velcro strap. Because of the discomfort, this projection must be completed as quickly as is possible. The base line of the skull is parallel with the table top, the sagittal and coronal planes are perpendicular to the table. The skull is surrounded with bolus bags and pieces of lead rubber moulded around the head.

Both petrous bones can be examined simultaneously centring the X-ray beam mid-way between the external auditory meati, or individually, centring 2–3 cm medial from the external auditory meatus of the side being examined. Measure the distance from the tragus to the table top for the level of the first 'cut'.

This view shows the facial nerve foramina, the jugular foramen, the foramen ovale and foramen rotundum. It demonstrates the petrous bones clear of the maxillae and their relationships with the base of the skull, external auditory canal, tympanic cavity, internal auditory canal and petrous tips, tempero-mandibular joints, semicircular canals and the relationship of the carotid canal and the jugular fossa. Malformation of the middle ear can easily be demonstrated in this view, the advantage being that it separates the attic from the labyrinth.

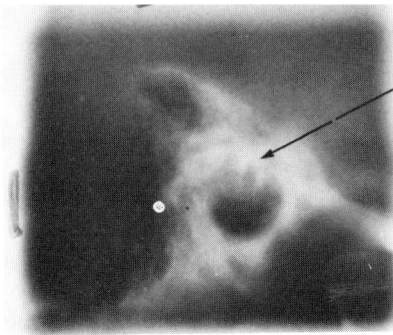

12.9

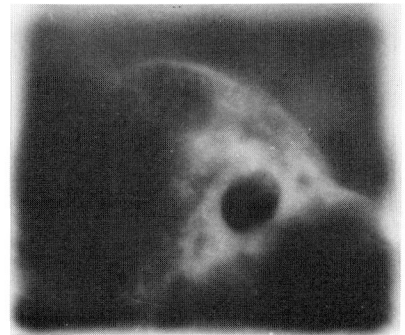

13.4

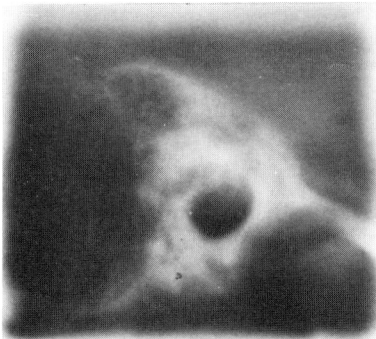

13.0

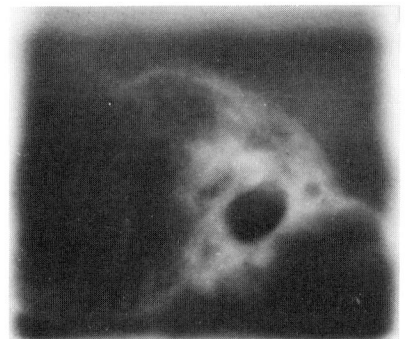

13.3

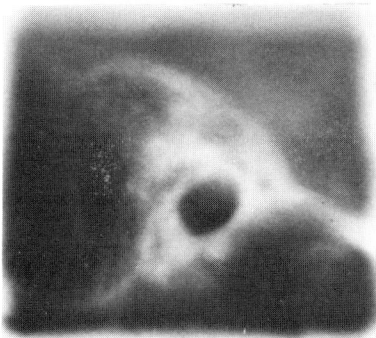

13.1

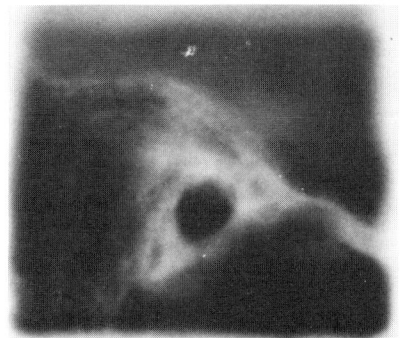

13.2

**Fig. 2.11** (D) Tomograms taken with patient in lateral position.

This view is used to demonstrate acoustic neuroma, cholesteatoma, fractures, otosclerosis (or ossicular dislocation).

*1st level.* Shows the floor of the external auditory canal, carotid canal and the jugular fossa.

*2nd level.* Demonstrates the continuation of the carotid canal and the jugular fossa, with the lower part of the hypotympanum just coming into focus.

*3rd level.* The jugular fossa can only just be seen, the middle ear shows well. There is a tiny spicule of bone on the anterior part of the most medial aspect of the external auditory canal which is part of the tympanic ring; the membrane is attached to this spicule. The carotid canal comes very near to the middle ear. The mastoid segment of the facial nerve canal is seen and its relationship to the posterior and medial aspects of the external auditory canal are also demonstrated.

12.0

12.2

12.4

12.6

12.8

13.0

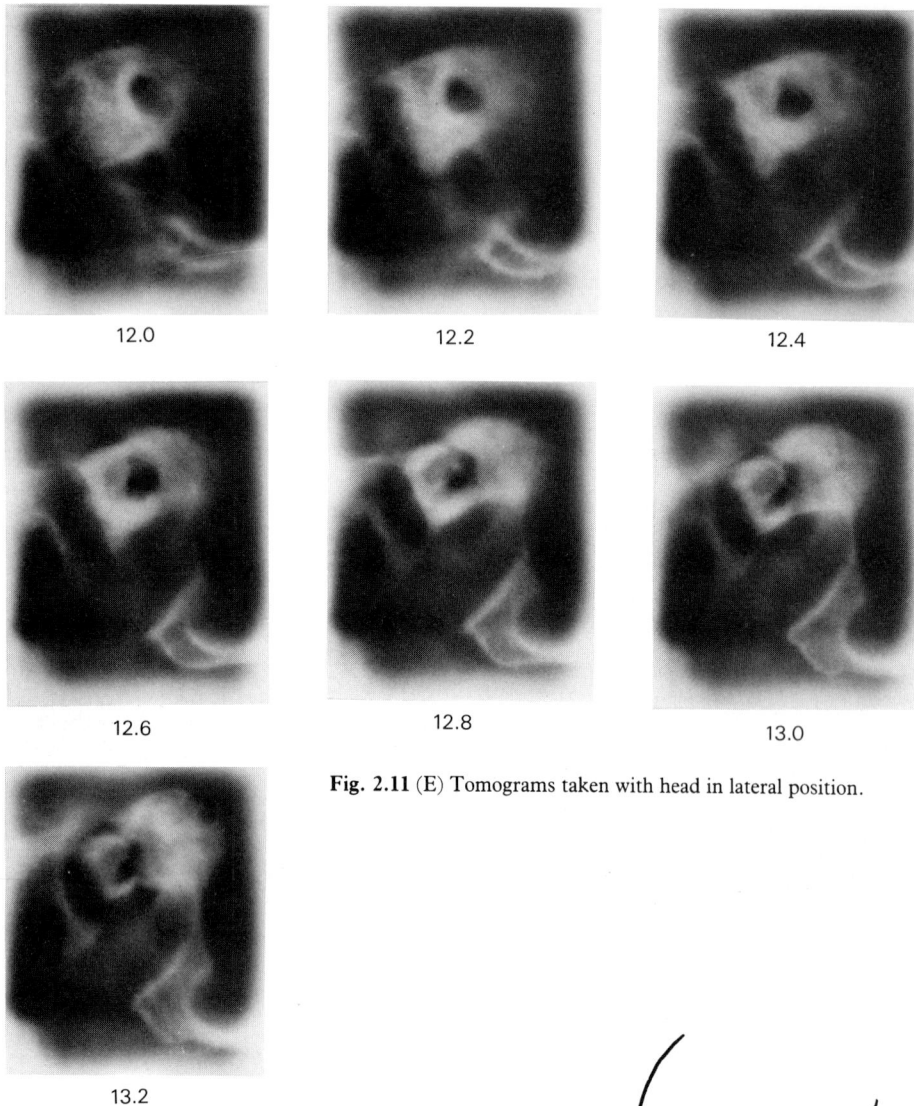

**Fig. 2.11** (E) Tomograms taken with head in lateral position.

13.2

*4th level*. It can be seen that the carotid canal has bent forward and continues antero-medially. The inferior part of the basal turn of the cochlea promontory protrudes into the middle ear. The Eustachian tube is an anterior extension of the middle ear. The carotid canal lies very close to the middle ear.

*5th level*. 1 mm cephalically to the previous level, shows both the internal and external auditory canals, and the coils of the cochlea. The long axis of the cochlea lies antero-laterally at an

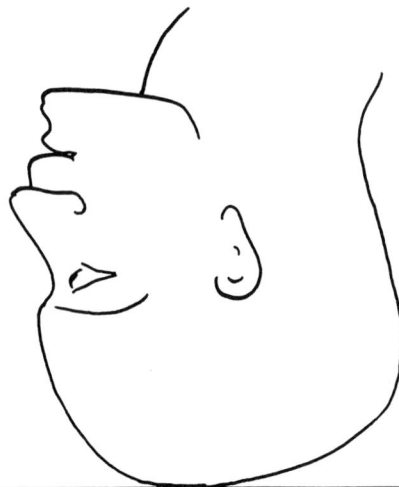

**Fig. 2.12** (A) Position for sub-mento vertical projection.

oblique angle to the axis of the petrous bone. The oval window shows the access into the vestibule. The lateral and posterior semicircular canals are also seen.

*6th level.* This 'cut' demonstrates the relationship between the middle ear, the mastoid antrum and its aditus. The aditus commences at the level of the most lateral aspect of the lateral semicircular canal. The crus commune (the junction of the posterior and lateral semicircular canals) shows as a lucency that opens into the posterior semicircular

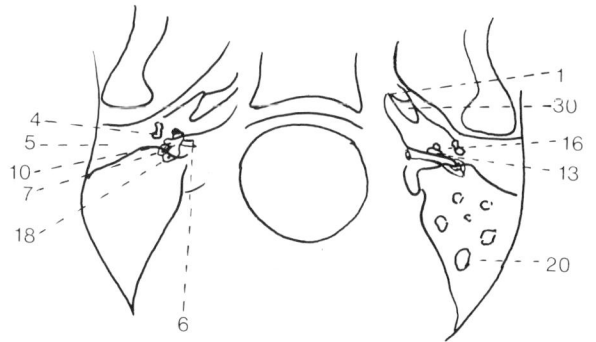

Fig. 2.12 (B) Diagram to illustrate radiographic anatomy of the petrous bone with sub-mento vertical position.

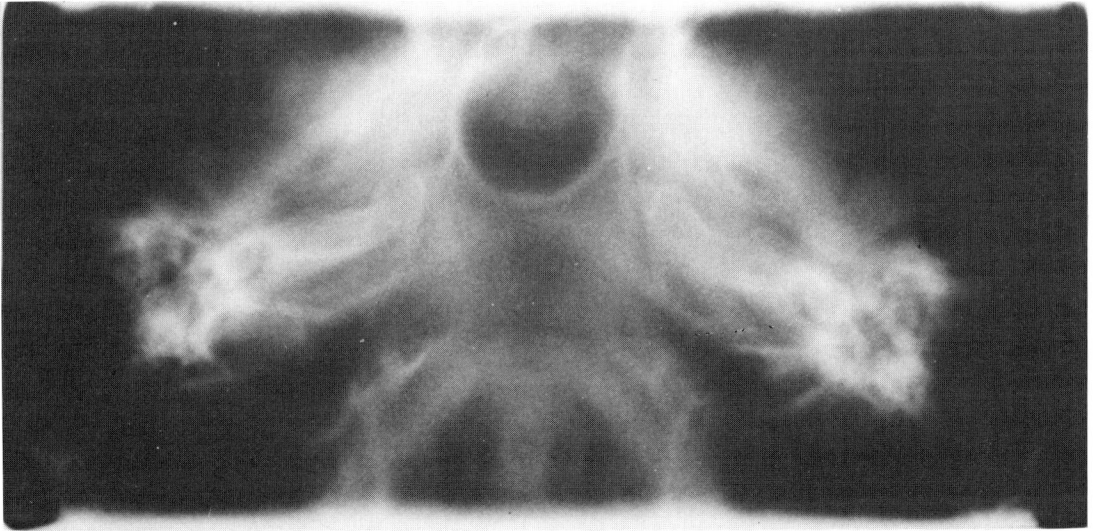

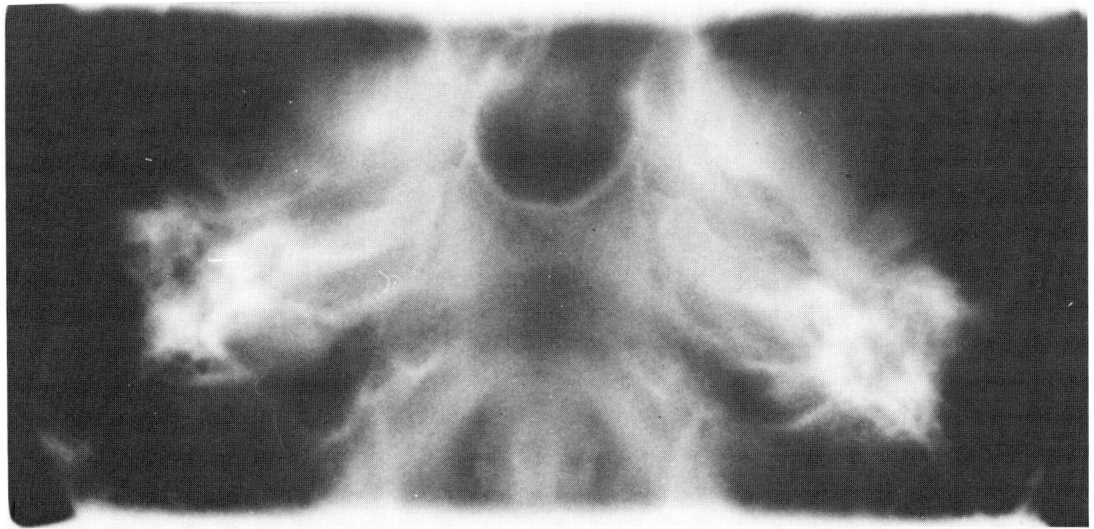

Fig. 2.12 (C) Tomograms of petrous bone in sub-mento vertical position (phantom).

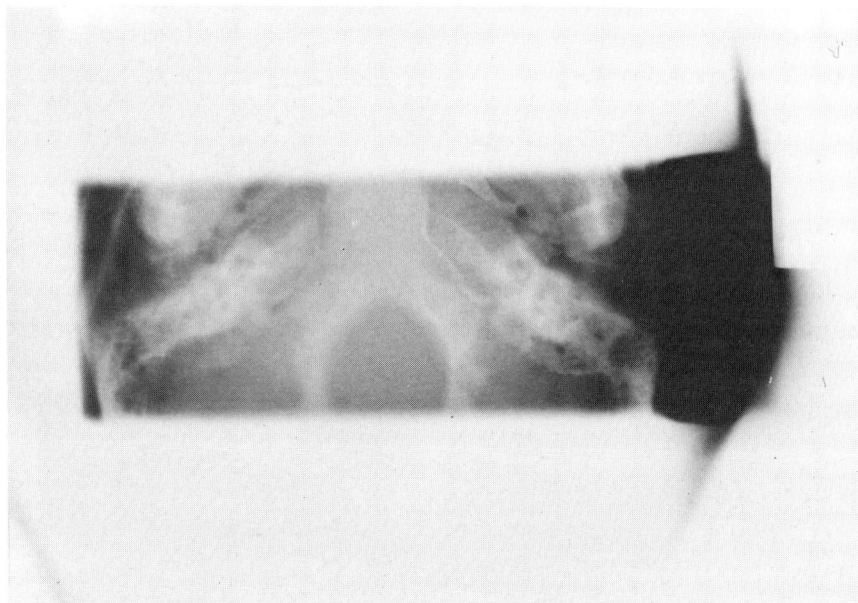

**Fig. 2.12** (D) Radiograph of petrous bones shown in sub-mento vertical position (phantom).

canal. The petrous and tympanic segments of the facial nerve canal are demonstrated; the sharp bend between these two is clearly seen. The geniculate ganglion is situated at this bend, or genu.

## CONCLUSION

Good plain films of the petrous bones are essential prior to tomography to enable assessment of the area to be made, e.g. the density and the angle of the petrous bones to the sagittal plane, as seen on the sub-mento vertical (SMV) projection, may not be uniform. These points can then be taken into consideration when positioning the patient and the selection of the correct exposure factors to obtain penetration of the part and produce good radiography. This requires a great deal of skill and knowledge on the part of the radiographer. The patient must be put at ease by an explanation of the examination, by placing them in a comfortable position and gaining their co-operation. Effective immobilization and accurate X-ray beam collimation are required to produce a series of radiographs of optimum quality. There is also a need for an X-ray tube which has high heat storage capacity, fast intensifying screens with lower light scatter, and X-ray films with a thinner base.

The modern trend which has been brought about by the Ear, Nose and Throat Centre in Brussels is that computerised tomography is now taking over from the conventional type. Tomography initially started in Brussels, but now there are very few examinations of this type being performed at that centre.

The advantages of computerised tomography are that the examination is easy to perform, the image is easier to interpret, and its contrast can be altered on the unit. The sloping surfaces of the temporal bone can be demonstrated more easily and the soft tissues are well demonstrated at the same time, e.g. in the temporal cavity — demonstrating adhesions and glomus tumours.

*Acknowledgements*

My thanks are due to Dr J.T. Lamb who started my interest and knowledge of tomography of the petrous bone, Mr R. Cadogan for the diagrams, Miss J. Wales for the radiographs and Mrs J. Papuga for her patience in typing this article.

REFERENCES

Binet E F, Moro J J 1972 A tomographic study of the base of the skull. Kodak Medical Radiography and Photography 48 (2): 30–35

Bonnard C Realisation pratique des incidences en radio-otologie. Manipulateur d'Electro-Radiologie. Editeur — Guy Ducros — Bordeau: 13–61

Jenson J, Rovsing H (ed) Fundamentals of ear tomography. Foreword by Gregers Thomsen. Contributor H Costa Davidsen: 5–17. Charles C Thomas, Springfield, Illinois

Korach G, Vignaud J, Lichtenberg R 1966 Selective employment of the polytome in accordance with type of examination. Medicamundi 11 (3): 82–92

McCallum A H 1963 The significance of the petrous bone in radiology. Ilford 'X-ray Focus' 4 (2): Part 1 Anatomy 24–26

McCallum A H 1963 The significance of the petrous bone in radiology. Ilford 'X-ray Focus' 4 (3): Part 2 Radiography 6–10

McCallum A H 1964 The significance of the petrous bone in radiology. Ilford 'X-ray Focus' 5 (1): Part 3 Pathology 14–21

McInnes J 1966 Tomography today — Part I. Ilford 'X-ray Focus' 7 (1): 6–9

Miller D, Naylor E 1972 Tomography of the ear. Ilford 'X-ray Focus' 7 (1): 6–9

Potter G D, Gold R P 1975 Radiographic analysis of the skull. Kodak Medical Radiography and Photography 51 (1): 2–15

Sangster J M 1966 Tomography Today — Part 3. The polytome in practice. Ilford 'X-ray Focus' 7 (3): 21–25

Schaefer R E 1972 Roentgen anatomy of the temporal bone. Tomographic Studies. Kodak Medical Radiography and Photography 48 (1): 2–22

KODAK DIAGRAM Radioanatomie du Rocher. Docteur J Vignauld — Service de Radiologie. Foundation A de Rothschild — Paris.

FURTHER READING

Grant J C B 1943 An atlas of anatomy. Williams & Wilkins. Baltimore

William P L, Warwick R 1980 Gray's anatomy, 36th edn. Churchill Livingstone, Edinburgh

## Key to numbers on Figures

1. Apex petrous pyramid
2. Apex of cochlea
3. Arcuate eminence
4. Auditory ossicles (a) malleus, (b) incus, (c) stapes
5. External auditory canal
6. Internal auditory canal
7. Lateral semicircular canal
8. Lateral wall of the attic
9. Promontory of cochlea
10. Superior semicircular canal
11. Temporal line of frontal bone
12. Vestibule
13. Cochlea
14. Hypotympanum
15. Mastoid antrum
16. Middle ear — tympanic cavity
17. Posterior surface of petrous pyramid
18. Posterior semicircular canal
19. Eustachian tube. Pharyngo-tympanic tube.
20. Mastoid air cells
21. Oval window — fenestra vestibule
22. Round window — fenestra cochlea
23. Epitympanic recess (attic containing malleus and incus)
24. Tympanic plate — ear drum
25. Styloid process
26. Tempero-mandibular
27. Inferior petrosal sinus
28. Occipital crest
29. Aditus ad antrum (aperture between mastoid antrum and epitympanic recess)
30. Carotid canal
31. Jugular fossa
32. Spur scutum
33. Facial nerve canal (petrous, tympanic and mastoid — descending segments)
34. Crus commune
35. Tegmen tympany — roof
36. Stylomastoid foramen
37. Crista transversa (crista falciformis)
38. Mastoid process

# 3

# Paediatric radiography

Children are our investment for the future and it is in our interest that they are dealt with properly. It is unforgivable to treat them as miniature adults; they really are different in proportions and in their methods of communication. The very first lesson a radiographer must learn is the different approach necessary to be understood by the paediatric patient. Certain techniques are the same or modifications of the equivalent adult examination, but some conditions apply only to infants. It has been stated that a successful paediatric examination is directly proportional to the quality of the radiographer.

A good paediatric department gives satisfaction when its patients leave smiling and willing to return and with a good diagnostic examination having been achieved. Conversely, a bad department leaves a child worried, distressed and nervous. To achieve good results, great skill and patience are necessary. As a profession and part of a diagnostic team, we ought to give doctors the best help available, whatever the time of day or night. Our patients are important — they are the citizens of the future.

## PLANNING A PAEDIATRIC X-RAY ROOM

Several reports have been published over the past 25 years on the welfare of children in hospital and numerous improvements have been made. The first report was instigated by the Ministry of Health under the chairmanship of Sir Harry Platt (Platt Report 1959). A further report 'The Welfare of Children in Hospital' was requested in 1973 by the Committee of European Societies of Paediatricians. The Earl of Halsbury chaired an en-

quiry into the pay and related conditions of the Professions Supplementary to Medicine (Halsbury Report 1974) and a report on the Child Health Services entitled 'Fit for the Future' was published in 1976, with Professor S. A. M. Court leading the investigation (Court Report 1976). There is a National Association for the Welfare of Children in Hospital, founded in 1961, which works constantly to improve the emotional well-being of children in hospital.

*All* these reports recommend that children should have an attractive and safe environment in which to wait, separated from sick adults, and that they should be examined by specialist staff who are trained and experienced in dealing with *sick* children. Unfortunately there are very few specialised Paediatric X-Ray Departments, although some District General Hospitals try to have at least one room dedicated to paediatric examinations.

Ideally, a Paediatric X-Ray Department should have its own entrance clearly signposted and a reception and waiting area separate from adult patients. Children can then avoid the possibly frightening and distressing sight of sick and injured adults. Moreover adults will be spared the harassing cries of children. The reception area should give very good first impressions and have a welcoming atmosphere which will go a long way towards establishing confidence in the parent and child. Delays can be interpreted as inefficiency but children cannot be hurried.

Play is an essential part of the waiting process in a children's department. The department should be as child-proof as possible, i.e. no sharp corners or access to electrical equipment or dangerous drugs. It should have an even temperature and plenty of suitable games and toys such as blackboards, books, crayons, rocking-horses and dolls houses. Anxieties can be forgotten in an enjoyable game and tensions are released with laughter.

## Equipment

The X-ray equipment should be dedicated so that it meets the requirements of children, i.e. permits very short exposure times and has an X-ray tube with small focal spot sizes to minimise unsharpness and reduce radiation dose by avoiding repeat examinations. The X-ray generator should be at least of the three phase or constant potential type but certainly not single phase. This is very important to provide high kW output and exposure times in micro-seconds. Several special X-ray tables are manufactured which give free access to a child, e.g. Siemens 'Infantoskope', Philips 'Junior Diagnost', and C.G.R. 'Pediatrix'.

The Infantoskope incorporates special support cradles which permits the examination of babies up to 270 cm tall. Siemens also manufacture 'Babix' cradles which can be suspended or used flat on a Bucky table. These are semi-circular troughs made from transparent radiolucent Cellon which not only immobilises the baby but also acts as a heat shield (Fig. 3.1).

The 'Junior Diagnost' consists of three units which share a high output X-ray generator and X-ray tube. These consist of a chest cassette stand, a flat table with an immobilising tray, a stationary grid with a very short object film distance and a tomography unit. The immobilising tray is fitted to the tomographic unit and its rapid movement means that even active children can have a successful examination carried out. Philips also manufacture paediatric attachments which can be used in conjunction with their adult tables.

The 'Paediatrix' is a remote control fluoroscopy unit which can be used for all paediatric examinations. It incorporates a patient support system and cradle support to fit any size of child.

Apparatus which reduces radiation dose is always useful in the department, e.g. small format (100 mm) film and ciné cameras with pulsed exposures attached to fluoroscopic units. Video-tape recorders reduce radiation dose if an examination can be carried out quickly and the tape

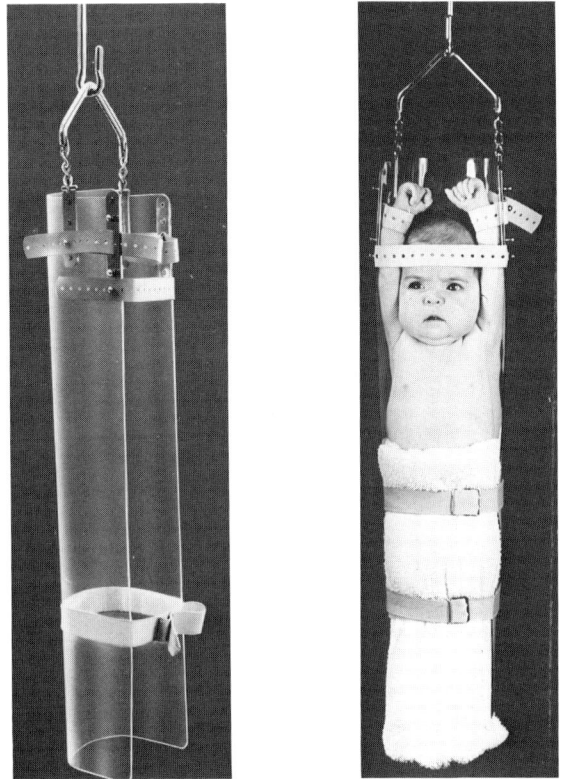

**Fig. 3.1** Siemens 'Babix' cradle

viewed several times at leisure instead of repeating parts of the examination. Systems which can freeze the image also reduce radiation dose, e.g. Siemens 'Memoscope' and Philips 'Vass' system. Here the image can be viewed, and then a permanent record made from the television monitor, using a multi-format camera.

Automatic exposure controls are in theory a great advantage when used for paediatric examinations and may reduce the number of repeat exposures due to exposure faults. However, they must be of the special paediatric type, and should preferably be solid state. Many departments have made their own immobilising devices to suit their specific needs and these generally work extremely well.

The ideal film/screen combination for paediatric use is very fine grain, high resolution intensifying screens to show soft tissue detail and yet 'fast' enough to avoid movement unsharpness whilst reducing radiation dose. This is impossible so a compromise has to be achieved and every department will have its own needs depending on type of equipment and X-ray generators. Rare earth screens are not suitable for neonatal examinations as there is too much 'grain' and quantum mottle at the kilovoltages required. For older children, however, a saving can be made when higher kilovoltages are used.

The kilovoltages used for abdominal and chest radiography in children varies from 40 kV for a neonate to 60 kV for a 12-year-old providing that the X-ray generator can provide mAs values sufficient to allow an exposure time of 0.1 second or less.

The mobile X-ray machine should be compact to allow for easy manoeuvring in the confines of a neonatal intensive care unit. Those of the condenser-discharge type are preferable as they give constant potential output, and shorter exposure times. Exposure times in the order of 0.01–0.003 second may be necessary since newborns may be breathing at a rate of 150 per minute. This type of mobile equipment is also conducive to easily maintained exposure charts, adjusting the kilovoltage to the baby's weight and keeping the X-ray tube current (milliamperage) and exposure time constant.

*Accessories*

Immobilising aids can be adapted for children, e.g. sand bags can be in the shape of toys, foam pads cut into interesting shapes. Dummies (comforters), feeding bottles and feeds, sweets, musical toys, glove puppets, and even lead rubber coats placed over the legs for protection immobilise a patient extremely well. Soft crêpe bandages make good restrainers for legs and arms, as does 'Tubigrip'. Velcro is used extensively as a restrainer and 'Dycem' placed on stools and under apparatus makes a good firm base. For babies up to 6 months of age wrapping them in a warm blanket makes them feel warm and secure when X-raying the skull. The same technique has the reverse effect on a 2-year-old who will object very strongly at being placed in a straitjacket, and can change a co-operative toddler into a frightened, impossible child. Helpers' hands are the recommended immobiliser for this age group with scrupulous

care being given to radiation protection. Children's eyes and ears must never be covered in an attempt to immobilise them, but playing a game and talking or singing to them as a form of distraction gives much better results.

**Preparation**

To achieve the greatest success in paediatric radiography, the most important rule is preparation. When an X-ray request is received by the radiographer, the X-ray room should be prepared down to the last detail. Cassettes are placed in the bucky tray with markers placed on them, the X-ray tube is centred and the exposure factor set. There should be as little movement of apparatus as possible once the child is positioned on the X-ray table. Trolleys and trays are prepared beforehand and syringes should not be filled in front of the patient. When everything is ready and the patient is called, the radiographer will then be able to devote all attention to the child. A tour of the X-ray room should be undertaken with a good look at pictures, toys and games as the patient settles down. At the same time explanations should be given as to what will happen; parents are told their role in the examination and by watching the child's eyes an assessment can be made as to whether the child is frightened. When the child responds to questions and starts talking, the examination can be quickly and efficiently carried out. If the child is co-operative there should be no need for repeat examinations, so a few minutes preparation will actually speed up the work flow. A smiling child returning to the waiting room, waving goodbye to the radiographer inspires confidence in those still waiting.

*Abdominal preparation*

It helps to improve the quality of radiographs and therefore diagnostic accuracy by making sure the abdomen is clear from gas and faeces.

In small babies up to 2 years old, glycerine suppositories placed in the rectum on the evening and morning prior to the examination are a safe and mild preparation.

From the age of 2 years a mild aperient may be given on two nights prior to the examination.

Suggested preparations are 'Senokot' syrup or granules. The dose is given according to the manufacturer's instructions. Children do not object to eating the chocolate flavoured granules or to taking a spoonful of the syrup. Granules are easier to send by post.

From the age of 6 years upwards a stronger aperient may be used, e.g. 'Bisacodyl', up to the age of 10 years the dose being one tablet of 5 mg for two nights.

Alternatively a low residue diet may be given for 4 days prior to the examination or a liquid diet, e.g. Triosorbon. This is manufactured in a variety of flavours and is especially suitable for children in hospital or for the physically handicapped. Preparation for a general anaesthetic will be arranged by the ward.

Non-opaque objects should be removed prior to X-ray examination as with an adult, but it is not always necessary to undress a child. From experience 9 out of 10 children prefer to keep their own clothes on, although some children do like dressing up. Children's clothes are usually of a cotton and mixture fabric which are radiolucent so there is no necessity to remove them for simple radiographic examinations. Metal fasteners can be manoeuvered out of the way or only one garment removed instead of two. Children feel vulnerable when undressed, being afraid that they are going to be put to bed or that a parent will leave them. Discretion is needed when there is a possibility that clothes could be soiled by barium spillage.

## RADIATION PROTECTION

Radiation protection is a vital aspect of paediatric radiography. Recent research has revealed that even very low dose radiation can affect growing bones and especially the epiphyseal centres. Gonad doses accumulate throughout life, but before puberty it is particularly harmful, as is radiation to gland tissue, e.g. thyroid, thymus, pituitary and mammary glands. Accumulated radiation may take 30 years to produce manifestation.

After extensive tests using a Diagnostic Dosimeter-Diamentor D, there is no doubt whatsoever that very tight, X-ray beam delineation reduces the amount of radiation dose to the child, and also reduces the scattered radiation to the parent, giving the added bonus of improved radiographic quality. For example a radiograph of a whole baby on a 35 × 43 cm film, centred on the abdomen with an appropriate radiographic exposure gives a Diamentor reading of 0.18 Gy × cm$^2$ × 10. Seven accurate radiographs tightly coned give a Diamentor reading of 0.06 Gy × cm$^2$ × 10, reducing the total radiation dose by one-third. Radiation protection must be very rigorously undertaken with escorts of paediatric patients. The Code of Practice (1972) limits their maximum permissible dose to one-tenth of that of a designated person.

Good record keeping *does* prevent unnecessary examinations, as does conscientious monitoring of radiation dose levels and fluoroscopic exposure times. Radiologists who have the fluoroscopic exposure times displayed on the television monitor become more radiation conscious when the seconds are 'mounting up' in front of their eyes. If fluoroscopic exposure times and radiation doses monitored by the Diamentor are recorded on the X-ray report:

Time: 1 minute 20 sec — Diamentor 0.10 Gy subsequent radiographic examinations promote the incentive to improve on the previous time.

Sensitive glands, if placed nearest the cassette, only receive the minimum 'exit' dose, compared to the skin 'entry' dose. For example if a radiographic examination of the thoracic spine is carried out in the postero-anterior position the mammary gland dose is reduced from 700 mSv to 66 mSv when recorded using thermo-luminescent dosemeter.

## PATIENT CARE

### Neonates

There is a worldwide agreement that a baby weighing 2500 g or less is a premature infant regardless of the period of gestation, therefore there are two types of low-weight infant, i.e. those less than 40 weeks' gestation and those who are small for dates. There are lots of reasons for the latter which include multiple pregnancy, pregnancy complications, ill health of mother, socio-economic reasons and smoking.

*Infant mortality*

A baby born before 32 weeks gestation has a 50/50 chance of survival. Eight out of ten deaths occur in the first week of life, and half of these after 24 hours. The major causes of death are due to Idiopathic Respiratory Distress Syndrome, congenital abnormalities, birth trauma or infection.

*Hazards with premature babies and neonates*

a. Infection
b. Dehydration
c. Hypothermia
d. Respiratory difficulties.

*a. Infection.* The neonate is prone to infection as it does not have many acquired antibodies (only those transferred via the placenta from the mother) and it has poor manufacturing properties for its own antibodies. There is no immunoglobulin present at birth. It is important therefore that the baby is kept in a stable environment and away from as many pathogenic organisms as possible. All personnel attending the baby must therefore obey the rules of isolation by handwashing, cassette wiping use of face masks and gowns and maintaining a good standard of personal hygiene and health. Certainly they should not attend the neonate if they have a respiratory tract or gastro-intestinal tract infection or any skin disorder.

*b. Dehydration.* This is the restriction of water in the diet or the removal of water from the body tissues resulting in reduced blood volume. Removal of tissue fluid causes an imbalance of the sodium, and potassium content of the blood and changes in the pH value. Severe changes in the electrolyte balance cause a low blood pressure, tachycardia, cold, grey skin, loss of skin turgor and severe shock which if not rectified results in death.

The main reason for dehydration involved with X-ray examinations is incorrect information given to parents regarding the starving and restriction of fluids prior to the examination. Every contrast medium which is injected is hypertonic and the amounts administered must be very carefully controlled. Although this is a medical responsibility, a radiographer must participate in this task. Charts indicating the dose applicable to the body weight should be clearly displayed since overdosage is dangerous.

*c. Hypothermia.* The baby's temperature regulating centre is underdeveloped and it is therefore unable to produce or conserve body heat, but rather takes up the temperature of the surroundings. Babies have a large surface area compared to volume and for this reason lose heat very rapidly. A baby lacks subcutaneous fat and does not have energy reserves, therefore it is nursed in the warm, humid atmosphere of an incubator. This has a pre-set temperature of 32°C (90°F) and can be raised to 35°C (95°F) to maintain the infant's temperature at 35–36.6°C (95–98°F). The incubator has an alarm bell which will ring if the temperature changes due to electric power failure, broken fuses or plugs or overheating due to fan failure etc. An infra-red heater, with an automatic adjustment according to baby's temperature can be used but radiant heat increases the water loss from the baby's skin and prolonged exposure would make the baby dehydrated. This heating method is useful when medical procedures need to be carried out necessitating the incubator to be open for long periods, or with the baby outside the incubator for improved access.

Failure to maintain the body temperature makes the baby lethargic and reluctant to wake or feed. The baby is cold to the touch and its heart rate slows down. Conversely, overheating or too rapid an increase in body heat is also detrimental.

*d. Respiratory difficulties.* Periods of temporary apnoea are quite common in full term infants, but more prolonged in premature babies and this can lead to cerebral anoxaemia. For this reason the neonate is nursed on a mattress which is connected to an apnoea alarm. Various types are available from several manufacturers but they are all based on the principle of an inflated mattress which has a built-in alarm system which sounds if there are no breathing movements for 15, 20 or 30 seconds. If the alarm sounds, the nursing staff can give external stimulation to get the baby to inspire once more. This equipment is expensive and delicate and staff have to treat it with care. A second type of apnoea alarm has small sensors which are fixed to the baby's chest and record movement of the chest wall electrically, and an alarm will sound if the movement stops.

Maintaining the baby's oxygen balance is complex but it is easier if the baby is in an incubator where the concentration and flow rate are controlled. The most effective way is in a head box or hood when up to 100% oxygen can be given. An oxygen analyser can be placed in the head box. The oxygen has to be warmed and humidified before it reaches the child. Too much oxygen will lead to oxygen blindness (retrolental fibroplasia) and too little to brain damage from oxygen starvation. The current method of monitoring oxygen concentrations apart from blood gas analysis is the use of a $PO_2$ monitor. In a newborn child blood gases are monitored by obtaining samples of arterial blood via an umbilical artery catheter. Capillary samples are not sufficiently accurate. There is an umbilical artery catheter available with an electrode at its tip which continuously measures the oxygen levels and these are displayed on a monitor beside the bed. When the baby has a good peripheral circulation a transcutaneous oxygen electrode can be used to monitor the $PO_2$ levels. Safety precautions must be observed when using a mobile X-ray machine around a concentrated oxygen source, i.e. avoidance of discharge of static electricity.

A premature baby has immature lungs and respiratory centres, a soft pliable rib cage and weak thoracic muscles. Babies stop breathing very easily, two causes being excessive heat and excessive handling. They should be nursed naked to make the observation of them easier. Radiographers have to aim for a very high degree of efficiency to minimise the amount of time the babies are disturbed.

Several indwelling tubes may be attached to a sick, premature baby (Fig. 3.2) and these create obstacles when trying to position a cassette. There may be a nasogastric tube, an intravenous 'drip' tube (often in a scalp vein), a catheter in the umbilical vein or artery or both, an endotracheal tube, or a central venous pressure line (for hyperalimentation) (or total parenteral nutrition), or a gastrostomy tube. Skin attachments would be electro-cardographic (ECG) leads, temperature probes, $PO_2$ monitoring leads. The fewer of these which appear on the radiograph the better and the medical and nursing staff will have to advise if they can be removed temporarily or moved to less strategic places, for example ECG leads on arms instead of over the lung fields. The phototherapy light treating a baby with neonatal jaundice will

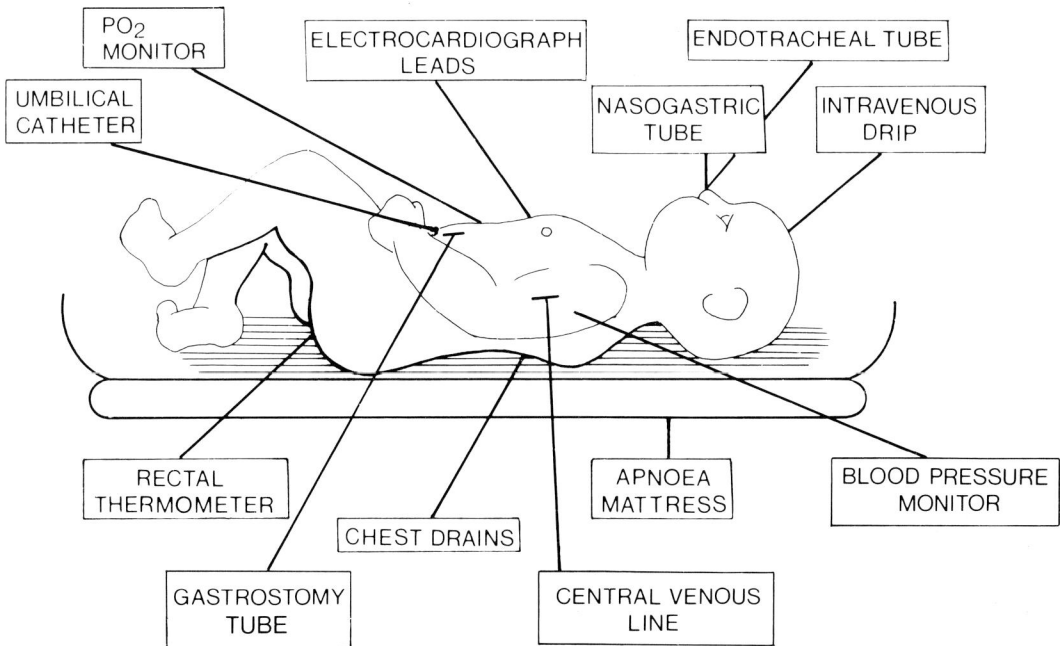

Fig. 3.2 Hazards when positioning a cassette under a sick neonate

also be in the way when trying to position the mobile X-ray machine.

Happily all the babies born with a low birthweight and without congenital malformations or cerebral damage should compare with a normal child by the age of 5 years. Those with a low Apgar score, infection, anaemia, apnoeic attacks, hyperglycaemia, hypocalcaemia may not be quite so fortunate. Virginia Apgar was an American anaesthetist who devised a system of assessing the condition of a newborn. This ranges on a scale of 1–10 and assesses heart rate, respiratory rate, muscle tone, colour and reflexes.

## Baby

When babies have outgrown the neonatal period the hazards become minimal and they are relatively easy to radiograph. Up to the age of 6 months they respond to anyone talking to them in a gentle, friendly tone, and providing they are kept warm, firmly held and regularly fed, they will lie still. Mother is the best person to hold them and talk to them as she is a familiar person. Care has to be taken to explain to the parent exactly how she can help to make the examination speedy and successful. Time taken to make the patient comfortable usually pays dividends. Music is appreciated and overhead musical mobiles are a great asset. The only time a parent should be asked to wait outside is during the insertion of some catheter or venepuncture, insertion of drip tubes. The sight of these procedures being undertaken can be extremely distressing and the parent is a greater support before and after the event. If the child sees its parent faint, a not uncommon occurrence, it would cause distress.

## 6 months–3 years

Children in this age group are very strong, need lots of patience and communication is difficult. The radiographer has to get down to the patient's level both in intellect and in size. Bend or kneel down until your eyes are on the same level and then an equal partnership without domination will exist. Children love being the centre of attention and many symptoms of misbehaviour are described as attention seeking. By talking to them

the radiographer should find a subject that interests them such as a pet or brother or sister.

Different toys, games and books will eventually produce some response even in a difficult child and should build confidence with the radiographer. If the examination can be made into a game so much the better, e.g. a hand will keep very still on the cassette till a count of three if a sweet is placed on the side and the child allowed to eat it on the number three. Children should always walk in and out of an X-ray room if they are able, and untruths should never be told. This is the age group which is very susceptible to psychological trauma. They are too young to understand explanations, but old enough to miss home. Some toddlers will make their unhappiness show audibly, whereas others will remain quiet and withdrawn and store up aggression. This shows itself in a reversion to childish behaviour and antisocial tendencies. To avoid this, it is recommended that they are nursed with other children of the same age. The staff have to be sympathetic to parents who may appear hostile and difficult, when underneath they are really worried and frightened because they do not understand fully procedures being undertaken.

## 3–6 years

This is the age group which expresses the greatest fear. They have vivid imaginations and the sight of the large equipment in an X-ray room can conjure up unreasonable doubts. Once again the antidote for fear is play, gradually building up confidence. To carry out a mock examination on either parent or Teddy Bear sometimes works, or if possible to watch an older, co-operative child undergoing an examination may help. A previous frightening experience can produce tears just at the sight of a white coat, therefore a lot of patience, kindness and distractions will be needed.

Two radiographers working together can speed up the examination, and often speed is the ingredient of success. Ask the child to 'smile' when taking the radiograph and this usually has the desired effect of keeping him still, even if it is a foot that is being examined.

Some examinations will require sedation but this will have to be sufficient and correctly administered by a paediatrician. A small quantity of sedation usually removes only the child's inhibitions and makes protests more volatile. There are some diseases whose symptoms produce a grumpy and miserable child. These children have to be treated normally but firmly. Mother can help to position them, but if she is not succeeding perhaps a nurse could replace the Mother and Mother be asked to wait outside. Alternatively, if the Mother and child are asked to wait outside for a little while until the child has settled, it gives the Mother a chance to reprimand, bribe or comfort. There are times when a child has had several other investigations performed before reaching the X-ray department and in this instance an appointment the next day may achieve better results providing that it is clinically acceptable.

*6–14 years*

Children of this age group attend school, have discipline other than parental, mix and respond to people other than family and should, therefore, be able to understand any instructions given to them. If they are frightened a simple explanation should allay their fears. They can be frightened and in pain, but they are more able to understand sympathy and help. Boys can be very self-conscious or fidgety so it is important to ease their discomfort or embarrassment.

**Handicapped children**

The disability of the child has to be assessed by the radiographer whether it is mental, physical, blindness or deafness and appropriate help is then offered. The person escorting such a child will always know the child's capabilities and can help in communication.

## CHEST RADIOGRAPHY

Radiography of the chest is the most common examination undertaken in children. Apart from heart and lung problems it is requested in cases of diagnostic uncertainty of abdominal pain and pyrexia of unknown origin. If the radiograph is taken on inspiration and providing that the patient is not rotated it is of great value; if rotated and exposed on expiration it is of no value whatsoever. Chest radiographs of children require slightly more penetration than adults to enable the space behind the heart to be seen. The use of fine grain intensifying screens is necessary because of the size of the lung structures to be visualised.

*Baby — antero-posterior-supine view — 100 cm f.f.d.*

The baby is placed directly on to the covered cassette. The baby is stretched out with arms extended over the head, the elbows being used to splint the head to keep it straight. A foam or polystyrene pad is placed under the shoulders to extend the chin and also to prevent the baby getting into the lordotic position (a very common fault). The central X-ray beam is directed above the sternal notch and the X-ray tube is then angled towards the feet to enable the radiation field to fit the size of the chest. This prevents the direct X-ray beam radiation reaching the eyes and neck. Exposure is made on inspiration by watching the baby breathe two or three times, and watching the movement of the abdomen, this being the reason for choosing the antero-posterior projection. If the baby is being X-rayed in an incubator, lead rubber can be placed on top of the perspex as added protection to the nurse's hands and the baby's head and abdomen. A skin fold is a common artefact on a supine radiograph due to the laxity of skin in a newborn — a fold can produce a density line.

*Toddlers — anterior posterior view-erect — 180 cm f.f.d.*

The child is seated on a stool against a cassette placed in a cassette holder. The escort stands directly behind the chest stand, behind a shield of 2 mm of lead equivalent and holds the child by the arms which are extended alongside the head keeping it straight. A foam pad is placed behind the shoulders to enable the chin to be raised; the shoulders are level. Centre above the sternal notch and again the X-ray tube is angled towards the feet and the field size adjusted to the area of the chest.

*Children — lateral view*

From either of the above positions the child is rotated through 90° and the central X-ray beam is directed to the midline of the axilla.

*Lateral view on a sick child.* When a patient cannot be moved because he is attached to a ventilator, a horizontal X-ray beam lateral view can be undertaken. The cassette is supported at the side of the patient making sure that there are no artefacts in the area. With a horizontal X-ray beam and a distance of 100 cm the central ray is directed to the mid-line at the level of the axilla. This is an extremely useful view to demonstrate fluid levels or a pneumo-mediastinum.

## Illness complications and requirements

All cardiac problems require good quality radiographs of the chest, e.g. non-rotated to visualise the cardiac outline, as even minor errors of centring, exposure and positioning can produce distortion of the heart size and shape.

Respiratory tract infections are numerous but a single good quality, straight, radiograph exposed on inspiration provides sufficient information for clinical management. Patients suffering from cystic fibrosis generally require both an antero-posterior and a lateral projection to monitor the progression of the disease. Asthmatic patients seldom require more than one projection to monitor progress and in a severe attack it is generally required to exclude a pneumothorax.

Babies are prone to aspirate food, meconium at birth or amniotic fluid. Inhaled foreign bodies particularly food or peanuts, require inspiration and expiration radiographs to demonstrate any obstruction. Respiratory distress syndrome in the newborn sometimes requires two radiographs when the first one does not demonstrate any aeration in the lungs. A repeat exposure would then be required to exclude a technical fault.

The signs of respiratory distress are tachypnoea, tachycardia, intercostal recession, sternal depression, grunting and cyanosis. A spontaneous pneumothorax occurs in 1% of all babies, other causes being infection and hyaline membrane disease. Hyaline membrane disease is most likely to occur in babies born before 36 weeks gestation and is due to a lack of pulmonary surfactant, the substance necessary to prevent lung collapse. The radiograph presents a 'ground glass' appearance and an air bronchogram. If a baby has its blood oxygen maintained for the first few days of life either by ventilation or using oxygen under pressure in a head box the surfactant levels should rise naturally. Babies also become very acidotic due to the excessive effort of breathing and this is treated by giving bicarbonate of soda.

Further complications of the newborn necessitating chest radiography include diaphragmatic hernia, which is a protrusion of the abdominal contents into the chest through a residual defect in the posterolateral segment of the diaphragm known as the foramen of Bochdalek. Other congenital abnormalities are tracheo-oesophageal fistula, congenital lobar emphysema, laryngeal clefts and lung cysts, terratomas and bronchogenic or duplications cysts.

*Oesophageal atresia* is the non-canalisation of part of the tube from the mouth to the stomach. When the baby is born it will choke on any fluid in its mouth and a nasogastric tube cannot be passed into the stomach. An antero-posterior and lateral view of the chest will be required using mobile X-ray equipment. An opaque tube will have been passed as far down the oesophagus as possible to demonstrate the extent of the blind pouch. The radiograph should include a small portion of the stomach to ascertain whether or not there is any air visible. If there is, this would indicate an associated fistula into the respiratory tract which occurs in over 70% of babies with an oesophageal atresia.

A further fistula between the oesophagus and the trachea without an atresia is an 'H' type. This manifests itself when the child is several weeks old and causes choking attacks with repeated chest infections which are associated with lobar collapse. To investigate this rare fistula a contrast examination has to be performed. The baby is placed prone and head down on a special box so that fluoroscopy can be performed with the child in the lateral position. A nasogastric tube is placed in the lower oesophagus and with the aid of an image intensifier, an injection of a contrast medium made. As fistulae arise from the anterior wall of the oesophagus this is the optimum position for their demonstration. Any contrast medium entering the respiratory tract, as the baby is head down should pass to the upper airway and can easily be removed by suction. This examination should only be undertaken when there is oxygen and a suction machine available.

For children with suspected croup (acute inspiratory stridor in the age group 6 months to 3 years), an antero-posterior radiograph of the chest and neck is required with a lateral view of the neck to exclude an inhaled foreign body, or retropharyngeal abscess. For the lateral projection, the cassette is supported at the side of the child's head, and with a distance of 180 cm and an airgap of 15 cm the central X-ray beam is directed to the cricoid cartilage. A high kV (e.g. 110 kV and 1 mAs) is used. It is important to record on the radiograph whether the child is quiet or crying, i.e. the position of the vocal chords when the radiograph was taken. This examination is frequently undertaken on the ward as any movement of the child could occlude the airway (Fig. 3.3).

The projection is very useful for all cases of stridor, e.g. subglottic cyst, vascular ring, bronchogenic cyst and epignathic cysts. Supplementary examination of a baby's airway could be made using 100 mm photofluorography, taking a series of radiographs over several seconds to record the different phases of respiration. Ciné fluorography may be undertaken as an alternative if this facility is available, giving a dynamic, rather than an anatomical examination.

A congenital anomaly of the nasopharynx occurs in the newborn, i.e. choanal atresia. This is a residual bone or membrane separating the nasal cavity from the nasopharynx, it can be incomplete, unilateral or bi-lateral. If it is complete the baby will have difficulty in breathing whilst it is crying or feeding therefore an oral airway is used until corrective surgery is performed. To examine a baby radiologically, a contrast examination is undertaken. The baby is placed supine on the X-ray couch and a cassette supported against its head. A horizontal X-ray beam technique is employed, with the central X-ray beam directed

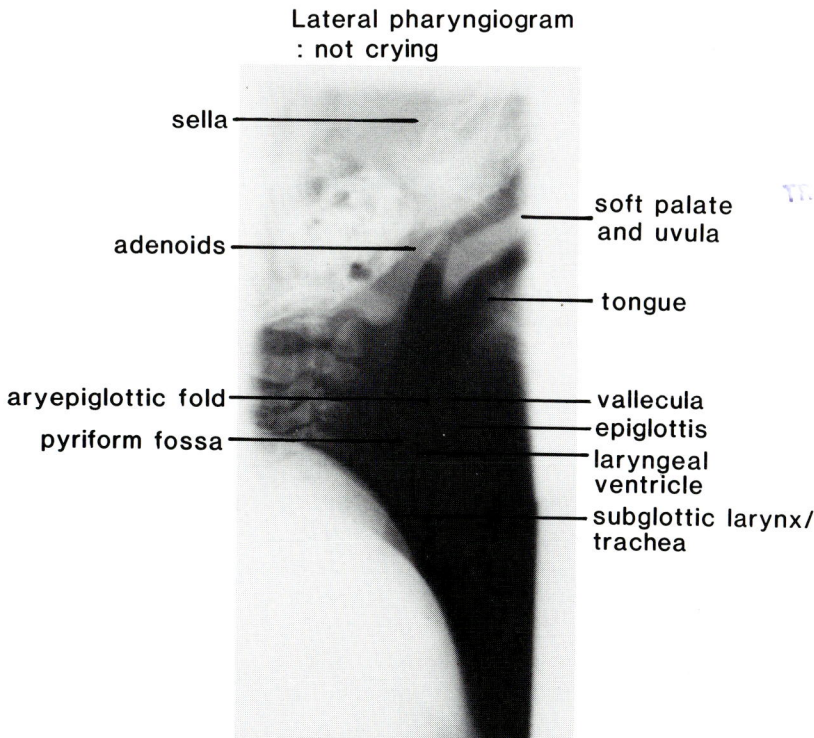

Fig. 3.3 Lateral pharyngiogram on a child with croup

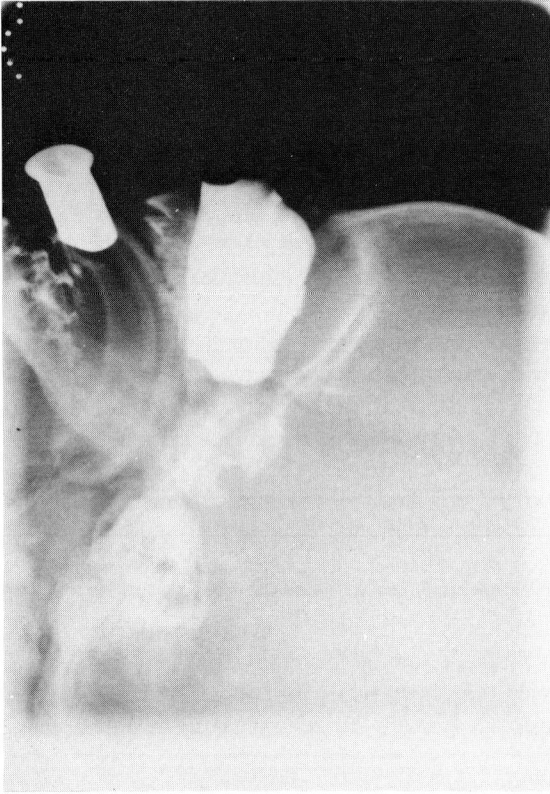

**Fig. 3.4** Contrast examination on a child with choanal atresia

to the condyle of the mandible as for a lateral projection of the paranasal sinuses. A small quantity of an oily contrast medium is placed in the nostril which is nearest to the cassette and an exposure is made. The cassette is then removed and the examination is repeated in the other nostril. Careful marking of the radiographs is necessary (Fig. 3.4).

*Magnification radiography of the chest*

Magnification radiography of the chest is used to help in the diagnosis of diseases which occur in premature and very small babies and to visualise the lung parenchyma and pulmonary vascularity. These are very simply carried out providing that there is an X-ray tube with a 0.3 mm$^2$ focal spot size, or less. A normal X-ray table can be used and intensifying screens with very fine grain together with the use of small focal spot size, taking tube loading and short exposure times into consideration.

The baby is positioned as for a supine projection of the chest at the end of the X-ray table (with bucky etc. moved out of the field) and the cassette placed on the floor directly under the table and baby. The X-ray tube focus will be centred the same distance above the baby as the cassette is below and the exposures made as for a normal chest X-ray. The exposure will be calculated as below:

$$\frac{\text{New exposure}}{\text{Old exposure}} = \frac{\text{New focus to film distance}^2}{\text{Old focus to film distance}^2}$$

The easiest unit to calculate is the mAs.

This system can be used for bones etc. when the detail required is too small to see on a conventional radiograph and the revelation of very subtle changes would give an indication of the diagnosis.

## RADIOGRAPHY OF THE ABDOMEN

The radiograph of the abdomen should show the amount and distribution of bowel gas, the presence of gas in unusual places, the soft tissue planes, the outline of solid organs, soft tissue calcifications and foreign bodies. It should therefore be taken with as low a kilovoltage as possible to give good soft tissue differentiation and the radiograph must include the whole abdomen from diaphragm to symphysis pubis.

*Baby*

The baby is placed supine directly on a covered cassette, a grid being unnecessary, the parents are asked to hold the child above and below the abdomen so one hand is over the chest and the other over the femora, or alternatively one of the restraining cradles is used. The central X-ray beam is directed to the umbilicus and the X-ray beam is collimated to the correct area. The exposure is made on expiration.

Sometimes a prone position is preferable if the infant will settle more readily and it has the added advantages of displacing bowel gas laterally and immobilisation is easier. The umbilical clip on a newborn has to be carefully positioned away from any area of interest or removed completely if possible.

### Toddler and older child

The patient is placed on a mattress on the X-ray table, the room having been prepared beforehand. The central X-ray beam is directed to the umbilicus and beam delineation is undertaken to include the diaphragm and symphysis pubis on the radiograph. If the child is not placed centrally it is necessary to make adjustments by moving the child on the mattress. The child can be gently held by its parents, or restrained by Velcro bands. A bucky band placed gently over the knees can assist immobilisation in a fractious toddler, but it is advisable to ask the parent's consent.

### Erect view of the abdomen in baby and small child

The cassette is supported in an erect cassette holder or chest stand and baby is suspended in the postero-anterior position either by parent or nurse holding him under the arms. Supporting or restraining bands are placed around the lower abdomen, and some of the weight taken by foam pads placed between the baby's legs. The central X-ray beam is directed to the umbilicus, making sure that the diaphragm and the rectum are included on the radiograph. The postero-anterior projection is preferable as the child cannot draw up its legs over the abdomen (this is a natural movement) and it is comforting for them to be looking at the person holding them. If a baby-cradle is available in the department this would make an excellent holding device for a child under 6 months of age. Differentiation between large and small bowel may be difficult in a baby due to lack of haustral markings, but erect lateral projections can sometimes help.

Occasionally an inverted lateral projection may be necessary using the effect of gravity to separate loops of bowel which may otherwise be difficult to interpret on the supine radiograph. Immobilisation is extremely difficult as a child cannot be suspended by the legs. Therefore two nurses are required to hold baby in this position, one supporting the child at the shoulders and the other around the lower abdomen. Strict limitation of the X-ray beam is required to exclude the nurse's hands.

When a child is old enough to stand at the erect bucky it only requires a nurse to make sure that he does not fall in order to achieve a satisfactory radiograph. If a patient is too sick to be placed in the erect position, or is in an incubator, a lateral decubitus projection would be the technique of choice to demonstrate free air and fluid levels. Using a horizontal X-ray beam with the central ray directed to the level of the umbilicus of the patient lying on its side, the X-ray beam is delineated to the area of the abdomen only. Care must be taken that the side of the patient next to the incubator is not obscured and it may be necessary to place the child on a folded sheet or foam pad to raise the abdomen from the floor of the unit. Indications of the technique used should be shown on the radiograph.

The following are the major indications for plain abdominal radiographs:

*Obstruction.* Erect and supine radiographs of the abdomen to demonstrate dilated loops of bowel with or without fluid levels. Additionally look for an absence of gas where it is normally expected to be present.

*Trauma.* Radiographs to give indications of bleeding, displacement of organs due to haematoma, loss of flank stripes or free air in the peritoneum. Further investigations may be necessary, i.e. urography or arteriography, CT scanning or ultrasound examination.

*Infection.* Acute gastroenteritis in infants may provide radiological evidence of paralytic ileus with air/fluid levels and perhaps gaseous distension of stomach.

*Appendicitis.* If the diagnosis is in doubt, the X-ray appearance may assist particularly when looking for a soft tissue mass, a faecolith or gas in an appendix abscess.

*Duodenal atresia.* A congenital developmental anomaly which is due to the non-canalisation of the gut in the duodenal portion. The radiographic appearance should give the 'double bubble' sign, the remaining abdomen being airless.

*Jejunal atresia.* This is treated as above except that the radiographic appearance would show more than two bubbles, but no air distal to the jejunum.

*Volvulus.* In fetal development, the gastrointestinal tract starts as a primitive tube. The middle portion gradually grows out and herniates out of the abdomen at the umbilicus. The foregut and hindgut rotate anticlockwise as growth proceeds and then returns into the abdomen. Any interruption to the growth pattern can result in an abnormal rotation causing a volvulus neonatorum which may give rise to obstruction.

*Pyloric stenosis.* Babies develop symptoms of pyloric stenosis a few weeks after birth. The cause is due to hypertrophy of the circular muscle fibres which cause thickening of the sphincter at the exit of the stomach thus producing signs of obstruction. This condition is nearly always diagnosed clinically, but a plain radiograph of the abdomen may be requested. A barium meal and ultrasound examination are occasionally requested in difficult cases.

*Plugs of meconium and milk.* Inspissated plugs of meconium or milk curds may give rise to an obstruction which will necessitate the child being examined in the supine position radiographically. These radiographs will confirm the clinical diagnosis and help to locate the site of the plug and differentiate between the meconium plug and meconium ileus or Hirschsprung's disease.

*Intussusception.* This condition may sometimes be diagnosed with plain abdominal radiographs. It is an invagintion or telescoping of the bowel due to enlarged 'Peyers patches'. It is most common in well-nourished babies but can occur at any age. The symptoms are abdominal pain which causes the baby to cry and draw up the legs, vomiting and possibly the passage of bloody stools ('redcurrant jelly' stool). A mass may be palpable and the radiograph will show the site of the intussusception by a crescentic edge to the pattern of gas or a coiled spring appearance. A barium enema may be requested to try to reduce the intussusception and a suggested technique is as follows:

An intravenous drip is inserted and adequate sedation given preferably hyoscine and papaveretum. The child will usually go to sleep as the pain is eased, and a large 30 ml balloon catheter can then be placed in the rectum. The balloon is inflated using a small amount of air and 1″ sticky tape is placed either side of the catheter to prevent it from slipping out. A barium reservoir containing a dilution of 2 parts normal saline and 1 part barium sulphate is connected to the catheter, the reservoir is placed 100 cm above the X-ray couch to produce the correct hydrostatic pressure. The barium is allowed to flow into the bowel and with fluorography the progress of the intussusception can be observed. Glucagon can also be administered as a smooth muscle relaxant at a dose of 0.02 mg per kilo of body weight. If the intussusception is not reduced with the barium enema, surgery may have to be performed.

*Hirschsprung's disease.* Radiographs of the abdomen are taken when a diagnosis of Hirschsprung's disease is suspected. This disease is an absence of nerve ganglion in the lower alimentary tract providing a lack of normal peristaltic action in the bowel. This occurs mainly in boys and the most common site for the aganglionic segment is in the rectum and distal part of the sigmoid colon but it can involve longer segments even in the small bowel. In older children where the disease will have progressed to dilatation or megacolon, the accumulation of faecal matter may be very evident on the abdominal radiograph. A severe case in the newborn will produce symptoms of obstruction. Barium studies will assist in the diagnosis as will a biopsy of the suspected area.

*Necrotising enterocolitis.* This disease of the premature baby has an unknown aetiology. The symptoms are abdominal distension and the passage of blood per rectum. The abdominal radiographs will confirm the typical sign of gas in the mucosa which forms a 'halo'. A left lateral decubitus radiograph will show free air in the peritoneum or air in the biliary tree.

*Ano-rectal malformations.* Anal anomalies are easily seen on inspection of a newborn baby, but the extent of any atretic portion cannot be assessed. 70% of rectal atresias have a fistula communication with some part of the genito-urinary tract. Radiographs can assist the surgeon in assessment of the type of treatment to be undertaken, either an immediate repair is possible or a temporary colostomy may have to be performed, providing that sufficient time has elapsed for swallowed air to reach the rectum. The baby should be placed in the Trendelenberg position for at least 15 minutes and then completely inverted for 5 minutes to allow the air to filter past the viscid meconium (Fig. 3.5).

*Inverted lateral view.* Baby is supported, upside

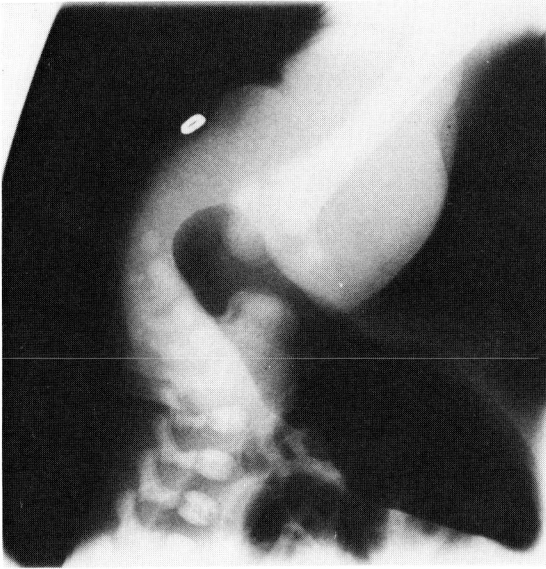

Fig. 3.5 Inverted radiograph on child with rectal atresia

down in the true lateral position, against a cassette supported in the erect cassette stand. The central X-ray beam is directed to the greater trochanter, so that the two ischial bones are superimposed and the hips are flexed over a foam pad so that the femora do not overlie the pelvis. A metal skin marker should be placed on the anal dimple or barium paste smeared in the natal cleft. The value of this radiograph is in assessing the position of the air-filled rectal pouch, relative to the anus and the pubo-coccygeal line. This line defines the level of the levator ani and pubo rectalis muscles.

*Alternative positioning.* This can be achieved by placing the baby prone for several hours, before the radiographs are taken with the patient prone on the X-ray couch, and using a horizontal X-ray beam, the patient's buttocks being raised on a 45° pad. This makes the holding of the baby much easier.

A small amount of contrast media may be injected through the anal membrane into the blind pouch and similar radiographs taken, making the measurement of the gap more accurate.

If a fistula into the urinary tract is evident a micturating cystogram can be performed to demonstrate possible reflux of the contrast medium into the blind rectal pouch.

*Jejunal biopsy.* Coeliac disease is a condition of malabsorption of the small intestine. This shows on a jejunal biopsy as a loss of normal villus structure of the upper jejunum. This abnormal structure returns to normal when the patient has a gluten-free diet. The diagnosis ideally requires three biopsies and if correctly carried out there is rarely any resistance to re-investigation by patients or their parents.

Patients are admitted fasting as day cases. They are given oral sedation such as trimeprazine tartrate 2 mg per kilo of body weight, on arrival, to allay anxiety. The child is taken to the X-ray department and placed on a fluoroscopy table, then the mouth and throat are anaesthetised with lignocaine spray (a metered spray of 10 mg). Intravenous diazemuls is administered (0.1–0.2 mg per kilo of body weight) followed by metaclopromide (0.5 mg per kilo of body weight) (Gaze et al 1974). The biopsy capsule is attached to a stiff catheter tube with a thicker, softer outer tube placed over part of its length. The two tubes are introduced orally and when the outer tube is in the oesophagus the inner tube can be advanced and retracted without producing a 'gagging' effect. With a minimum of fluoroscopy (less than 5 seconds) the capsule is advanced until it reaches the 3rd/4th part of the duodenum or duodenal/jejunal flexure and the capsule is fired.

If the specimen is satisfactory the patient is returned to the ward and discharged the same day. As the diazemuls has an amnesic effect, the child has no recollection of the experience and quite happily returns 6 months later after treatment on a gluten-free diet for a further biopsy, and again 6 months later after a gluten challenge. The only memory the child will retain will be of the reception received in the X-ray department before sedation.

## RADIOGRAPHY OF THE SPINE

To demonstrate fractures, congenital anomalies, secondary deposits or bone infection, antero-posterior and a lateral projections of good technical quality are usually all that is needed. Limitation of the X-ray beam to the area in question is most important. By keeping the kilovoltage to a minimum, the whole of the spine in a small child

**Fig. 3.6** Aluminium filters for whole spine radiography

can be visualised on one radiograph. Unless for specific orthopaedic purposes a coned radiograph of lumbar V/sacral I junction should not be taken as this gives a gonad dose of radiation equal to that of a chest radiograph every day for several years.

*Scoliosis.* Radiography plays a vital part in the evaluation and treatment of scoliosis, by determining the type of spinal deformity and monitoring its progress. The radiographs must be clearly labelled, showing the technique and position used in order that future examinations can be comparable. There is a standard series of radiographs used for initial assessment, i.e. postero-anterior and lateral views of the whole spine standing, plus a supine antero-posterior and lateral view of the affected area plus a radiographic assessment of the bone age.

*Postero-anterior view of the whole spine — standing*

This radiograph should include the whole spine from the level of the external auditory meatus to the level of the anterior superior iliac spines of the pelvis. Gonad protection should be used and radiographs taken in the postero-anterior position to give vital glands, e.g. thyroid and mammary, as low a dose of radiation as possible. The patient stands erect (shoes are removed) with feet together and knees straight. The pelvis is checked to ensure there is no rotation, i.e. anterior superior iliac spines are equidistant from the cassette, arms are placed level with the shoulders and the exposure is made on full inspiration. Because of the variation in densities of the spine, it is advisable to employ an aluminium filter (Fig. 3.6).

*Lateral view of the whole spine*

The patient adopts a similar position but at right angles to the cassette with arms raised to shoulder level, and the exposure is made on full inspiration. A long cassette containing rare earth screens is employed. If a grid of sufficient length is not available an alternative technique would be to use an airgap. A 30 cm airgap gives a very acceptable quality of radiograph if used in conjunction with a 360 cm f.f.d.

*Supplementary views.* These are undertaken as necessary in the supine or oblique position. Additionally views may be required of selected areas and incorporating flexion, hyperextension and traction techniques.

## RADIOGRAPHY OF THE PELVIS

To accomplish a quick, successful examination of the pelvis the usual room preparation must be carried out. The child is placed towards the bottom end of the X-ray couch. Mother stands at the foot of the couch and holds the child's knees to keep the legs straight, knees and heels together, toes pointing to the ceiling, and the pelvis straight. Mother looks directly at the child, and should talk to distract his attention. If the cassette was ready before the child came into the room, the exposure can be very quickly made. No lead protection is used on the first attendance, only on follow-up examinations (Fig. 3.7). There is generally no need to undress the child (except remove nappies) as the children's panties are radiolucent. Pads can

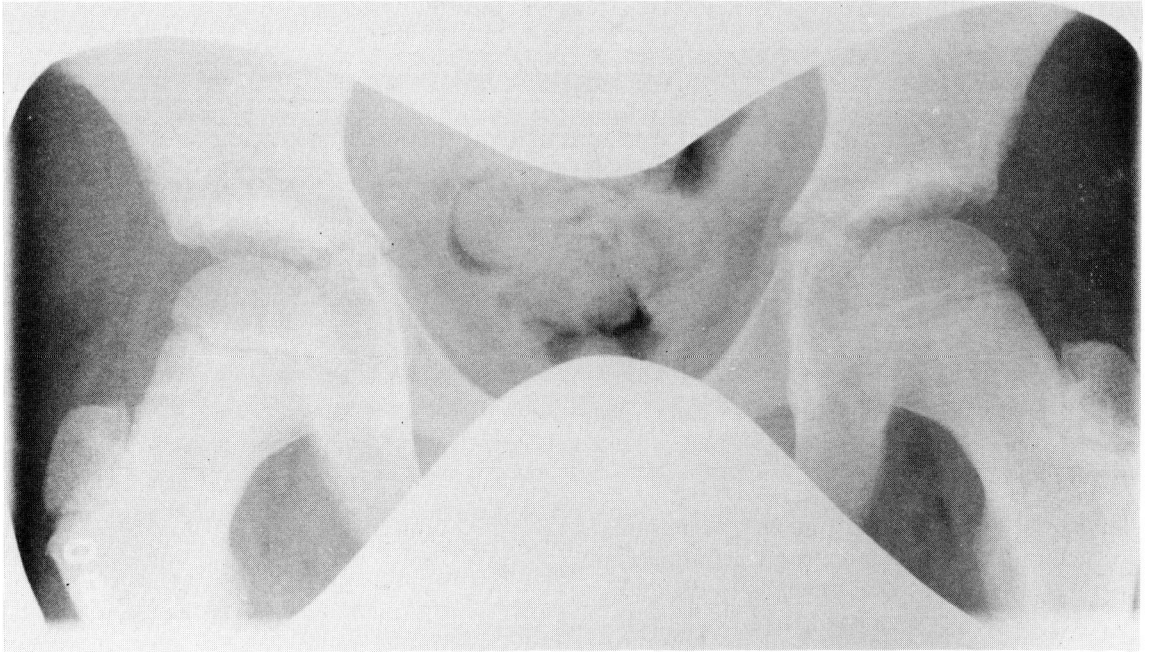

**Fig. 3.7** Lead gonad protection as used in Austria

be placed on the couch so that shoes do not soil the sheet.

### Frog lateral position or view

From the supine position the thighs of the child are gently flexed and then the legs are allowed to fall outwards and the soles of the feet are placed together. Sandbags can support the outstretched knee to keep them at an equal angle. Mother holds the child's feet keeping them together all the time. Lead protection is put in place and the exposure made.

### Andrén-Von Rosen view

From the antero-posterior position the femora are abducted to an angle of 45° and the feet inwardly rotated. This radiograph has no value at all unless the femora are at equal angles of 45°. The central X-ray is directed over the symphysis pubis.

### Congenital dislocation of the hips (CDH)

In 1976, 60 children per 10 000 were treated surgically for this complaint. To demonstrate congenital dislocation of the hips the radiograph must be exposed when the pelvis is non rotated in the antero-posterior position. However, before commencing treatment orthopaedic surgeons like to visualise the whole of the femoral head, therefore a frog lateral projection is performed. A few centres still perform the Andrén-Von Rosen projection. Lead protection should be used as accurately as possible for all follow-up examinations as children with a CDH will endure many examinations often including an arthrogram.

### Perthes' disease (osteochondritis of the femoral head, a disease occurring mainly in boys in the ratio 4:1)

To diagnose this disease antero-posterior and frog lateral views of both hips are required. The lateral projection generally gives the most information, but the comparison of both hip joints is important. Radionuclide imaging is a very useful diagnostic procedure to undertake in Perthes' disease as increased uptake of the isotope in one femoral head would indicate bone disease even before it is visible radiographically.

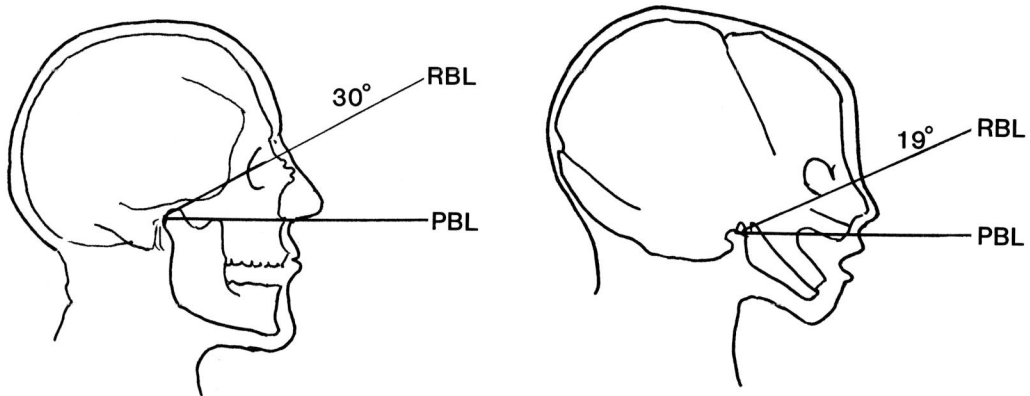

**Fig. 3.8** Difference between paediatric base line (right) and orbitomeatal line

*Slipped capital femoral epiphysis*

Slipping of the femoral head can be produced any time due to trauma, but more commonly it occurs in teenagers who are overweight (Pickwick's fat boy). An antero-posterior radiograph of the pelvis and hips together with a frog lateral view is required. Radiation protection should be rigorously adhered to as these patients will have numerous examinations during the operative and postoperative treatment.

## RADIOGRAPHY OF THE SKULL

Good quality radiographs of the skull may be very difficult to obtain on a small child, and sedation may be required. Immobilisation is best undertaken by the parents holding the head in position with foam pads. Tight restraining bands are not recommended on the baby's head where sutures are open. Foam pads should *not* be placed over the ears. Babies can be wrapped in a cotton blanket to resemble an Egyptian Mummy as this will make them feel warm and secure, but this would have the reverse effect on a toddler, since they resent having their hands tied. A 2-year-old will require immense patience and skill to obtain accurate positioning and to hold that position long enough for the exposure to be made. Experience has shown that distraction is most effective in this age group by the use of toys, mobiles, pictures at the critical moment and making the examination into a game.

A baby's head is not the same as an adult head and the ratio of the size of facial bones to vault alters considerably during the first 10 years of life. For this reason the paediatric base line is taken as that line which extends from the external auditory meatus to the anterior nasal spine. The angle between the paediatric base line and the normal radiographic base line (orbitomeatal line) varies from a newborn of approximately 19° to an adult of 30° (Fig. 3.8).

For a routine examination of the skull on a child three basic views should be taken: antero-posterior, occipital (Towne's) and a lateral. It is rarely possible to persuade a small child to lie with the face down on the table.

### Antero-posterior view

The patient is placed on a flat table or skull unit in the supine position and the head flexed by a small sponge pad behind the neck to bring the paediatric base line 5° caudally from the perpendicular. The sagittal plane is at 90° to the cassette and the central X-ray beam is directed 2 cm above the nasion in the mid line (Fig. 3.9).

### Occipital (Towne's) view

The patient is placed in the supine position with a wedge-shaped foam pad under the occiput to bring the paediatric base line 15° caudally from the perpendicular. The medial sagittal plane is at right angles to the cassette and the central X-ray beam

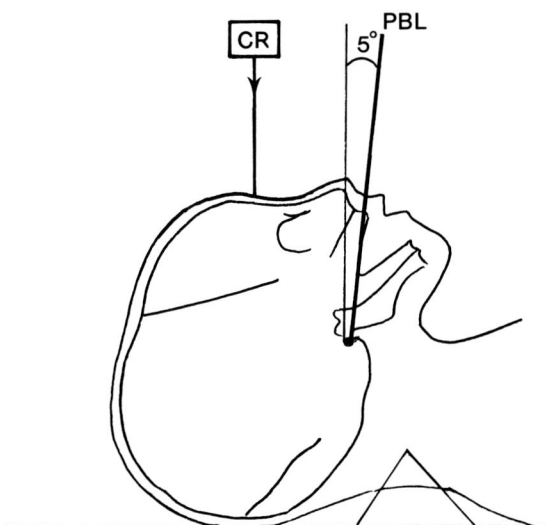

**Fig. 3.9** Antero-posterior position for skull

is directed 35° caudally, through the midline on a level with the external auditory meatus.

## Lateral view

The patient is placed supine with a pad under the occiput to raise it to the centre of the cassette. The cassette is supported on one side of the head parallel to the median sagittal plane and the X-ray beam is directed 2 cm anterior and superior to the external auditory meatus. A toy or mobile is positioned above the child or he may be given a teat to suck to direct his gaze vertically. If the child is extremely restless a supine view can be undertaken. The patient is placed in a postero-oblique position with the head turned into the lateral position. A sponge pad of the correct size is positioned under the head to keep the median sagittal plane parallel to the cassette. A compression band can be used to keep the head in the correct position.

## Postero-anterior 15° view

When a child is co-operative he can be placed prone with the paediatric base 20° caudally from the perpendicular. The median-sagittal plane is at right angles to the cassette and the central X-ray beam is directed 15° caudally through the mid-line to emerge 2 cm above the nasion.

## Submento-vertical view

Due to the suppleness of a child's neck, this view is quite easy to obtain in the normal manner. By lowering the head from the supine positon until the paediatric base line is parallel to the cassette the central X-ray beam is directed perpendicular to the cassette in the mid-line between the angles of the jaw.

## RADIOGRAPHY OF THE SINUSES

Examinations of the paranasal sinuses in babies have very little value. The sinuses are under-developed and as the films will have to be exposed in a supine position, three views are adequate: a Caldwell's view, a Waters (reverse) projection of the maxillary sinuses and a lateral. The lateral projection should be centred to the condyle of the mandible (to include the whole of the postnasal space) with the mouth closed. In an older child a submentovertical projection would demonstrate the ethmoid and sphenoid sinuses. When a child reaches an age to be able to co-operate the routine adult views would be applied.

By using the paediatric base line most projections for facial bones, temperomandibular joints Stenver's mastoid views and optic foramina can be modified to be taken supine.

## Stenver's view

The patient is supine and the head is flexed to bring the paediatric base line 20° caudally from the perpendicular. The median sagittal plane is rotated 45° from the perpendicular and the central X-ray beam is directed 12° caudally to pass through the petrous portion of the temporal bone, i.e. at a midpoint between the external auditory meatus and the outer canthus of the eye. (Note: this projection demonstrates the mastoids on the side away from the film.)

### Views of the optic foramina

For the right side: the patient is placed supine with the paediatric base line perpendicular to the film. The head is rotated towards the left side until the median sagittal plane forms an angle of 35° from the perpendicular. The central X-ray beam is directed straight through the right orbit. Repeat with head rotated to the right for the left eye.

## RADIOGRAPHY OF THE FEET

To obtain a dorsi-plantar view, babies and toddlers can be seated on a small box or pad and supported from behind by Mother. To demonstrate pes planus a weight-bearing lateral view of the foot would be required. This can simply be carried out with the child standing on a polystyrene box against the cassette and held by the escort.

For congenital talipes equinovarus (club foot) a routine set of radiographs would include weight-bearing dorsi-plantar views of both feet, a lateral view of the foot with true lateral projection in relation to the ankle joint, an axial view of the calcaneum and an oblique projection of the affected foot. In infants the weight-bearing effect is achieved by the parent applying pressure against the sole.

## RADIOGRAPHY TO ESTABLISH BONE AGE

The maturity of bone development (bone age) is compared to the child's actual age. This is done by comparing the patient's hand and wrist with a standard set of radiographs such as those published by Greulich & Pyle (1966). More accurate charts are those prepared by Tanner & White-house (1959). It makes little difference to the assessment if the right or left hand is used, but by convention the non-dominant hand is chosen. Final height of children may be important in career choices for example, that of ballet dancers. For assessment of bone age in a baby of less than 40 weeks gestation, a knee is radiographed. The distal femoral epiphysis appears at 36 weeks of fetal life and the proximal tibial epiphysis between 38 and 40 weeks.

### Radiographic technique

It is important that the radiograph is correctly exposed. The palm of the non-dominant hand faces downwards in contact with the cassette. The axis of the middle finger is in direct line with the axis of the forearm. The fingers should be straight and not touching and the thumb should be placed in a comfortable position. The hand can be secured in this position with adhesive tape, or clear film or a baby's hand can be held firmly by Mother or nurse and released a split second before the exposure is made to avoid exposure to themselves.

Radiographs with high quality definition are needed and a non-screen or single screen type system is preferable. A film focus distance of 100 cm is used with the central X-ray beam directed to the head of the third metacarpal. For radiation protection a lead rubber coat must be worn by the patient and the table top covered with lead if the patient is seated. It is easier if babies are placed prone with the arm extended.

### Leg length measurement

It is sometimes necessary to measure the length of the lower limb bones with a greater accuracy than can be achieved by routine radiography. This will determine whether it is necessary to treat a discrepancy in the length of the two limbs by operative means. One technique would be to make six exposures on one long cassette (115 cm × 35 cm) accurately centred and exposed over the six joint spaces of the hips, knees and ankles. The feet are placed approximately 14 cm apart to keep the joints in longitudinal alignment. The long bones are shielded from the X-radiation as much as possible to protect the bone marrow, and immobilisation with sand bags has to be undertaken carefully. Any movement of the patient would make the examination inaccurate therefore two radiographers working together make this a more speedy and efficient procedure, with one radiographer adjusting the exposure and the other changing the X-ray tube position.

Measurements can then be made from each joint space. To achieve an even density of the various joints on the radiograph several suggestions are made, including use of graded screens, graded filters, and a stationary grid placed under the pelvis. An alternative method is to use a Scanogram procedure which gives a very small slit beam of X-radiation which is exposed to the entire length of the leg. The disadvantages of this procedure would be the X-radiation dose and the capital outlay for such a special type of examination.

## NON-ACCIDENTAL INJURY

This term covers a complicated mixture of abuse, neglect, physical violence, emotional trauma and nutritional deprivation which result in soft tissue and bony injury, malnutrition and wasting. The main types are:

Drug abuse: intentionally neglecting or giving drugs which cause harm

Medical abuse: intentionally failing to obtain medical treatment

Sexual abuse: assault or molestation

Safety neglect: wilful neglect of normal safety precautions taken in the average home

Emotional abuse: belittling, humiliating the child and depriving it of love and affection

Child neglect: lack of basic survival needs, i.e. food, clothes, warmth and love

Physical abuse: shaking or battering the child.

The causes of the above are many and varied but the main reasons are a lack of discipline, rejection by the parent, hyperactive child, parental stress, family crises, illness, parental abuse of alcohol or drugs, abuse from child minder and lack of mother and child contact in an unwanted child.

### Signs of non-accidental injury

These are most common in children under the age of 3 years. The general obvious appearances show neglect, nappy rash and malnutrition. There may be fractures of bones, and evidence of old fractures, soft tissue injuries, and bruising at different stages of healing. There may be a history of several episodes of trauma and frequent visits to the casualty department. The explanation given for the injury may be unconvincing.

Radiological investigations would be either a skeletal survey or a radionuclide bone scan, which would present signs of increased uptake of the isotope in damaged bone or epiphyses.

A suggested skeletal survey would include:

lateral view of the skull
postero-anterior view of the chest (including shoulder joints)
antero-posterior view of the abdomen (including hip joints)
antero-posterior projection of all long bones.

All the above projections should be correctly centred and all exposed separately. Further projections would be taken if found necessary. From the X-radiation aspect a skeletal survey carried out as above gives the patient approximately one-third of the X-radiation dose of a 'babygram'.

## BARIUM EXAMINATIONS

There is very little difference in barium examinations for children compared with adults other than the basic approach. Barium sulphate can be made more dilute as the patient is relatively slim, and for barium meals it can be flavoured with milk shake solutions — banana, strawberry or chocolate — to make it more palatable. For babies, sterile water and sterile bottles would be used, or a catheter dummy (Fig. 3.10). A bottle steriliser with Milton should be kept in the department. Toddlers prefer a feeding cup, or even their own special cup from home. Barium enemas require dilution with normal saline if there is any risk of large amounts of fluid being absorbed via the colon causing a fluid imbalance. As with all examinations a minimum amount of fluoroscopy and radiographs to minimise the radiation dose to the patient.

Barium meals and swallows demonstrate gastro-oesophageal reflux, hiatus hernia, unco-ordination of swallowing, congenital abnormalities and strictures following surgery. Small bowel meals reveal a malrotation of the gut. Metaclopromide can be given to speed up the examination. Bowel preparation is not required for a barium enema for Hirschprung's disease, and a 24 hour radiograph will be taken after the administration of barium.

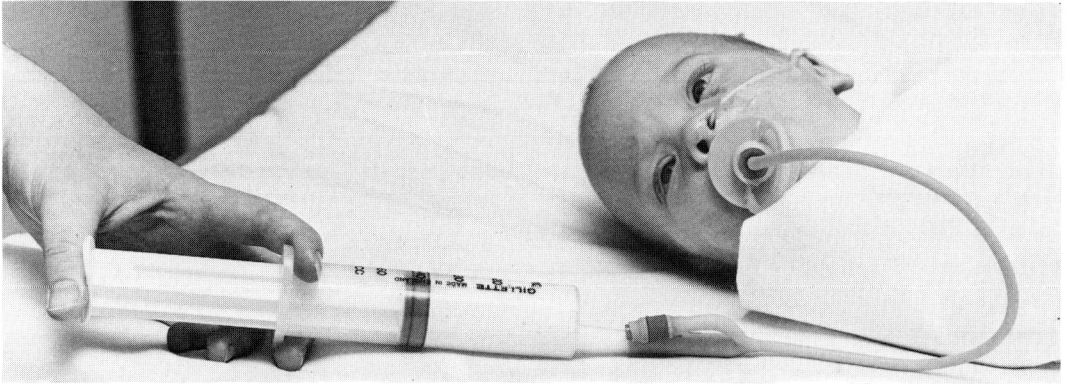

**Fig. 3.10** Baby with catheter dummy

When searching for polyps, Crohn's disease and ulcerative colitis, bowel preparation would be administered according to the patient's condition. For a therapeutic enema for babies with suspected meconium ileus, a Gastrografin solution can be used, but as this contrast agent is very hypertonic it is neccessary to have an intravenous drip in situ, the osmolality being 2150 milliosmols per kg of water.

## RADIOGRAPHY OF THE GENITO-URINARY TRACT

### Intravenous urography

Apart from chest radiography, the IVU is as frequently performed as any other procedure. The indications for the examination range from urinary tract problems, assessment of congenital abnormalities unrelated to the urinary tract to investi-

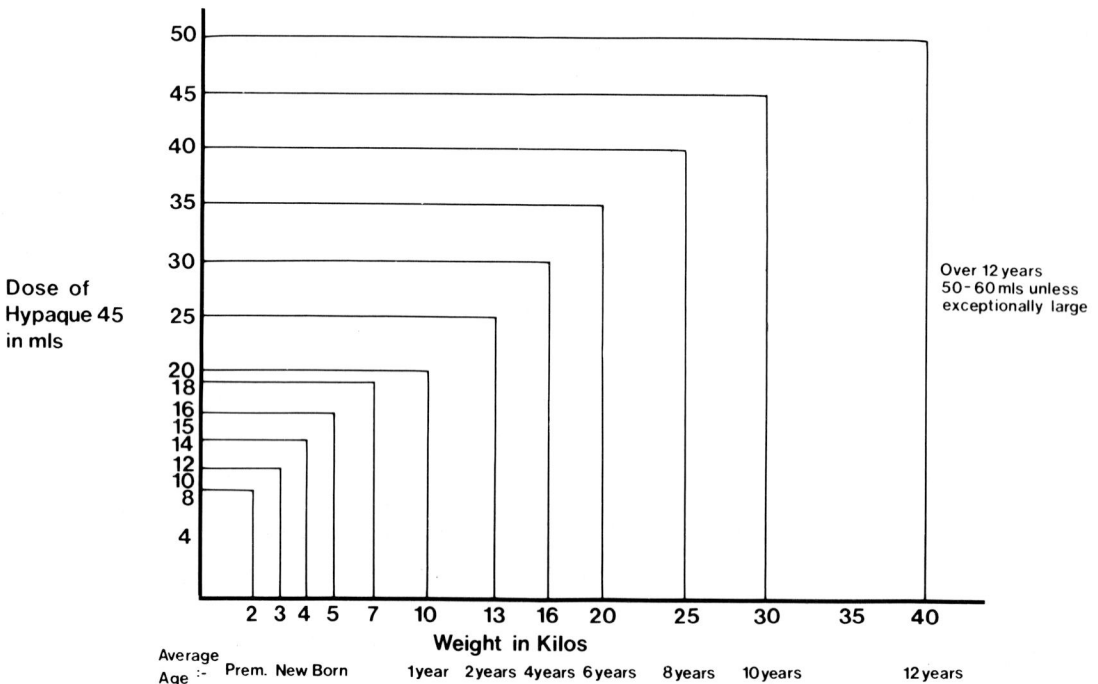

**Fig. 3.11** Doses of 280 mg/ml contrast for children's urography (Wessex Paediatric Radiology, suggested dosage for IVP)

gations of pyrexia of unknown origin. Parents should be told beforehand what the examination involves and they can decide whether to tell the children that an injection will be given. Abdominal preparation is important so that gas and faeces do not render the examination useless. Rigorous fluid restriction is not advised, but nothing is given by mouth for 4 hours and this is sufficient preparation. An empty stomach is an advantage so that the child does not vomit during the injection. A fizzy drink or a feed is given on completion of the injection of contrast media to produce a 'window' of gas through which to view the left kidney and to settle the patient after the venepuncture.

A suitable dosage of contrast media is shown in Figure 3.11. The choice of contrast is one having an iodine content of 280 mg/ml. The non-ionic contrast media are now first choice for neonates because of their lower osmolality which causes less change in the physical and chemical properties of the blood than do ionic contrast media. In a small

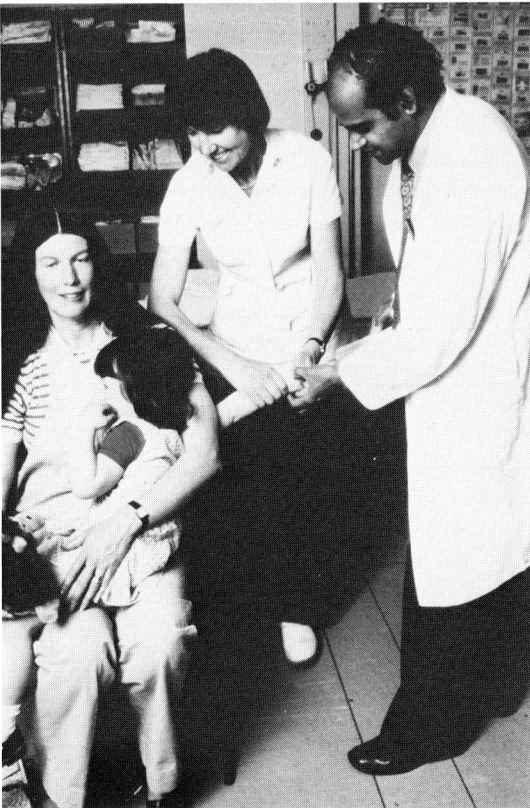

**Fig. 3.12** Toddler on mother's lap for venepuncture

child the easiest injection site to immobilise is the hand and a toddler can sit on Mother's lap (Fig. 3.12). In the neonate a scalp vein may prove to be the best site. The antecubital fossa is used in older children as it is slightly less painful than the hand. A butterfly infusion needle is the needle of choice and can be obtained in sizes ranging from 25G to 21G which should be ready on the trolley.

**Radiographic technique**

This will vary from centre to centre but a minimum number of radiographs should be taken. Control radiographs are rarely needed for follow-up examinations. For a first time examination, a radiograph coned to the kidney area immediately after the injection and a further radiograph 10 minutes later after a fizzy drink is a common sequence. At 20 minutes a radiograph including the kidneys, ureters and bladder completes the examination. A post-micturition radiograph of the bladder gives very little useful information and may be omitted especially as there are better tests of bladder disorders. The aim of the examination is to see the whole of the renal outlines, the thickness of the cortex and a sharp picture of the collecting system and ureters.

If the renal outlines are not clear due to gas and faeces, a 'renal window view' can be undertaken by directing the central X-ray beam 35° caudally through the xiphisternum. This enables the right kidney to be visualised through the liver substance and the left kidney through the lower part of the heart and gas-filled stomach.

Follow-up examinations require only two radiographs, one at 10 minutes after the injection and a 20 minutes radiograph to show the entire renal tract except in cases of hydronephrosis where a delayed radiograph at 1 hour following a water load will show the efficiency of drainage. If an assessment of bladder residue is required then ultrasound would be the technique of choice.

The findings on children's intravenous urograms would include duplex kidneys and ureters, ureterocoeles, neurogenic bladder, Wilms' tumours, neuroblastomas, hydronephrosis, pelvi-ureteric junction stenosis, congenital urethral valves, bladder neck obstructions, bladder diverticulae and nephrocalcinosis.

During the injection an oxygen supply and suction apparatus should be available and an emergency tray with the following drugs: adrenaline, calcium gluconate, diazepam, hydrocortisone, atropine, propranolol, phytomenadione, sodium bicarbonate and lignocaine, with the recommended paediatric doses.

**Micturating cystograms**

This is a safe, efficient, accurate examination of the lower urinary tract. The main indications for performing this test are suspected urethral abnormalities and vesico-ureteric reflux. The recording methods can be by cinéfluorography, fluoroscopy with spot films and photofluorography. The choice of method is influenced by the equipment available, the radiation dose and the purpose of the examination.

Sodium diatrizoate 20% solution, or a meglu-minediatrizoate 30% solution are suitable contrast media being of adequate density and non-irritant. After emptying the bladder the child is catheterised on the X-ray table.

With Mother standing at the top of the table and the nurse at the foot the child is readily restrained. A 6 or 8 FG feeding tube, or a Foley catheter lubricated with 1% Lignocaine gel is suitable. It is a sterile procedure so a catheterisation tray has to be prepared beforehand. The tray includes sterile towels or drapes, swabs, 2 ml and 50 ml syringes, feeding tubes or catheter.

The bladder can be filled through a drip in-fusion bottle placed 50 cm above the table top or with a 50 ml syringe and hand pressure. The filling of the bladder is observed by intermittent fluoroscopy. When it is judged that the child is ready to micturate, the catheter is removed. Over the age of 4 years most children can micturate erect, even girls. The table is brought to the erect position and a low box placed on the footstep. The child turns to the right and the right foot is placed on the box. This separates the femora and gives better access to the Mother who holds a receptacle to catch the micturition stream. For smaller children the table remains horizontal and they micturate whilst lying on their side with absorbent pads. Spot radiographs are taken during mictur-

ition. After micturition a final radiograph is taken to show the bladder area and ureters.

Speed is the key to success as it is an uncomfortable experience for the child and they will not tolerate delay or hesitation. It is preferable for two radiographers to be present, one for the position of the patient and the other for setting the exposures and changing the cassettes. The whole procedure should not take more than 5 minutes after catheterisation and sedation is better avoided. The patient is told to have a warm bath to remove any sticky residues and also encouraged to drink more fluid than usual. A lollipop or some sweets given encourages the child to be cooperative in the future.

## ULTRASOUND IN PAEDIATRIC PRACTICE

Ultrasound is being used increasingly in paediatric imaging because:

1. It is non-invasive.
2. No harmful effects are known to occur at the power levels used.
3. Several different organ systems can be investigated at the same examination.
4. Imaging does not depend on organ function.
5. Little or no patient preparation is required.
6. The child does not usually have to be restrained in difficult positions as in radiography.
7. Follow-up after treatment can easily and safely be performed.

Babies can be fed during the examination to help settle them. Toddlers however can prove the more difficult age group to deal with but a real-time facility can enable a rapid assessment without any great loss of detail. The older children often enjoy watching the probe movement and seeing their 'insides' on the TV monitor.

The modern high quality compound B-scanners and real-time machines now available give very good detail. Most young children are scanned using a 5 MHz probe and the older ones, from 5–6 years upwards usually need a frequency of 3.5 MHz. Many real-time systems are fully mobile and thus can be used on the Intensive Care Unit, obviating the need to move very sick children.

## CONCLUSION

In conclusion, many techniques have not been described due to lack of space, but this does not mean to say that they are not successful or even better. However, the techniques that have been described are well tried and tested. Modern technology and equipment may contribute to a successful result but of greatest importance is the paediatric radiographer's own skills. When dealing with children the radiographer needs lots of patience, wisdom and integrity. Patience at all times, integrity to keep all promises and wisdom never to promise the impossible.

*Acknowledgements*

The author would like to thank Dr Ivan Hyde, Consultant Radiologist of Wessex Paediatric Unit for all his help and encouragement, and Mrs Sandra Brown for typing the manuscript.

REFERENCES

Code of Practice for the protection of persons against Ionizing Radiations arising from medical and dental use. 1972 HMSO, London
Court S A M 1976 Fit for the future. HMSO, London
European Society of Paediatricians 1953 The welfare of children in hospital, London.
Gaze H, Rolles C J, Signer E, Sagaro E 1974 Premedication for jejunal biopsy in childhood using intravenous diazepam and metaclopramide. Archives of Disease in Childhood 49:4
Greulich W W, Pyle S I 1966 Radiographic atlas of skeletal development of hand and wrist. Stanford University Press, USA
Halsbury H 1974 The Halsbury report on pay and conditions of nursing and paramedical staff. HMSO, London
Hyde I, Danby B 1980 An aid to barium meal examinations in infants. British Journal of Radiology 53: 997–998
Platt H 1959 The Platt report. Welfare of children in hospital. HM(59) 43. HMSO, London
Tanner J M, Whitehouse R H 1959 Standards for skeletal maturity. International Children's Centre, Paris
Tanner J M, Whitehouse R H, Healy M J T 1962 A new system for estimating skeletal maturity. Department of Growth, Institute of Child Health, University of London

BIBLIOGRAPHY

Aspects of Sick Children's Nursing 1981 General Nursing Council for England and Wales
Ardran G H, Coates R, Dickson R A, Dixon Brown A,

Harding F M 1980 Assessment of scoliosis in children. British Journal of Radiology 53: 146–147
Bates M, Practical paediatric nursing. Blackwell Scientific, Oxford.
Brunner S B, Suddard D S 1981 Lippincott manual of paediatric nursing. Harper & Row, London
Caffey J 1978 Paediatric X-ray diagnosis, 7th edn. Year Book Medical Publishers, Chicago
Clark K C 1973 Positioning in radiography, volume 1, section 6. Heinemann, London
Cremin B J 1971 Radiological assessment of ano-rectal anomalies. Clinical Radiology 22: 239–250
Fundamentals of Radiographic Photography 1973 Kodak International Commission on Radiological Protection 1977 ICRP No. 26
Krepler P, Vana N, Havranek C 1977 Gonad protection in radiological examination of hips. Pediatric Radiology 5: 231–235
Moe J H, Winter R B, Radford D F, Lonstein J E 1978 Scoliosis and other spinal deformities. Saunders, Philadelphia, ch 3
Mole R H 1963 Possible consequences of radiographic tissue dose. British Journal of Radiology 424: 241–246
Ozonoff M B 1979 Pediatric orthopedic radiology. Saunders, Philadelphia
Pollen A G 1973 Fractures and dislocations in children. Churchill Livingstone, Edinburgh
Stephens D F, Smith E 1971 Ano-rectal malformations in children Year Book Medical Publishers, Chicago
Whitson B J, McFarlane J M 1980 Pediatric nursing skills. Wiley, New York

# 4

# Accident and emergency radiography

## INTRODUCTION

The Accident and Emergency Department in any hospital acts as the public relations department to the local population, in that the reputation and efficiency of the hospital tends to be judged by the populace when attending for diagnosis and treatment of his or her trauma. Patients have come to 'expect' that an X-ray is mandatory to confirm or disprove any underlying injury or malady, therefore the radiodiagnostic department is an integral part of all Accident and Emergency organisations.

All injuries are major to any patient who is suffering. However, the majority referred to the Accident and Emergency Department have sustained relatively minor injuries affecting only one part of the body. A small percentage of patients arriving in the department suffer from multiple injuries, usually as the result of a road accident, and this group demands attention from the most experienced professionals, including the radiographic team. It is essential that there is always a senior, experienced radiographer available in the radiodiagnostic department attached to the Accident and Emergency Department.

Radiography of the accident and emergency patient tends to be the most popular section amongst the radiographers. This is probably because:

1. there is a large variety of examinations ranging from minor injuries to the acutely ill patient or the multiple injury patient.

2. the radiographers are part of a small team and they are working alongside the referring clinician as opposed to being completely remote from the clinical situation.

3. in most instances their professional expertise

is relied upon more than in other sections of the radiodiagnostic department.

Unfortunately, all too often, the pressure of the workload does not allow the radiographer to spend sufficient time with the patients and standards are apt to fall. This must not be allowed to continue in any department, especially in the accident section, as the majority of patients will be relatively young and fit and a wrong diagnosis due to poor radiographs can result in a permanent disability. In order to maintain a high, uniform radiographic standard in the Accident and Emergency Department, the following points must be followed:

a. Senior experienced radiographers should be available (on a 24 hour basis).

b. Sufficient radiodiagnostic rooms and radiographers are available at all times so that optimum time can be allotted to each patient.

c. Radiographers must be meticulous in the production of each radiograph and check that all the details are correct.

d. All multiple and severely injured patients must be radiographed in the main radiodiagnostic room — mobile radiography in the resuscitation room has little or no value.

Most Accident and Emergency Departments are open 24 hours each and every day. It is essential, therefore, that this heavily used X-ray apparatus is properly and regularly maintained and that any malfunction is rectified immediately.

## INITIAL PLANNING

The number of radiodiagnostic rooms to be

provided for the workload from the Accident and Emergency Department can be determined:

(i) from local and national statistics — the number of examinations currently arising per year per thousand of the hospital's catchment area.

(ii) the average length of time each examination requires

(iii) the length of the working day.

Where the accident and emergency radiography takes place in the main radiodiagnostic department, the total number of examinations arising per year per thousand of the catchment area, can be used to determine the total size of the department, but where it is separated (and this is not to be encouraged in the planning stage of a new hospital), then the number of accident and emergency patients arising per year per thousand has to be determined.

The number of radiodiagnostic rooms can be calculated from the formula:

$$\frac{N \times E}{D}$$

when N = number (in thousands) of people in the catchment area.

E = number of examinations per 1000 per year of existing catchment area (from departmental statistics).

D = number of examinations per radiodiagnostic room.

D is calculated from the working time available (T); the average time per examination (M); and the room utilisation rate. An average for the room utilisation rate is 70% so the formula becomes

$$D = 0.7\frac{T}{M}$$

LOCATION

Generally two radiodiagnostic rooms will be required for a District General Hospital. This

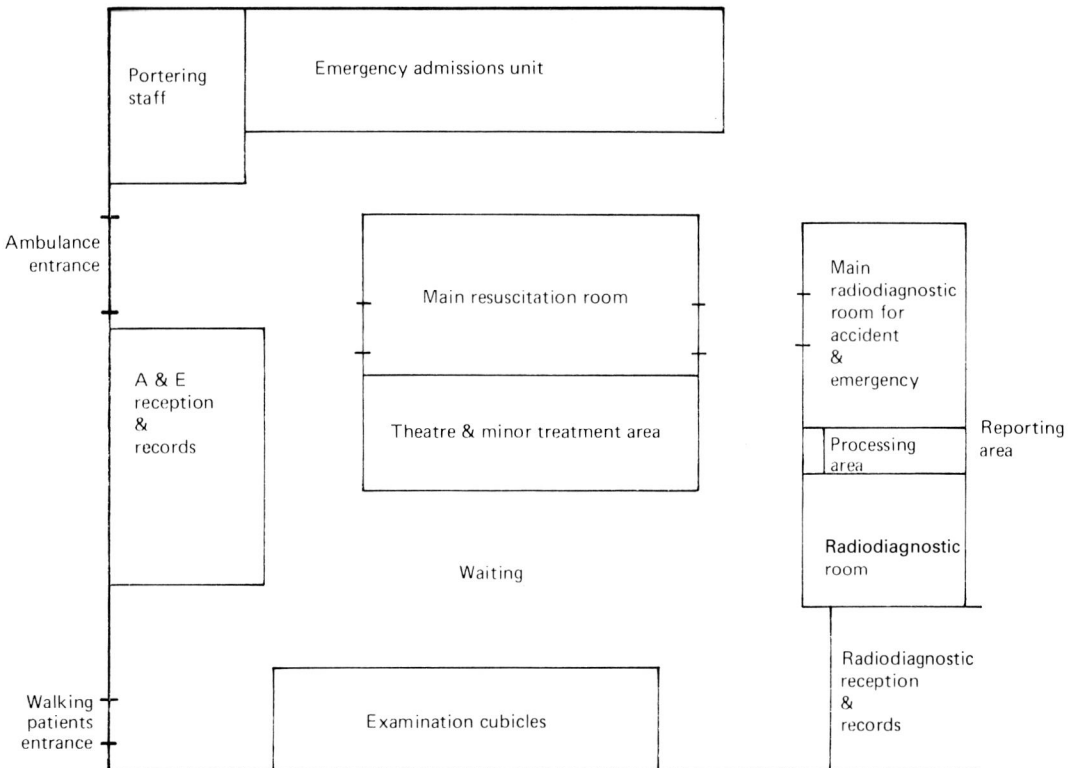

Fig. 4.1 Site of the radiodiagnostic rooms in relation to the Accident and Emergency Department.

allows one to be equipped to radiograph all major and minor trauma cases with the additional room being equipped only for minor work. The main X-ray examination room should be sited adjacent to the resuscitation room of the Accident and Emergency Department, with processing facilities as an integral part of the room. In this way, emergency medical care can be continued in the radio-diagnostic room on the severely injured patient (Fig. 4.1).

## RADIODIAGNOSTIC ROOM

The main examination room should comply with the recommendations given by the DHSS in the Health Building Note No. 6. The main criteria is to have a large room — minimum of 36 sq. metres and 3.1 metres in height. The room not only has to accommodate all the X-ray equipment and accessories, but there must be available space for the medical personnel to care for, monitor and resuscitate, if necessary, the severely injured patient.

Piped oxygen and vacuum are essential in these radio-diagnostic accident rooms, together with a good level of illumination. The lighting system should be capable of being dimmed for the purpose of seeing the centring lights of the X-ray apparatus.

Additional facilities in the room will be:

(i) hand basins and soap dispenser
(ii) radiographic illuminators
(iii) cupboard and working surface
(iv) lead protection panel with large lead glass area (15 mm lead equivalent is usually adequate)
(v) lead protective apron hanger
(vi) trolley containing all the necessary items for resuscitation and emergency treatment (e.g. airways, ambu-bag, suction catheters, intravenous infusion sets, normal saline etc.).

### Equipment requirements

The choice of equipment will vary considerably from one department to the next, but some basic guidelines should be followed to take into consideration the condition of the patients and the examinations to be performed. Patients ranging from small babies to the elderly can be expected through the Accident Room and have to be satisfactorily examined whether they are able to co-operate or not. The multiple injury patient should be moved as little as possible to avoid discomfort and possible worsening of their condition and all patients should be examined quickly and with as little inconvenience as is practical. The capital cost of the equipment should be reasonable and maintenance cost should be minimal.

Equipment required will be:

(i) Patient support/transport system
(ii) X-ray tube supports
— general radiography
— skull radiography
(iii) X-ray tubes
(iv) Erect bucky/cassette holder
(v) High tension generator
(vi) Accessories
(vii) Processing facilities.

*Patient support/transport system*

Generally patient support/transport trolleys should have the following characteristics:

a. Simple and easy to use.
b. To allow the patient to be moved as little as possible (to reduce likelihood of shock).
c. Easily manoeuvrable, being strong yet light, and manageable by one radiographer.
d. Variable in height with tilt available at both ends.
e. Acessories available to be added for the convenience of the patient such as a suction device, oxygen, equipment for blood transfusion or electrolyte infusion, and even a basket for the patient's clothing.
f. Reasonable capital cost.

There are various trolleys commercially available which offer most of these characteristics. There is one which has all the requirements but is expensive. This patient transport trolley is nearing the 'ideal' and ought to be considered by the staff of both the Accident and Emergency Department and the radiodiagnostic department when equipping their respective units. The system involves the use of a special table with bucky support in the main radiodiagnostic room, which in turn enables a trolley top to be used as the floating table and thus eliminates the need to move the patient during a complete X-ray examination. A total of two of these special trolleys would be required in most resuscitation rooms of the Accident and Emergency Department for the management of patients with multiple injuries.

The somewhat cheaper, yet more basic, trolleys can be used for patients who are less severely injured and are able to move from the trolley to the X-ray table. Some patients who only require chest X-ray examinations can be radiographed on the basic trolley. Most radiographers prefer the X-ray table to be of the floating top variety for speed and convenience in examining the patient. Various trolleys with cassette tray, whilst being adequate, are somewhat inconvenient to use, and difficulty in aligning the cassette and X-ray tube accurately is often a problem.

Radiographers are often working alone, especially during unsocial hours, so it is an advantage if the table is adjustable in height. It is easier and safer for patients to sit on to a lowered table than for them to climb on to steps in order to sit on a fixed height table.

### X-ray tube supports

All the main Accident and Emergency radiodiagnostic rooms need to be equipped with a ceiling supported X-ray tube. This support allows full movement of the tube in most directions, covering a large area of the room. This versatility is essential for the radiographic examination of the injured patient, as many standard techniques may have to be adapted for both the comfort of the patient and the production of accurate radiographic projections.

Head injuries account for approximately 10% of new admissions to Accident and Emergency departments. In order to examine these patients, an isocentric skull unit is an advantage in the radiodiagnostic room allocated for the accident patients. Ceiling-suspended skull units allow patients to remain supine and the unit to be positioned around the patient. This is especially important, for example, in radiography of the facial bones when it is vital to produce a postero-anterior projection. The skull unit will require an X-ray tube with a fine focus, be able to withstand high loadings, but will only need to give limited coverage and not have a very high repetitive rate. To this end a 0.3 mm square fine focus with 0.6 mm square large focus on a high speed rotating anode with a 7° angle is required.

The ceiling suspended tube, used for general radiography, will require a fine focus for extremity radiography but will also have a broad focus which can cope with high ratings for abdominal radiography. An X-ray tube with 0.6 mm square and 1.0 mm square foci on a high speed rotating anode with a 16° target angle is required.

An erect bucky and cassette support is essential in the radiodiagnostic room for the examination of the patient in the erect posture, necessary for chest radiography, certain abdominal radiography and patients suffering from particular injuries which make lying flat difficult.

### High tension generator

Patients attending the Accident and Emergency Department are invariably in pain and so are sometimes unco-operative due to their injury, their age, or alcohol abuse. For these reasons the shortest exposure time is required, thus necessitating a high output, preferably constant potential generator.

The accident radiodiagnostic department will usually have to be staffed throughout each 24 hours, resulting in frequent changes of radiographic staff. Therefore, an anatomical exposure programme is recommended as this will give initial guidance in exposure factors for the new radiographers.

### Accessories

Various accessories should be available to aid positioning and for the comfort and safety of the patient. The table should be equipped with arm supports, removable side supports and fixation bands for the agitated patient. Positioning pads, lateral cassette holder and two fine line grids (one

24 cm × 30 cm and the other 35 cm × 43 cm) will be needed for particular radiographic projections.

## Processing

Without any doubt, daylight automatic processing is an advantage in any Accident and Emergency radiodiagnostic department. It allows the radiographer to be in contact with the patient at all times, as opposed to being locked away in a darkroom. Careful initial planning can accommodate a daylight processing system in the main radiodiagnostic room (Fig. 4.2). Some daylight processing systems limit the user to five sizes of film but this number is usually sufficient if careful choice of cassette size is made.

## Alternative equipment

With the advent of large field size image intensifiers giving reasonable image resolution, some departments have installed their units in conjunction with a 100 mm cut film camera. Although an

initial survey of the patient can be obtained with a reduction in radiation dose (to the patient) compared with conventional large radiographs, accurate positioning in more than one plane without moving the patient is difficult.

In those Accident and Emergency Departments where radiography in the resuscitation room is considered essential, some departments have installed a ceiling suspended X-ray tube with its own high tension generator. This is beneficial for the production of good radiographs, but is not very cost effective.

When medical personnel request a chest radiograph on a critically ill patient to be taken in the resuscitation room, then a condenser discharge mobile should be the unit of choice in order to obtain the shortest exposure factors.

Cassetteless radiography systems are now commercially available and the manufacturers are recommending them for the Accident and Emergency Departments. Undoubtedly, as the films are not handled, but are fed automatically through the intensifying screens and into the processor, there

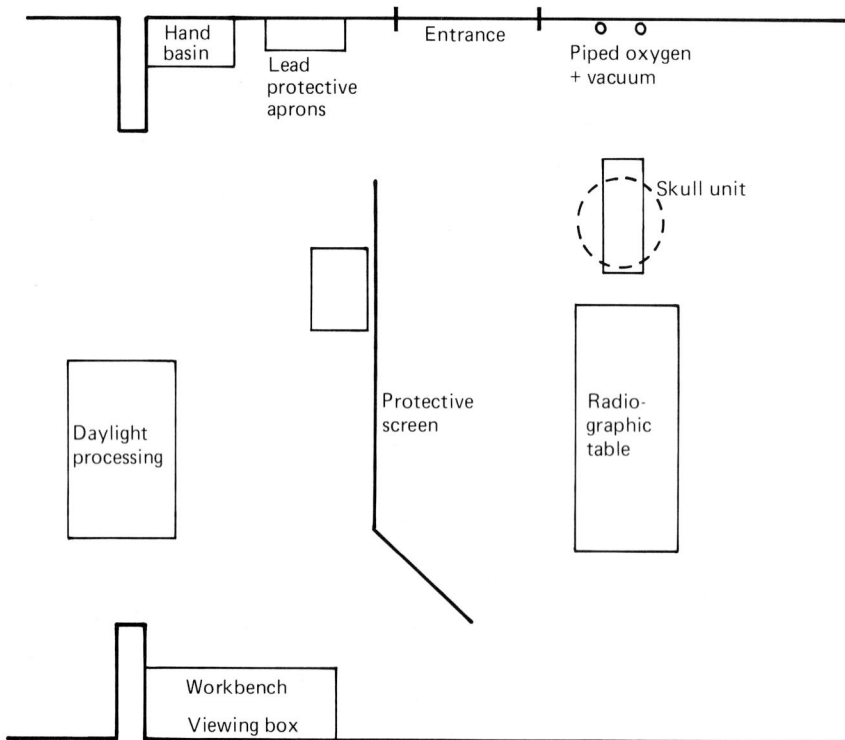

**Fig. 4.2** Layout for the main radiodiagnostic room for the Accident and Emergency Department.

are fewer artefacts and in theory the final image is available in a shorter time than the conventional processing. If any problems arise, however, no part of the radiodiagnostic room may be used until the faults are rectified. Accident and emergency radiodiagnostic equipment should be easily and readily maintainable.

## RADIOGRAPHIC TECHNIQUE

### General considerations

It is the responsibility of the radiographer in the Accident and Emergency team:

  a. to produce radiographs which demonstrate fully that part of the body requested by the referring clinician.
  b. to ensure the patient is properly cared for whilst in the radiodiagnostic room.
  c. to ensure the recommendations of the Code of Practice for the Protection of Persons against Ionising Radiations arising from Medical and Dental Use (1972) are followed. This will ensure that patients are exposed to the minimum amount of radiation, that no other personnel are unnecessarily irradiated and the unborn fetus is not accidentally irradiated.
  d. to produce radiographs that are correctly labelled with the patient's name, date and indication which is the Left or Right aspect of the body.

This latter point is easy to overlook when the radiographer is busy and too often mistakes occur.

Two elementary points to be stressed in accident and emergency radiography are:

1. All the patient's clothing must be removed from that part of the body which is being examined to reduce the possibility of artefact.
2. Collimation of the X-ray beam is essential for the production of good radiographs, i.e. reduction of the radiation dose to the patient.

When radiographing children, special care is required to eliminate patient movement and to obtain the true radiographic projection. Many departments stipulate that it is not only the injured limb which is to be radiographed, but also the opposite one for comparative purposes.

As a general guide, a fast intensifying screen film combination should be used throughout, except for radiography of the hand, wrist, forearm and humerus, ankle and lower leg, when a 'detail' intensifying screen should be utilised with the fast film.

Finally, it is important that the patient has a full clinical examination followed by a complete X-ray examination. This should eliminate delay in the patient's treatment, and for those seriously ill or injured patients, this process should negate further radiographic examination on the ward.

### Upper limb

Examination of the upper limb is regarded as elementary, but great care and precision with projections must be taken to give a complete and accurate picture of any injury. Even minor disabilities to the upper limb need to be correctly and quickly treated as they may affect the patient's capacity as a wage earner.

Details of routine basic projections of the skeleton should be well known to the radiographer (Clark 1979), but additional and modified techniques will be required by the Accident and Emergency radiographer.

### Hand

To demonstrate fully fractures of the 5th metacarpal, a 40° antero-posterior oblique and lateral projection of the hand are most adequate.

### Wrist

When radiographing the scaphoid, ulnar deviation is essential in the postero-anterior projection. Both obliques and lateral projections will demonstrate the bone completely.

Injury to the carpal bones will necessitate the sound side being radiographed for comparison, so that casualty officers will be aware of dislocations or small fractures on the injured side.

### Forearm

It is frequently necessary to radiograph the radius and ulna using a horizontal X-ray beam or have the patient in various acrobatic positions. Injured limbs are usually splinted and they have rarely been immobilised in the position required by the radiographer to demonstrate the radius and ulna in their true positions.

## Elbow

Injury to the elbow is often extremely painful. Most patients will be unable to straighten their arm completely so additional antero-posterior projections must be made; one to demonstrate the head of radius and ulna with the forearm on the film and a second projection with the upper arm on the film to demonstrate the lower end of humerus. Occasionally, an axial projection will be the only position possible (Fig. 4.3). Demonstration of the head of radius requires an externally rotated antero-posterior projection of the elbow and a lateral projection with the hand in the lateral and then pronated position.

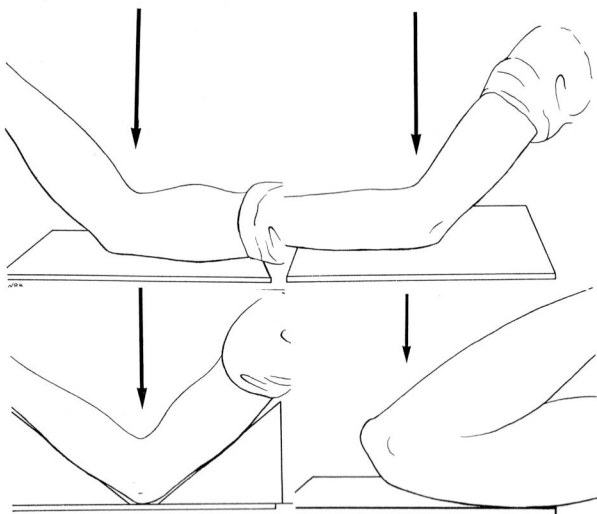

Fig. 4.3 Radiographic examination of the injured elbow often requires multiple views for full demonstration of any injury.

## Humerus

Obtaining two projections at right angles may be difficult. Moving the arm forward or backwards, allows the humerus to be radiographed clear of the chest wall.

## Shoulder

Following the antero-posterior, a lateral projection of the upper humerus and joint capsule is necessary. As the patient may not be able to move the arm (and it may be dangerous to move the arm in trauma cases), a lateral projection of the scapula is best and it is achieved most easily with the patient in the erect position. Where a suspected dislocation is to be examined, the true antero-posterior projection of the shoulder joint is taken, followed by a lateral projection of the scapula to demonstrate whether or not the dislocation is anterior or posterior.

If the patient has multiple injuries and has to remain supine, the scapula may be imaged in the antero-posterior position with abduction of arm, and then the lateral projection being obtained by raising the affected side between 30°–40° supported by pads and then pulling the shoulder and elbow forward slightly whilst keeping the back of the patient's hand against the side. The centring point is through the head of humerus (Fig. 4.4).

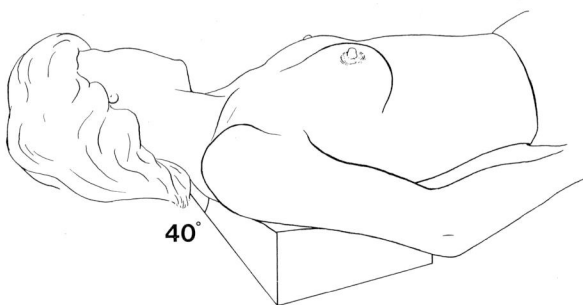

Fig. 4.4 Modified position for the lateral scapula projection on a supine patient.

## Clavicle

To demonstrate a fracture of the clavicle fully and show the degree of displacement, an inferosuperior projection should be taken in addition to the routine postero-anterior.

## Lower limb

Injuries to the lower limb usually mean that the patient is in a wheelchair or on a trolley. For radiography of the knee and lower leg the patient need not necessarily be placed on the X-ray table.

Patients requiring radiography of the femur must be examined on the X-ray table using the bucky.

## Foot

When requested to demonstrate the calcaneum, axial and lateral projections must be taken of both feet.

## Ankle

Demonstration of the ankle joint should include the antero-posterior lateral and the medial oblique projections, so that any small fractures of the lower tibia and fibula will be demonstrated.

## Tibia and fibula

They should be radiographed in their entirety so that both the ankle joint and knee joint are demonstrated.

## Knee joint

The lateral projection should be taken using a horizontal X-ray beam, so that any fluid levels may be detected. This is important for the detection of small fractures involving the joint. Fractures of the patella may be more clearly visible using the 'skyline' patella projection. Where a patient's knee has 'locked', intercondylar projections may be requested to look for loose bodies or even fractures. If a curved cassette is not available, the orthopantomogram cassette may be used.

## Femur

Four projections are usually required in examining the femur of an adult. The exposure factors required for the proximal portion are more than those required for the distal part.

## Hip joint

It is a common request to radiograph the hip joints of the elderly following a fall. Generally an antero-posterior projection of both hip joints, followed by localised antero-posterior and lateral projections of the injured hip are necessary. These patients are usually frail old ladies and must be handled with kindness. If a fracture of the upper end of femur is confirmed, a chest radiograph should be taken. This enables the treatment for the patient to be planned at an early stage.

## Pelvis

A full demonstration of the pelvis will be required at the first examination, so it is always advisable to omit the use of any radiation protective device for the gonads. When fractures are shown they must be fully demonstrated, so that the orthopaedic surgeon can plan the management of the patient. As the pelvis is a three dimensional ring structure, it may be necessary to perform either oblique, inlet and outlet, or Judet's projections to demonstrate the full extent of any injury.

## Thoracic cage and contents

The chest radiograph taken at 180 cm, erect on full inspiration is the most common examination in any imaging department. It is also probably one of the most important and must be performed with meticulous attention to detail in positioning, collimation and exposure. Following this initial radiograph, further views may be necessary depending on the pathology suspected or demonstrated. Any localised lesion will require a lateral projection; any pathology demonstrated in the apices — an apical projection and dense shadows or pleural effusion — a penetrated postero-anterior projection. If a pneumothorax is suspected but not demonstrated on the initial radiograph exposed on inspiration, a further postero-anterior projection should be performed with the patient holding his breath in expiration. The most satisfactory method of demonstrating the movement of the diaphragm is to examine the patient under fluoroscopy when the patient is asked to breathe in and out deeply. However, if the facility and a radiologist are not available, the range of movement of the diaphragm may be demonstrated on one radiograph with the patient being exposed on inspiration and then repeated with expiration without movement. Exposure values will need to be halved for each exposure. When patients have suffered trauma to the ribcage, an erect postero-anterior chest radiograph should be taken if possible with oblique projections of the affected area as required.

The sternum tends to be difficult to demonstrate, but if the examination is performed erect with the use of a bucky and short exposure time, reasonable images can be produced both in the postero-anterior, oblique and lateral projections.

Dislocation of the sterno-clavicular joint is usually easily detectable clinically, and may be confirmed radiographically by a postero-anterior projection of both joints, using a short focus film distance, 50 cm, and then an oblique projection of each joint. Note the postero-anterior projection should not be repeated due to the high radiation skin dose.

## Abdomen

Most requests for examination of the abdomen will be from the emergency admissions unit and will tend to be requests for pathology of the abdomen, rather than trauma. It is especially important that the radiographer is made aware of the pathology suspected before the examination, so that the relevant projections may be taken. It is usual practice to perform a chest X-ray when abdominal radiographs are requested, as chest/abdominal pathology can be closely linked.

The most common pathology for the request of abdominal radiography is intestinal obstruction and renal colic and both these conditions will require different projections and different exposure factors. Demonstration of a suspected intestinal obstruction requires both supine and erect projections of the abdomen showing the abdomen from the symphysis to diaphragm in both positions. A radiograph of an obese patient sitting over the side of the trolley is of very little value if the lower abdomen is obliterated by the thighs. In these instances, if the patient cannot stand, then a lateral decubitus radiograph ought to be performed. A short exposure time is imperative.

A low kilovoltage must be used when a supine abdominal radiograph is exposed for a patient suffering from renal disease. This radiograph must include the symphysis pubis and both renal areas. Additional projections may be required to confirm, or otherwise, that small opacities are within the renal, ureteric or bladder areas. Inspi-ration and expiration radiographs may help to identify opacities in the renal areas, and oblique projections are advocated for the ureteric and urinary bladder areas. Patients suspected of having a perforated duodenal ulcer should have a chest radiograph performed in the erect position and this must include the diaphragms. An abdominal radiograph in the supine position will be required and if it is not possible to radiograph the chest in the erect position, a lateral decubitus projection of the abdomen (patient lying on the left side) should be taken.

Patients suspected of suffering from paralytic ileus, strangulated hernia and intussusception (in infants), must have a radiograph of the chest in the erect position and abdomen in the erect and supine positions. Patients with hepatic or biliary pathology ought to have the abdomen X-rayed in the supine position and in the prone position with the X-ray beam collimated to the right upper abdomen.

For patients suspected of having pancreatitis or ruptured spleen, a radiograph of the upper abdomen in the supine position is all that is necessary, together with a chest radiograph with the patient erect. A lateral projection of the abdomen as well as one exposed in the supine position, are the projections to demonstrate best a suspected aortic aneurysm.

## Skull and facial bones

Radiographers are all aware of the large numbers of minor head injuries which are sent to the X-ray department for skull radiography, mainly for medico-legal purposes. The controversy about whether too many examinations are requested for skull radiography is still not resolved. However, the referring casualty officer in requesting these radiographs, must be given perfect projections each time to assist him in his diagnosis of any injury to the patient.

Skull radiography requires a high degree of skill and expertise to produce the required radiographs. Unfortunately, all too often the radiographs produced are of a poor standard due either to incorrect or poor positioning (rotated antero-posterior projections), or to the incorrect exposure. There is very little excuse for the latter two prob-

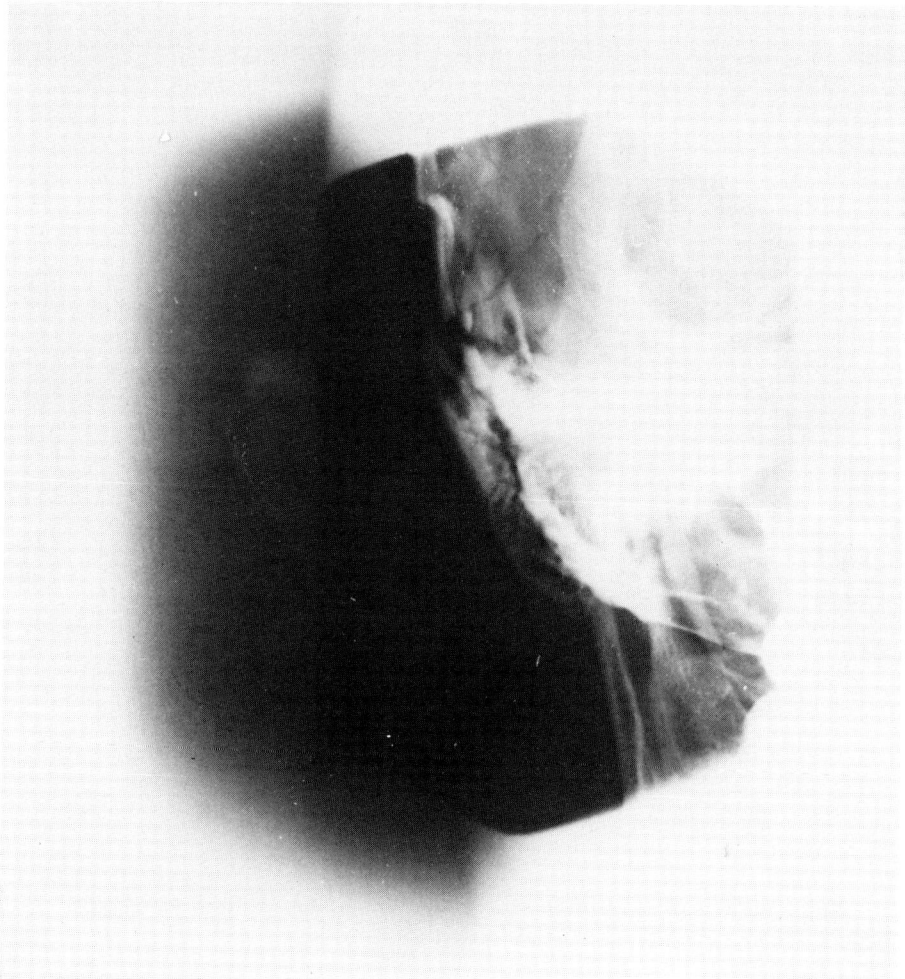

**Fig. 4.5** The importance of the tangential projection for further demonstration of a fracture in the skull.

lems, but incorrect positioning usually occurs because the radiographer is junior and/or the equipment provided is inadequate for good skull radiography, e.g. mobile X-ray unit. Specialised skull units must be provided, together with adequate staff with sufficient experience so that time can be taken to obtain the correct projections. For patients with a suspected fracture of the skull or intracranial haematoma, four projections should be performed:

1. 30° half axial (Townes projection) to demonstrate any displacement of the pineal gland (shift of midline), or fracture in or around the occiput and temporal bones.

2. Postero-anterior 20° to demonstrate the frontal and parietal bones.

3. Left and right lateral projections to demonstrate each side of the skull and base of skull. One of the laterals should be brow up and a horizontal X-ray beam used in order to show any fluid levels, especially in the sphenoid sinuses.

4. Tangential projections of any depressed fracture area will be necessary (Fig. 4.5).

Radiography of the facial bones is an even more complex subject than skull radiography. Before radiographers are allowed to cover accident and emergency radiography, they should be given

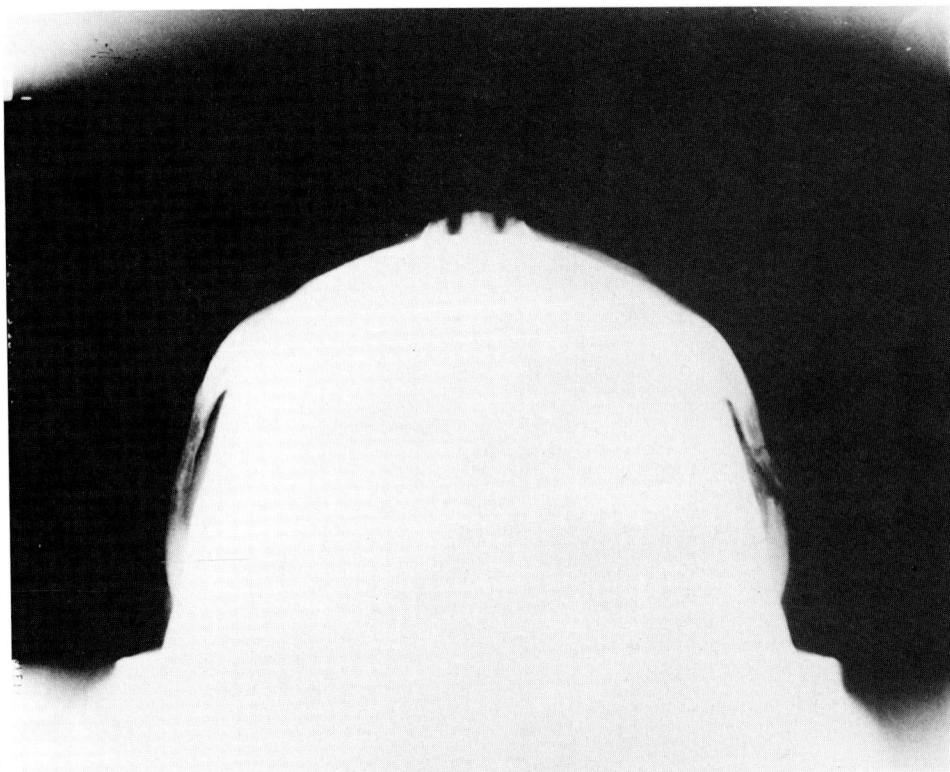

**Fig. 4.6** Axial projection to demonstrate the zygomatic arch.

specialist instruction in the anatomy of the face and facial bones, where fractures can and do occur, the projections necessary to demonstrate the facial structure and the normal appearance of the completed radiographs. In this way the standard of radiography would be improved and thus the service to both the patient and casualty officer.

Ideally patients who are requested for facial bone radiography ought to have a full skull series. Too often departments have insufficient staff available in the out of hours time to ensure complete examinations.

The standard basic projections are:

1. an occipito-mental to include the whole facial structure, including the zygomatic arches.
2. a 25° occipito-mental which is virtually a tangential projection of the facial skeleton.
3. a lateral of the facial structures.

These radiographs may be performed erect or supine. The occipito-mental projections are of very

limited value if they are taken in the reverse direction.

Additional radiographs may be a lateral of the nasal bones, and for the adequate demonstration of the zygomatic arch a 100° axial centred through the zygomatic arch (Fig. 4.6) or a 35° half-axial centred again through the zygomatic arch, both these projections requiring reduced exposure from the standard exposure.

Another useful projection to demonstrate the zygomatic arch is 45° rotated occipito-mental projection, centring through the zygomatic arch thus projecting the zygomatic arch over the opposite parietal bone but free from other facial structures (Fig. 4.7).

**The mandible**

An orthopantomogram (OPG) is the ideal way to image the mandible. However, there are not many Accident and Emergency radiodiagnostic depart-

**Fig. 4.7** Rotated OM projection for the demonstration of the zygomatic arch.

ments that have an OPG unit either attached or nearby.

The standard projections are a postero-anterior and both lateral obliques. The lateral oblique projections should be performed with:

a. the patient prone so that the body and symphysis menti are nearer the film
b. the patient supine so that the ramus, angle and body of the opposite side are nearer the film.

In this way, together with true lateral and occlusal projections, any fractures of the mandible will be demonstrated.

**The spine**

Patients with injuries to the spine need to be handled with care. Yet it is essential that a total demonstration of the requested area is achieved. Too often when examining the cervical area, cervical vertebra VII is missed in the lateral projection.

Generally antero-posterior and lateral radiographs are necessary of the cervical, thoracic and lumbar regions. Additional information regarding the articular facets and neural arches can be obtained with the 45° oblique projection.

The cervical spine needs special mention. All

the cervical vertebrae must be demonstrated in both planes. Cervical vertebra I and II can usually be imaged with the patient's mouth open in the antero-posterior projection, but tomography must be considered if the patient's mouth cannot be opened (or even supported open). If a dislocation at the cranio-vertebral junction is suspected (or shown), full demonstration can be achieved by taking two supine oblique projections.

The lateral projection of cervical vertebrae VI and VII and thoracic vertebra I is often difficult due to the patient's shoulders and, even more so, if the patient has a shoulder injury. Yet experience shows that this projection is attainable in most patients if the radiographer gently eases the shoulders down and has them held down whilst the radiograph is exposed. In the rare case cervical vertebrae cannot be demonstrated adequately in the lateral projection, a lateral oblique will be necessary with one arm raised and one lowered. The upper thoracic vertebrae may be examined in a similar way, but often tomography is the only true way of demonstrating these vertebrae.

## Multiple injured patient

Wherever possible these patients should be placed on a special patient transport trolley directly from the ambulance. If this is done then the patient need not be moved significantly again, thus causing no further aggravation to their injuries. These patients should be examined in a specially equipped radiodiagnostic room so that medical staff can provide emergency treatment while radiographs are being taken. If the patient has to remain in the resuscitation room, then only radiographs which are absolutely necessary for the emergency treatment of the patient should be taken. These will usually be a supine chest radio-graph and a lateral radiograph of the cervical spine.

Only senior experienced radiographers should examine these patients so that the radiographs will be correct first time and few repeat radiographs will be required. It is important to be methodical in the radiography and so the chest radiograph should be taken first, then a lateral cervical spine radiograph. These films should be viewed before continuing. Then radiographs of the rest of body are taken, starting with the antero-posterior projections from the skull downwards, followed by lateral projections. It is essential that all radiographs have the correct demographic data and are marked with the correct side of the body.

## Foreign bodies

Radiography to demonstrate foreign materials in the human body requires different thoughts and techniques. The basic rule is that two views at right angles must be taken so that the referring clinician is able to pinpoint the exact location of the foreign material and is thus able to remove the foreign material with the least amount of trauma to the patient. Depending upon the site of the foreign material, the radiographer may put a metal marker on the skin surface of the patient, either at the suspected point of entry, or some other convenient location. The radiographic exposure needs to be chosen very carefully in order to enhance the suspected foreign material. When a patient is suspected of swallowing a foreign body, then it is important that the radiographer demonstrates the whole of the alimentary tract, from the mouth to the anus, including both. These latter requests are often for children and it is essential to take an antero-posterior projection of the chest and abdomen and also a lateral projection of the naso-pharyngeal area.

REFERENCES

A Study of the Utilisation of Skull Radiography in 9 Accident and Emergency Units in the U.K. The Lancet, 6 December 1980.
Clark K C 1979 Positioning in radiography, vols 1 and 2. Heinemann Medical, London
Code of Practice for the Protection of Persons against Ionising Radiations arising from Medical and Dental Use. 1972 DHSS, London
Grech P Casualty radiology. Chapman and Hall, London
Griffiths D M, Freeman N V 1984 Expiratory chest X-ray diagnosis of inhaled foreign bodies. British Medical Journal 288(April)
Health Building Note Radiodiagnostic Department 1981, DHSS, London
Health Equipment Note No. 6 Radiodiagnostic Department 1981 DHSS, London

# 5

# Neuroradiography — new techniques

## INTRODUCTION

Neuroradiology has fascinated the medical profession since 1896, when the first skull X-ray was taken. After 75 years of progressive discoveries we are now experiencing an explosion of information which challenges every radiologist, physicist and radiographer.

Since 1972 it has been possible to reconstruct images showing the differentiation of brain tissue instead of just variation in bone densities; now the whole brain can be mapped out in horizontal sections in a matter of minutes. This means that abnormalities can be detected in incipient stages when early treatment can be undertaken, often with highly successful results.

At the time of writing, further developments are taking place in nuclear magnetic resonance which are complementing the recent advances in computed tomography. With today's technology it is impossible to forecast what new ideas will revolutionize the field within a few years.

## HISTORY

It is interesting to step back into the history of neuroradiology and look at some of the milestones.

The first X-rays of the nervous system are thought to have been taken in April, 1896, in the town of Nelson, near Manchester, England in the presence of Dr Stanton, the deputy of physicist Arthur Schuster, together with the Mayor of Nelson and the Town Clerk. Two views were taken of a woman's skull; the first took 1 hour and the second 10 minutes longer. The X-rays revealed the presence of four bullets. As a result of this gruesome experience, Dr Stanton is said to have suffered a nervous breakdown from which he never fully recovered! (Schuster, 1962).

Professor Arthur Schüller, who graduated the same year that X-rays were discovered, worked with Holzknecht in the Viennese School of Röntgenology. He later became known as the father of neuroradiology.

In 1919 Walter Dandy performed the first encephalogram by introducing air into the lumbar theca after removing an equivalent amount of cerebro-spinal fluid, having performed ventriculography on infants with hydrocephalus the previous year. Dandy predicted the procedure of air myelography. The first case was performed by Jacobaeus in Stockholm.

In the same year positive contrast was used for the first time in neuroradiology by Sicard. He used Lipiodol for the treatment of sciatica and discovered that it tracked down the nerve root sheath. A student of Sicard's accidentally penetrated the subarachnoid space when treating a patient with Lipiodol; the contrast was seen on fluoroscopy and observed to have collected at the bottom of the subarachnoid space with the patient in the standing position. Sicard had the idea of tilting the patient to observe the movement of the contrast; thus myelography came into being.

Egas Moniz, the Portuguese neurologist, struggled for some time to find a suitable and safe contrast media to demonstrate the cerebral arteries. It was he who described the brain as 'mute to X-rays'. Many of the substances he tried were too toxic or not well visualized until he used 5 ml of 2% sodium iodide. He finally succeeded in 1927 by performing a cut down procedure of the carotid artery.

Erik Lysholm is best known to radiographers for the equipment he designed in conjunction with the engineer George Schönander and his book on the positioning of the skull.

In 1935, Ziedses des Plantes, who also was a prominent figure in the development of tomography, made the important contribution of subtraction, which is so valuable in vascular radiography.

In 1939, Leksell perfected the technique of stereotaxy, which is now used increasingly in neurosurgery.

The discovery that has caused the greatest advance in neuroradiology since Wilhelm Conrad von Röntgen discovered X-rays in 1895 was made by Godfrey Newbold Hounsfield. Several groups in different parts of the world were developing the fundamental theory of computed tomography (CT) at the same time, but it was Hounsfield who succeeded in turning his theory into practice. The first CT scan was performed in 1972. The scanning time for two contiguous horizontal sections taken simultaneously was only 4 minutes. The results sent ripples of excitement around the world, as the medical profession realized the immense potentialities of the new discovery.

In the last few years, nuclear magnetic resonance imaging techniques have been perfected. This is an imaging technique which uses non-ionising radiation. Nuclear magnetic resonance has been used for chemical analysis for the last 30 years, but it is only since 1981 with the improved quality of the images that it has begun to make a serious impact on radiology.

The last two developments will now be discussed in some detail.

## HIGH RESOLUTION COMPUTED TOMOGRAPHY OF THE SPINE

High Resolution Computed Tomography (HRCT) of the spine was developed in 1978 by David King, formerly of EMI, at the instigation of Dr Romeo Ethier of the Montreal Neurological Hospital. This was to improve visualisation of the spinal cord and achieve higher spatial resolution.

### Development

Initial attempts at improving the quality of spine

### High Resolution Scanning (2nd Generation)

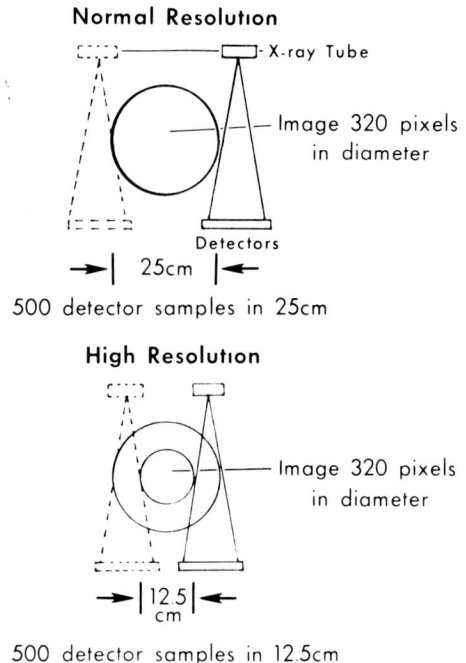

Fig. 5.1 Twice as many detector readings sampled during each traverse. (Reproduced from Canadian Journal of Opthalmology Vol. 16, July 1981 by kind permission of Dr D A Nicolle & the Editors.)

images as produced by regular CT simply involved averaging a number of slices (typically four) of the same area of the spine. However, it was quickly realized that higher spatial resolution was also required.

The need for further improvement brought the addition of a hardware-software package known as 'High Resolution Sector Scanning' (Ethier et al 1980). King (1980) stated the following were requirements needed to bring this about:

1. additional collimation
2. twice as many detector readings in the area of interest sampled during each traverse (Fig. 5.1).
3. modified software to collect the high resolution data and a 320 × 320 matrix.

When viewing a small structure such as a vertebra or the spinal cord, it was more convenient to magnify part of the image for easier visualisation. This was achieved by using a 160 quadrant, which magnified the image twofold (Fig. 5.2).

A                                                    B

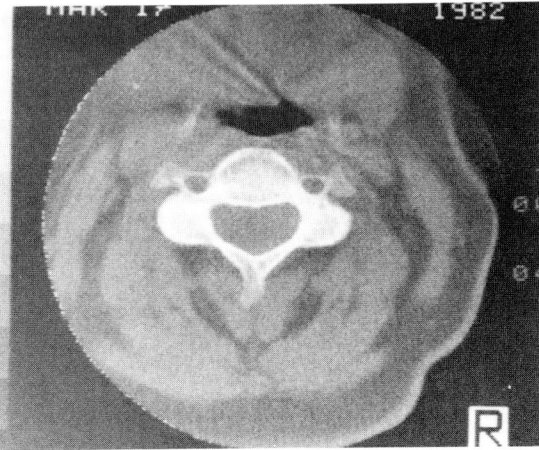

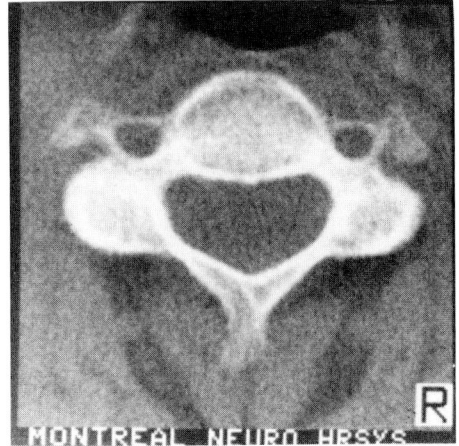

**Fig. 5.2** A Normal sized image 320 × 320 pixel matrix. B 160 × 160 quadrant.

King reported that a series of 8 mm high resolution slices will result in a peak dose of less than 10 rads. Wall et al (1979) report 6 rads, the dose being approximately the same as that delivered during a barium enema.

## Application

Using an 8 mm slice, a 120 mm diameter scan field, a 70 second scan time, and a 320 × 320 pixel matrix, this new field technique of scanning made it possible to examine the size and shape of the spinal canal and spinal cord with and without the aid of contrast media. It is important to remember, when planning the sequence of tests, that the introduction of oil-based contrast media into the subarachnoid space or the central canal will seriously interfere with the results of the CT scan if the myelogram is done first. However, with the increase in the use of water-soluble contrast this problem has virtually been resolved.

A discussion on the spine and its contents and associated conditions follows.

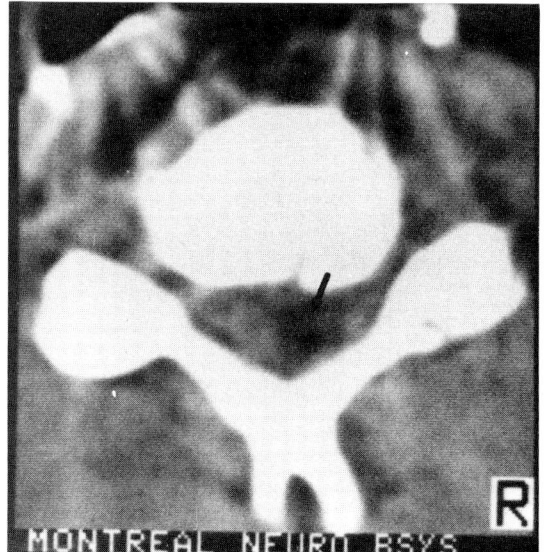

**Fig. 5.3** Oedema of the spinal cord and fracture of the vertebral body. (Reproduced from M.J.D. Post's *Evaluation of the Spine* ch. 14 by kind permission of Dr R. Ethier, Dr M J D. Post & Masson Publishing U.S.A. Inc, New York.)

### Trauma

Myelography in traction following severe trauma can be a difficult and risky procedure. Therefore, the preferred examination, following good plain films, should be a CT scan. For the latter the patient should remain supine in traction; overlapping scans should be done at the appropriate level, using high resolution technique as previously described.

The scout view available on the new total bodyscanners may, in certain cases, rule out the necessity of plain radiographs, thus eliminating the pain, risk and delay involved in moving the patient for a separate examination.

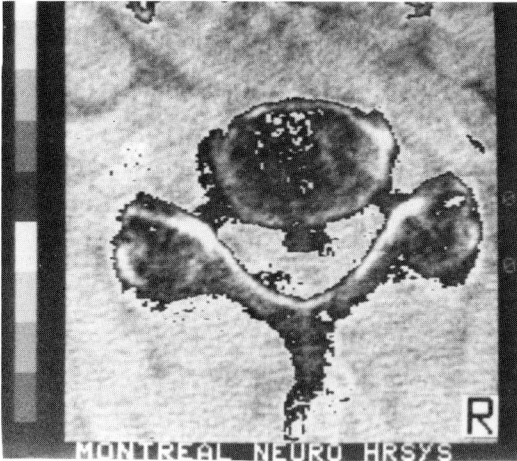

**Fig. 5.4** Dual windowing. The edge enhancement delineates the protrusion of the disc into the spinal canal. Note dual grey scale.

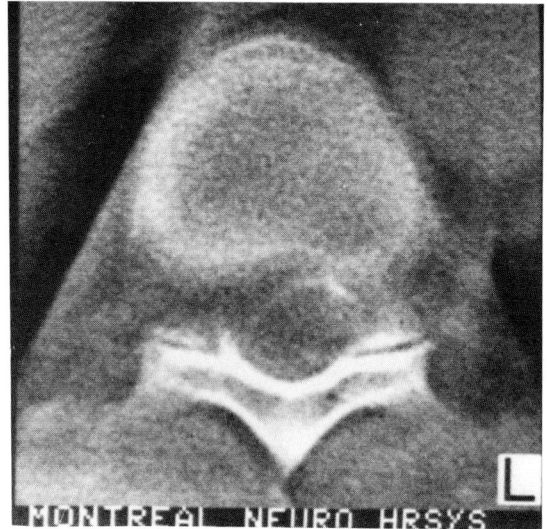

**Fig. 5.5** Thoracic meningioma discreetly outlined with metrizamide.

The CT scan will not only demonstrate the bony fracture, but will also differentiate between a haematoma or oedema of the spinal cord (Fig. 5.3) (Ethier et al 1980).

*Disc disease*

Herniation of a disc may be demonstrated by a plain CT without the aid of myelography. Ethier et al (1980) stated that in cases of severe disc protrusion the subarachnoid space may not fill at that particular level during myelography. This could lead to misdiagnosis; in such cases CT using overlapping 8 mm slices of the level in question should be performed.

Dual windowing is essential to delineate the bony structures while retaining good visualization of the disc (Fig. 5.4); by filtering and averaging it should be possible to differentiate between a diagnosis of tumour or of disc disease.

This technique is a marked improvement and has virtually brought about the discontinuance of epidural venography (Shapiro 1975, Smith 1976, Gargano 1980). It gives better information, is more comfortable for the patient, and is a safe procedure. Epidural venography was a high risk with the possiblity of complications such as venous thrombosis and embolism; there was also the added risk of allergic reaction to the iodinated contrast media.

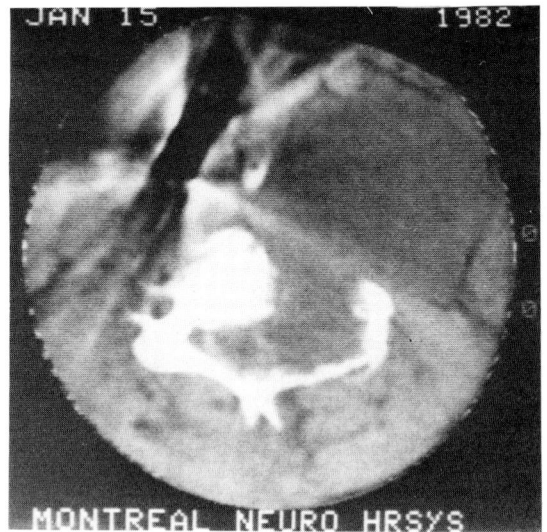

**Fig. 5.6** Neurofibroma is seen expanding the intervertebral foramen. The spinal cord, outlined with metrizamide is displaced to the left.

*Tumours*

Tumours of the spine and spinal cord may be divided into the following categories:

*Intracanalicular*
1. Primary
   Consists mainly of meningiomas and neuro-

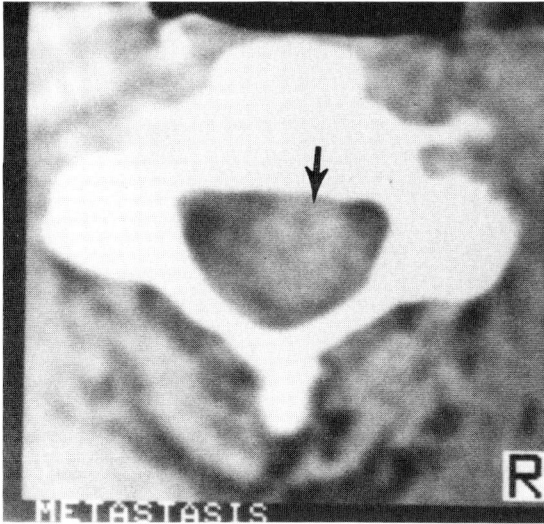

Fig. 5.7 Intramedullary metastasis with intravenous enhancement.

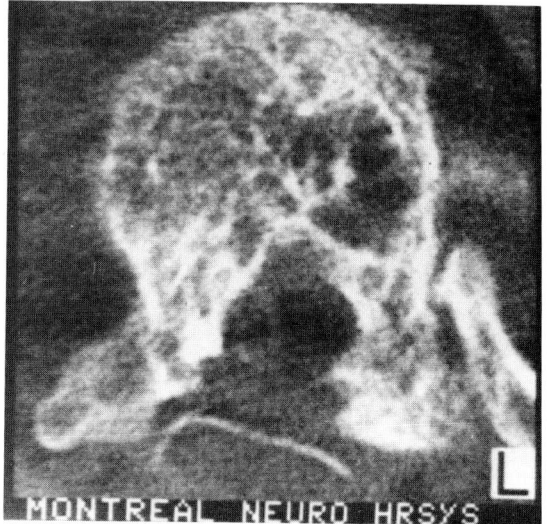

Fig. 5.8 Bone tumour (haemangioma) of 10th thoracic vertebra.

fibromas. It is necessary to use contrast media to delineate the lesion (Figs. 5.5,5.6).

2. Secondary

Even a small metastatic lesion can be seen which could be missed on routine myelography (Fig. 5.7).

*Extracanalicular*

1. Bone tumours generally have an isotope scan done first, to establish the level or area by

demonstrating the 'hot' spot. HRCT would then follow in order to determine the type of lesion, and sometimes even the presence of a lesion (Fig. 5.8).

2. Inflammatory diseases, such as osteomyelitis or septic arthritis, can be diagnosed in the early stages, as CT has the capability of demonstrating early changes in soft tissue (Fig. 5.9).

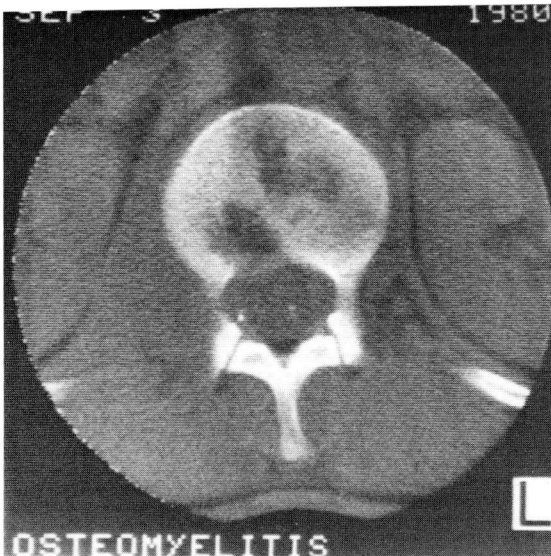

A

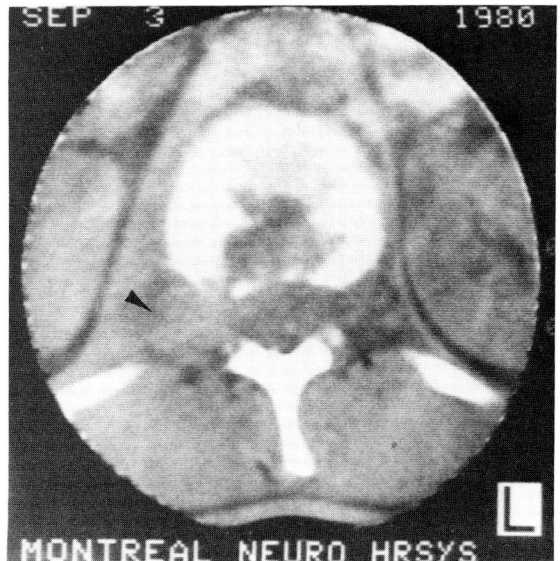

B

Fig. 5.9 (A) Osteomyelitis of the vertebral body.  (B) Osteomyelitis demonstrated in the soft tissues.

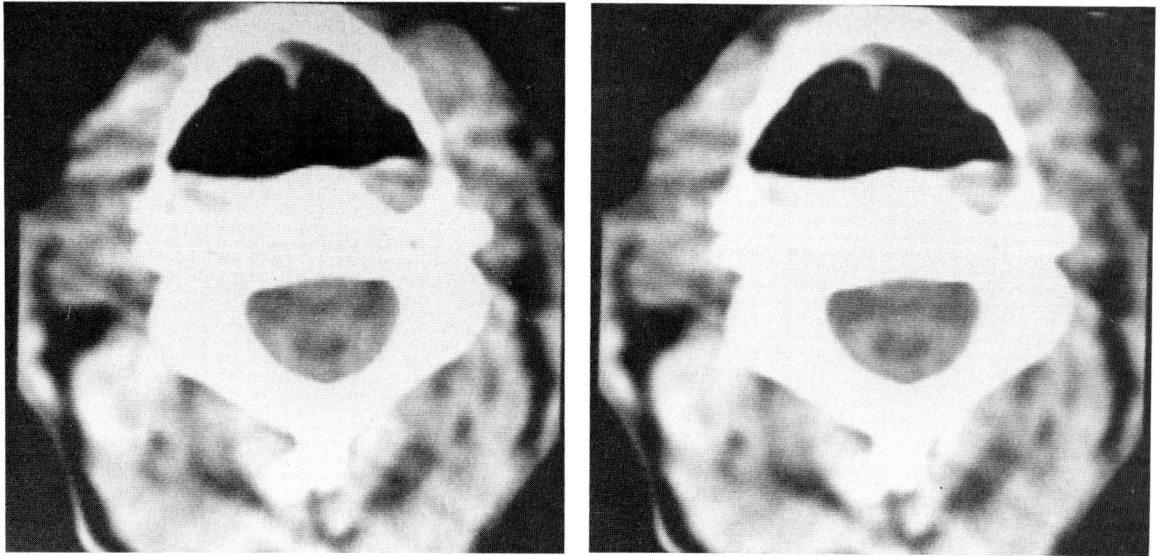

**Fig. 5.10** Flat atrophic cord (A) before averaging; (B) after averaging.

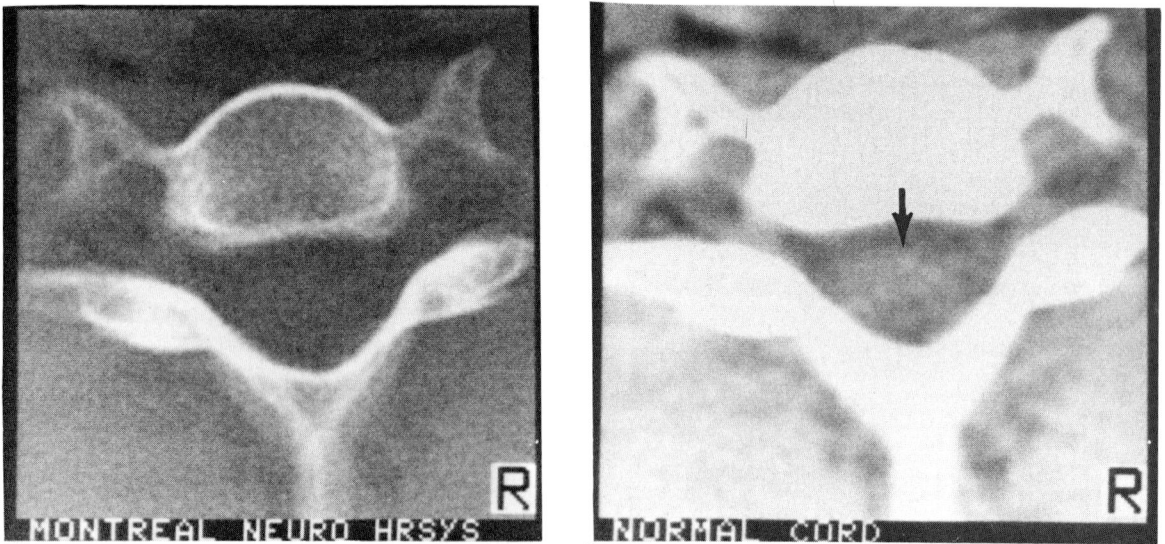

**Fig. 5.11** Normal spine (A) before filtering; (B) after filtering, demonstrating normal cord.

The improved spatial resolution afforded by HRCT, together with the various viewing options, such as averaging (Fig. 5.10), filtering (Fig. 5.11), dual windowing (Fig. 5.4), as well as coronal (Fig. 5.12) and sagittal reconstruction (Fig. 5.13), make it possible to demonstrate erosion of the bone, soft tissue masses or swelling, and displacement of nerve roots.

*Spinal stenosis*

HRCT is essential to assess the size and shape of the spinal canal and intervertebral foramina. These structures cannot be visualized well on plain films or routine myelography. Claims have been made by Gargano (1980) that transverse axial tomography plays a role here. Experience has shown the superi-

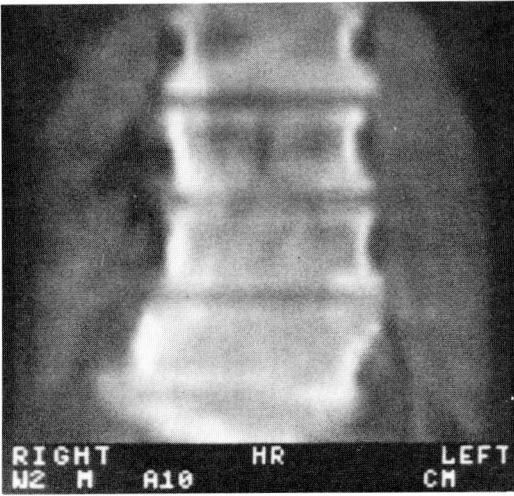

Fig. 5.12 Coronal re-formation of lumbar spine.

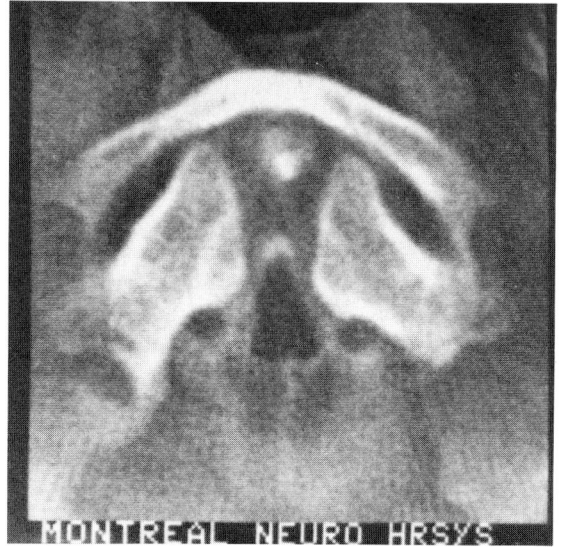

Fig. 5.14 Developmental spinal stenosis. Intrathecal metrizamide demonstrates triangular canal between the occipital condyles. (Reproduced from M.J.D. Post's *Evaluation of the Spine*, ch. 14 by kind permission of Dr R Ethier, Dr M J D Post & Masson Publishing U.S.A. Inc., New York.)

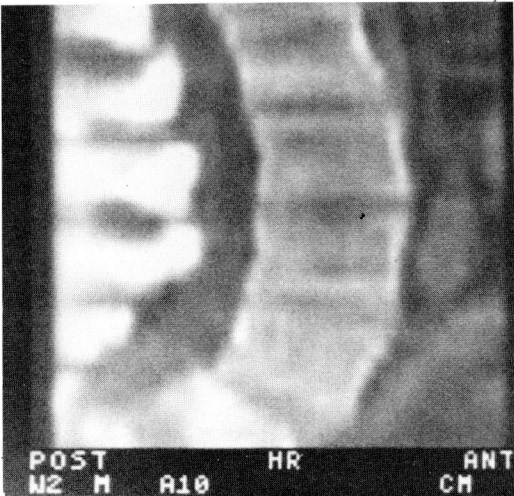

Fig. 5.13 Sagittal re-formation of lumbar spine.

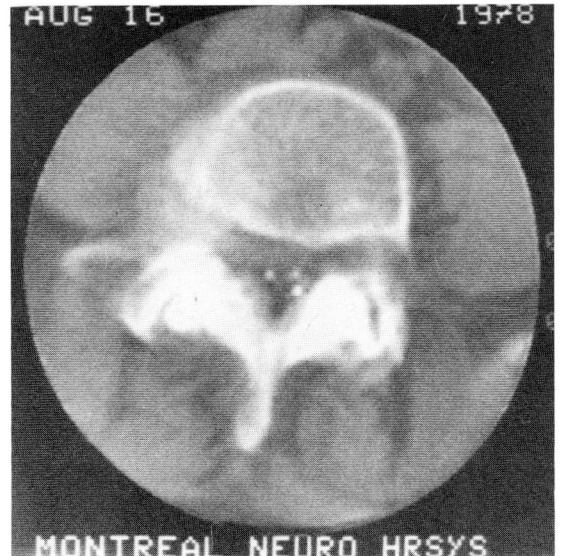

Fig. 5.15 Acquired spinal stenosis. Disc herniation impinges on spinal canal. Note droplets of contrast media. (Reproduced from M.J.D. Post's *Evaluation of the Spine*, ch. 14 by kind permission of Dr R. Ethier, Dr M.J.D. Post & Masson Publishing U.S.A. Inc., New York.)

ority of the HRCT techniques (Ethier et al 1980).

There are two types of spinal stenosis: developmental and acquired (Fig. 5.14 & Fig. 5.15).

*Developmental.* In general this is a narrowing or distortion of the canal, present since early life. This constriction is usually seen starting in the dorsal area, but is generally more marked in the lumbar region as seen in achondroplasia (Gargano 1980).

*Acquired.* This category includes disc protusions, spondylolisthesis, thickening of the laminae and ligamenta flava, hypertrophy of the facets and osteophytes. Any of these, alone or combined, may cause narrowing of the canal and possibly compression of the cauda equina.

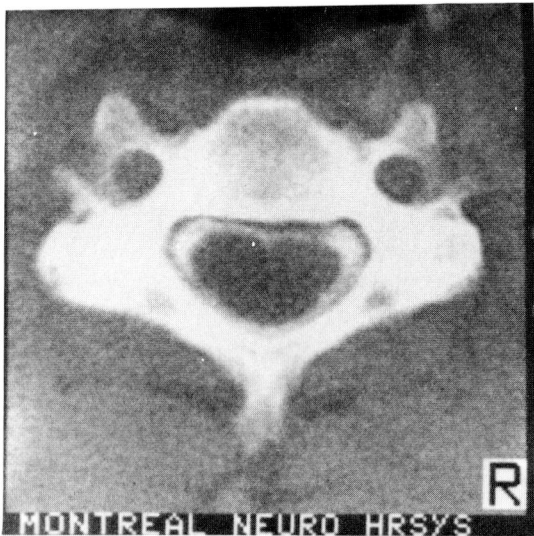

Fig. 5.16 Round large cord outlined with metrizamide.

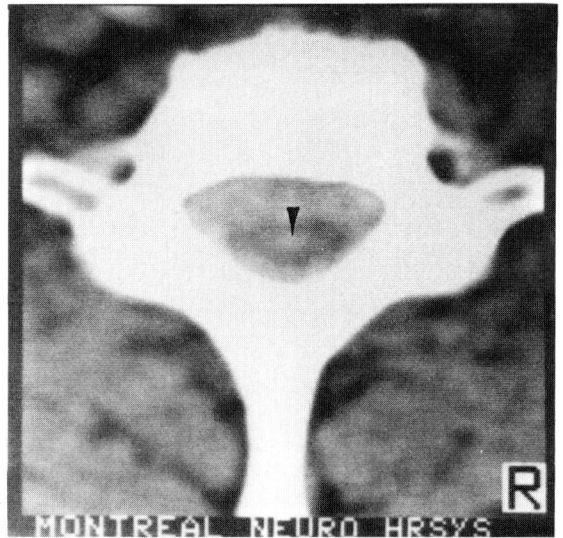

Fig. 5.18 Syringomyelia. Flat atrophic cord with metrizamide demarcating the cavity.

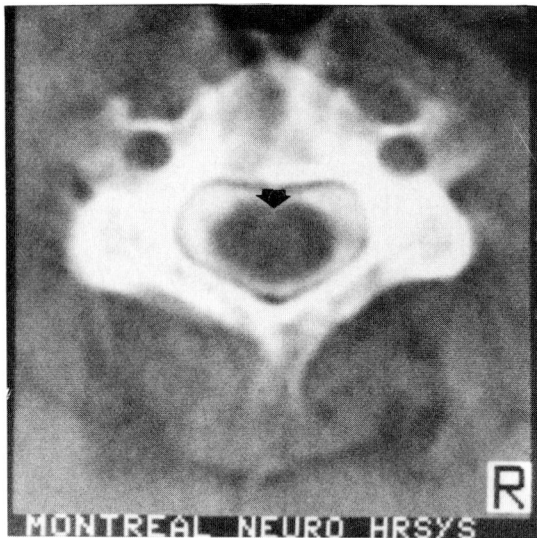

Fig. 5.17 Syringomyelia. Normal cord but with cavity faintly opacified with metrizamide.

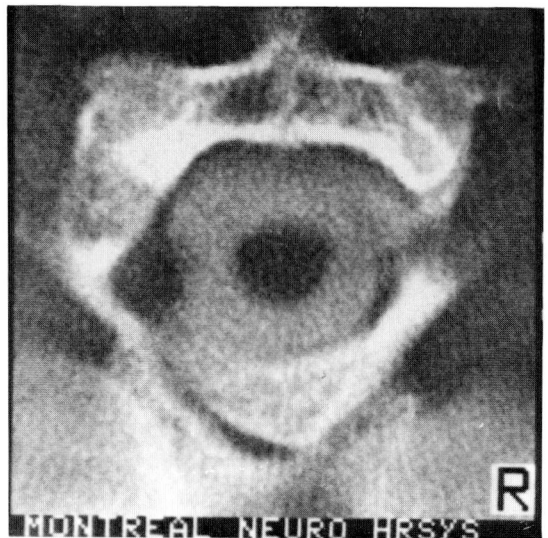

Fig. 5.19 Hydromyelia. Metrizamide outlining atrophic cord.

### Syringomyelia and hydromyelia

Four types of spinal cord have been described by Ethier et al (1980) in these conditions:

*Round large cord.* This fills the spinal canal; the central canal is distended with fluid (Fig. 5.16).
*Flat cord.* Flattened front to back and resting against the posterior border of the canal (Fig. 5.17).

*Normal size.* But with cavity of the central canal of varying size (Fig. 5.18).
*Atrophic or small cord.* This may contain a cavity (Fig. 5.19).

Once again HRCT proved to be of great value in these cases. In an early study of 32 patients by Bonafé et al (1980), 75% showed indirect evidence of syringomyelia on oil myelography and 25% were negative. All 32 cases showed cavitation on the CT studies. This was a significant breakthrough in the investigation of this condition. In 1973 Foster & Hudgson reported only 79 myelograms were positive out of 92 cases of clinical cervical syringomyelia.

### Cord puncture

Transcutaneous puncture is sometimes used to demonstrate the upper and lower limits of a cystic or hydromyelic cavity. In the latter it is also possible to ascertain whether there is communication with the fourth ventricle. Oil-based and water-soluble contrast have been used, sometimes combined with air.

As cord punctures are performed infrequently the method will be described briefly. The patient lies in the prone position; local anaesthesia is administered; a short 22 gauge lumbar puncture needle is advanced as far as the dura; then a 25 gauge needle is inserted through the first needle in the midline. The position is checked with biplane fluoroscopy (a Mimer III or Neuro-diagnost-N is useful in such cases). Following the aspiration of fluid for analysis, 0.25 or 0.50 cc (220 mg I per ml) of metrizamide is injected. Plain radiographs or scout films are taken, followed by HRCT.

### Use of metrizamide as a contrast media

It is obvious from the foregoing that a water-soluble contrast media has proven to be the more effective to use in conjunction with computed tomography because of its lower attenuation factor.

Metrizamide was first used in Norway in 1972 (Sortland et al 1979). Despite the fact that seizures were among the reactions reported (0.1–1.0% of cases), its use has been generally acclaimed, with the exception of patients having a known history of seizures and those needing neuroleptic medication. Phenothiazine drugs are used in some forms of anaesthesia; this drug reduces the seizure threshold to metrizamide.

The superiority of metrizamide over an oil-based contrast is easy to recognize. It will outline fine structures such as nerve root sheaths, which may not be clearly defined or not seen at all with the oil-based contrast. Another reason for discontinuing the use of the oil-based contrast was the realisation that cases of thecal scarring and chronic arachnoiditis were being seen many years following myelographic studies whereas no cases of thecal scarring were observed following myelography with metrizamide (Kimber 1980).

### Reaction

In order to avoid the possibility of reactions, 10 mg diazepam were administered prior to the procedure. As this was considered to be ineffective after 15 mins, the premedication was changed to 200 mg phenobarbital by mouth the night before the examination.

Even with this premedication and careful screening for history of low seizure threshold, a recent 48 hour follow-up study at the Montreal Neurological Hospital showed that out of 218 cases, 119 (56%) had minor reactions and 7 (1%) major reactions. The minor reactions consisted mainly of headache, dizziness, nausea and vomiting, but visual as well as personality disturbances have been recorded. The breakdown of the major reactions was as follows:

1 — seizure
3 — aphasia
1 — dysphasia
1 — paralysis
2 — urinary retention.

None of these reactions resulted in a permanent deficit.

A survey of the literature on reactions is quite alarming, although in some instances the report may describe an isolated case. It is still debatable whether or not permanent damage is being done to the nerve cells, and whether this risk is offset by the improved visualisation.

Gonsette (1977) reports five patients who described a feeling of dual personality, one of whom suffered an attack of paranoia. He also states that 50% of patients after cervical myelography demonstrated EEG abnormalities (bursts of slow waves).

Skalpe (1977) reports transient hallucination and disorientation in three patients. Smith & Laguna

(1980) report 'Confusion, dysphasia and asterixis following metrizamide myelography'. Bergvall et al (1981) report two cases of confusion, myoclonus and speech arrest. Bertoni et al (1980) report 'Asterixis and encephalopathy following metrizamide myelography'.

Kelley et al (1980) describe one case reported to have Guillain Barré syndrome following metrizamide. Russell et al (1980) report on two patients with complex partial status epilepticus following myelography with metrizamide. However, these cases could be explained by a relatively high dose of iodine and the fact that one of the patients lay down for 15 minutes soon after the procedure was finished.

Major reactions can be avoided in the majority of cases by means of optimum concentration of iodine, and proper pre- and post-examination care (screening for history of seizures, premedication, adequate hydration before and after, and elevation of the patient's head following the procedure).

It is already known that metrizamide is less toxic than other forms of water-soluble contrast. There are also new water-soluble contrast agents which are said to be less toxic, which are in the process of being tested. It is to be hoped that all centres will document any adverse effects observed during the 48 hours immediately following each procedure. This should be the responsibility of every radiographer in collaboration with the radiologist and hospital ethics committee (Table 5.1).

**Table 5.1** The advantages and disadvantages of metrizamide

| Advantages | Disadvantages |
| --- | --- |
| Good visualisation of fine structures | Dilutes rapidly due to its miscibility with CSF |
| Lower kV required | Patient must remain sitting or semi-recumbent following examination |
| Does not have to be removed, therefore less pain for the patient | Many unpleasant side effects |
| Computed tomography may follow several hours after myelogram | |

## Summary

1. It has been shown that the radiation dose for one examination using HRCT is no higher than that of a barium enema. The radiologist must always assess the benefit to the patient when doing special studies; the seriousness of the problem often outweighs the question of the amount of radiation being delivered.

2. Despite all the literature describing serious reactions, it has been shown that the non-ionic water-soluble contrast (metrizamide) is far superior to that of oil-based contrast. With careful handling, reactions can be kept to a minimum and serious adverse effects do not appear to excede 1% of cases.

3a. It has been shown that HRCT is a useful, if not indispensable, adjunct to plain film radiography, and will often rule out the necessity of conventional tomography.

It has been proven that lesions not previously seen may be well demonstrated even in the very early stages. These added benefits to the patient more than compensate for the expense involved in adding the hardware-software package.

b. The higher cost of water-soluble over oil-based contrast media is more than compensated for by the improved visualisation which leads to more accurate diagnosis.

## NUCLEAR MAGNETIC RESONANCE IMAGING

### Introduction

Nuclear magnetic resonance imaging, or zeugmatography, is a technique which stimulates and records the effect of resonances of atomic nuclei when subjected to a magnetic field. By placing the patient within such a strong field, applying radiofrequency pulses at a frequency appropriate to stimulate resonance, and analysing the radiofrequency signals emitted by the nuclei with a computer, images of the nuclear density (usually hydrogen) can be obtained (Kaufman et al 1981). This produces an image showing grey and white matter, cerebrospinal fluid, the ventricles, the subarachnoid space, the fissures, sulci and blood vessels. It appears also that it will differentiate between normal and abnormal tissue. All of this can be shown more clearly and with fewer artefacts than with CT.

## History

Nuclear magnetic resonance (NMR) has been used for the past 30 years in the analysis of matter in chemistry and physics.

Damadian (1971) proposed a theory for imaging solids and in 1972 Lauterbur succeeded in producing the first images and reported his results in 1973. This technique used in imaging is now generally known as Magnetic Resonance Imaging (MRI). The principles of MRI have already been well documented in the literature (Kaufman et al 1981, Andrew & Worthington 1981, Lerski 1983). There are currently two approaches to the generation of the high magnetic fields:

a. Resistive. This water-cooled system consumes 50 kW of power and generates a great deal of heat. It involves the considerable cost of the power consumption and the heat dissipation.

b. Superconductive. This type of unit is cooled by liquid helium (−269 °C). It involves the cost of the latter and a higher initial outlay.

c. A further approach which uses permanent magnet technology is being evaluated by some companies.

## Application

Sagittal, coronal or transverse planes may be scanned by manipulating the magnetic field. This means that the patient has only to be positioned once. This is particularly valuable in cases of trauma to the spine and spinal cord, as the patient will remain supine throughout the entire examination.

The scanning time for one level with an approximate slice width of one centimetre is 4–8 minutes; however, total scanning time for 8–10 sequential slices is approximately the same. Thicker sections of more than 1 cm may be recorded; this would facilitate three dimensional reconstruction. The resolution is in the region of 0.5 mm (Andrew & Worthington 1981).

Images are normally displayed in black and white in a similar manner to CT scans. The results may be viewed either on a 256 × 256 or 512 × 512 matrix. Hard copies are made using a multi-imager.

## Discussion

In contrast to CT bony structures do not give any appreciable NMR signal and thus appear black in the image. This is of particular significance in the area of the skull.

Mansfield & Maudsley (1977) stated that the NMR signal coming from a localized region of the sample is directly related to the mobile proton density at that point. For example, a glioma has a high content of mobile protons and shows as a homogeneous high density. The advantage is that infusion of contrast media is not necessary to demonstrate this lesion. Bone contains relatively few mobile protons and, therefore, gives a low signal, whereas soft tissues which contain more mobile protons give a relatively high signal. This contrasts sharply with conventional X-rays, which give poor soft tissue definition.

The subarachnoid space is well visualized and subdural haematomas may be demonstrated clearly, as stationary blood gives a higher signal and will appear more intense than blood in motion. The sylvian and interhemispheric fissures are clearly seen; there is also good differentiation between the grey and white matter. It is this strong sensitivity to differentiation between the grey and white matter which has already made MR Imaging an important factor in the early diagnosis of multiple sclerosis.

MRI has the advantage over CT in the area of the posterior fossa. The cerebellum is clearly outlined. This is due to the fact that, as already stated earlier, there is no NMR signal produced by the bony structures. It is possible to identify the fourth ventricle, the cerebellar fissure and the cerebellar tonsils.

In cases of Arnold Chiari malformation herniation of the tonsils has been demonstrated. With this malformation, syringomyelia and hydromyelia have been better visualized with MRI than CT with contrast. Other diseases involving grey and white matter, such as Wilson's & Binswanger's, have also been demonstrated.

The cerebrospinal fluid in the ventricles gives a signal when static or in motion, that is, the motion caused by peristalsis and pulsation of the brain. It is thought that blockages of the ventricular system may be demonstrated.

Blood vessels are identifiable by their position in the brain and show as tortuous black lines. Static blood and blood in motion give different signals. It should be possible to detect arteriovenous malformations and thrombosed vessels (Moore et al 1980). Dynamic studies have been carried out in order to demonstrate acute cerebral infarction. Occlusion of the common carotid has been visualized during the first 24 hours.

Abnormalities of the spinal cord and canal, such as stenosis, syringomyelia, hydromyelia, tumours of the cord and discs, have all been demonstrated better with MRI than CT with contrast. Brant-Zawadski et al (1983) summarize the demonstration of cerebral abnormalities well by stating that 'the depiction of brain anatomy and its major components (i.e. the grey and white matter) is superior to that offered by CT especially in regions where beam-hardening artifact occurs on CT images. NMR is more sensitive to the presence of brain abnormalities than is CT. This sensitivity is especially evident in foci of demyelination and/or edema of the white matter.'

## Care of the patient

Andrew & Worthington (1981) point out three possible sources of hazard:

1. high static magnetic fields
2. the radiofrequency pulses
3. The time dependent magnetic field gradients.

On the other hand, Moore & Holland (1980) as well as Hinshaw et al (1980) state that to date MRI appears to be hazard free. However, it is important to monitor patients closely during the procedure for possible arrythmia or other reactions. Seizure patients are not recommended for MRI.

Pavlicek states that six pacemakers were tested; all demonstrated a response to a magnetic field. Some units will switch from demand to fixed rate mode in a magnetic field as low as 5 gauss. The torque on at least two resulted in significant movement within the chest wall. The amount of movement appears to be dependent on the amount of fibrotic tissue present. It is important that attention be paid to patients with pacemakers before submitting them for an NMR study of the brain or spinal column. Personnel or public wearing pacemakers must be eliminated from the MRI suite.

Surgical clips may rotate, with the possibility of the dislodging of the clip from the site of an aneurysm, allowing haemorrhage to occur. Precautions must be taken to prevent accidents when loose objects made of magnetic metal are brought into the room. Such items as scissors, cigarette lighters, or instruments used with life support systems could fly several feet and cause damage to the apparatus or injury to the patient.

It is important to have an extensive follow-up programme for a considerable number of years. Stringent regulations have already been laid down in 1981 by the British National Radiological Protection Board at Harwell.

## Artefacts and other complications

In more recent literature extensive tests have been performed on dental amalgam, stainless steel, 14 karat gold, a variety of surgical clips. Artefacts from some dentures and orthodontic braces were seen on the NMR image in the area of the face. No artefacts were seen with amalgam, gold or steel suture wire (New et al 1983). Tape recordings will be spoiled if brought into the magnetic field. Credit cards can be invalidated and the function of mechanical watches disturbed.

## Summary

1. NMR imaging to date has not presented any serious side-effects to the patient. But further tests have to be made in the light of recent experiments. The fact that non-ionising radiation is used is of particular importance in the investigation of children, although the need to remain still for 10–20 mins may be a limitation.

2. The necessity of the use of contrast media with MRI has not yet been determined. Experiments are now being conducted to find a paramagnetic contrast media.

3. As with the new generation of CT scanners, the cost factor of MRI poses a significant problem to hospital administration and health services (Table 5.2). The resistive unit is less expensive than the superconductive, but the costs of power consumption and heat dissipation are high. On the

**Table 5.2** The advantages and disadvantages of magnetic resonance imaging

| Advantages | Disadvantages |
|---|---|
| Non-invasive | Cost factor |
| Non-ionising (important for examination of children) | Space factor & installation requirements |
| No moving parts | Length of examination |
| Patient remains in one position | Confining effect of apparatus for patients with tendency to claustrophobia |
| Potential capability of discriminating neoplastic from normal tissue | Lack of definition of bony structures |
| Free from artefacts caused by dense bony structures, especially in the base of the skull | Problems associated with metallic objects, especially pacemakers and life support equipment |
| Good visualization of anomalies of the spinal cord | At the present is complementary to CT scanning |
| Improved visibility over CT soft tissue structures | |
| Demonstrates demyelinating diseases | |
| Evaluation of occlusive stroke in early stages | |
| Possibility of dynamic studies of CSF and cerebral blood flow | |

image quality (i.e. resolution and signal to noise ratio) is also greater. The superconductive unit can be used for spectroscopy as well as imaging.

## CONCLUSION

These great technological developments raise some serious questions:

1. How much radiation can safely be delivered to the patient during these procedures?
2. Can the use of contrast media with possible serious side-effects be justified?
3. Are the medical benefits worth the very high cost of purchasing and operating sophisticated new equipment?

Each person in authority must search in his own mind for the answers to these and other questions raised by today's advances in technology.

*Acknowledgements*

I would like to thank my peers and colleagues for their help, encouragement and patience; especially Dr R. Ethier, Dr D. Melanson, and Dr T. Peters. I would also like to thank Margaret Allison for technical assistance and Charlotte Payette for preparing the manuscript.

other hand, the superconductive carries a much higher initial cost. However, to offset the added expense, it should be pointed out that it has ten times the field strength of the resistive unit. The

## REFERENCES

Andrew E R 1979 Nuclear magnetic resonance imaging: Zeugmatography. In: Kreel L (ed) Medical imaging CT U/S IS NMR. HM & M, Aylesbury, Section 1, ch 7, p 38–43

Andrew E R, Worthington B S 1981 Nuclear magnetic resonance imaging. In: Newton T H, Potts D G (ed) Radiology of the skull and brain, vol. 5 Technical aspects of computed tomography, ch. 132: 4389–4404

Bergvall U, Brismar T, Lying-Tunell U, Valdimarrson E 1981 Confusion myoclonus and speech arrest: epileptic manifestations after metrizamide myelography. Acta Radiologica Scandinavia 63: 315–322

Bertoni J M, Schwartzman R J, Van Horn G, Partin J 1980 Asterixis and encephalopathy following metrizamide myelography: Investigations into possible mechanisms and review of the literature. Annals of Neurology 9: 366–370

Bonafé A, Ethier R, Melançon D, Bélanger G, Peters T 1980 High resolution computed tomography in cervical syringomyelia. Journal of Computer Assisted Tomography 4(1): 42–47

Brant-Zawadski M et al 1983 NMR demonstration of cerebral abnormalities: Comparison with CT. American Journal of Neuroradiology 4: 117–124

Di Chiro G 1980 Improvement in computed tomography spatial resolution. In: Caillé J M Salamon G (ed) Computerized tomography. Springer Verlag, Berlin, p 12–15

Ethier R Personal communication

Ethier R, King D G, Melançon D, Bélanger G, Thompson C 1980 Diagnosis of intra- and extramedullary lesions by CT without contrast achieved through modifications applied to the EMI CT 5005 Body Scanner. In: Post M.J.D. (ed) Radiographic evaluation of the spine. Masson Publishing USA, New York, ch 14, p 377–393

Fischgold H, Bull J 1967 A short history of neuroradiology. VIII[th] Symposium Radiologicum, Schering

Foster J B, Hudgson P 1973 The radiology of communicating syringomyelia. In: Barnes H J M, Foster J B, Hudgson P (ed) Syringomyelia. Saunders, Philadelphia, vol I p 50–63

Gargano F P 1980 Transverse axial tomography of the lumbar spine. In: Post M J D (ed) Radiographic evaluation of the spine. Masson Publishing USA, New York, ch 19, p 475–490

Gargano F P 1980a Extradural venography. In: Post M J D (ed) Radiographic evaluation of the spine. Masson Publishing USA, New York, ch 27, p 579–591

Gonsette R E 1977 Cervical myelography with metrizamide by sub occipital puncture. Acta Radiologica Supplementum 355: 121–126

Grainger R G 1977 Technique of lumbar myelography with metrizamide. Acta Radiologica Supplementum 355: 31–37

Grainger R G 1979a Further developments in water soluble myelography. In: Lodge Sir T, Steiner R E (ed.) Recent advances in radiology and medical imaging 6. Churchill Livingstone, Edinburgh, ch 10, p 173–193

Haughton V M, Williams A L 1980 CT anatomy of the spine. In: Caillé J M, Salamon G (ed) Computerized tomography. Springer Verlag, Berlin

Hawkes R C, Holland G N, Moore W S, Worthington B S 1980 Nuclear magnetic resonance (NMR) tomography of the brain: A preliminary clinical assessment with demonstration of pathology. Journal of Computer Assisted Tomography 4 (5): 577–586

Hinshaw W S, Andrew E R, Bottomley P A, Holland G N, Moore W S, Worthington B S 1980 Current progress and future prospects in NMR imaging. In: Caillé J M, Salamon G (eds) Computerized tomography. Springer Verlag, Berlin, p 271–274

Holland G N, Hawkes R C, Moore W S 1980 Nuclear magnetic resonance (NMR) Tomography of the brain: Coronal and sagittal sections. Journal of Computer Assisted Tomography 4 (4): 429–433

Holland G N, Moore W S, Hawkes R C 1980 Nuclear magnetic resonance tomography of the brain. Journal of Computer Assisted Tomography 4 (1): 1–3

Holland G N, Moore W S, Hawkes R C 1980 NMR neuroradiography. British Journal of Radiology 53 (627): 253–255

Hounsfield G N 1980 Computed medical imaging: Nobel lecture, December 8, 1979. Journal of Computer Assisted Tomography 4 (5): 665–674

Hounsfield G N 1978 Potential uses of more accurate CT absorption values by filtering. American Journal of Roentgenology 131: 103–106

Kaufman L, Crookes L E, Margulis A R (ed) 1981 Nuclear magnetic resonance imaging in medicine. Igaku-Shoin Ltd, Tokyo

Kelley R E, Daroff R B, Sheremata W A, McCormick J R 1980 Unusual effects of metrizamide lumbar myelography. Archives of Neurology 37: 588–589

Kimber P M 1980 The new myelographic agent and its spin-off. Radiography 542 (XLVI): 29–34

King D G 1980 Computed tomography of the spine: Resolution requirements and scanning techniques to achieve them. In: Post M J D (ed) Radiographic evaluation of the spine. Masson Publishing USA, New York, ch 13, p 366–376

Kreel L, Osborn S 1976 Transverse axial tomography of the spinal column; a comparison of anatomical specimens with EMI scan appearances. Radiography 946 (XLII): 73–78

Lerski R A 1983 Physical principles of nuclear magnetic resonance imaging. Radiography 49 (180): 85–90

Mansfield P, Maudsley A A 1977 Medical imaging by NMR. British Journal of Radiology 50: 188–194

Moore W S, Holland G N 1980 Nuclear magnetic resonance imaging. British Medical Bulletin 36 (3): 297–299

Moore W S, Holland G N, Kreel L 1980 The NMR CAT Scanner — a new look at the brain. Journal of Computed Tomography 4 (1): 1–7

Naidich T P, Pudlowski R M 1980 High resolution CT of the cervical spinal cord. In: Caillé J M, Salamon G (eds) Computerized tomography. Springer Verlag, Berlin

National Radiological Protection Board 1981 Exposure to nuclear magnetic resonance imaging. Radiology 563 (XLVII): 258–260

New P F J et al 1983 Potential hazards and artifacts of ferromagnetic and nonferromagnetic surgical and dental materials and devices in nuclear magnetic resonance imaging. Radiology 147: 139–148

Newton T H, Potts D G (eds) 1981 Radiology of the skull and brain, vol. 5. Technical aspects of computed tomography. Mosby, St Louis

Oldendorff W H 1980 The quest of an image of the brain. Raven Press, New York

Partain C L, James A E, Watson J T, Price R R, Coulam C M, Rollo F D 1980 Nuclear magnetic resonance and computed tomography. Radiology 136: 767–770

Pavlicek W et al 1983 The effects of nuclear magnetic resonance on patients with cardiac pacemakers. Radiology 147: 149–153

Peters T Personal communication

Russell D, Anke I M, Nyberg-Hansen R, Slettines O, Sortland O, Veger T 1980 Complex partial status epilepticus following myelography with metrizamide. Annals of Neurology 8: 325–327

Sackett J F, Strother C M 1979 New techniques in myelography. Harper and Row, Hargerstown, Maryland

Shapiro R 1975 Myelography, 3rd edn. Year Book Medical Publishers, Chicago

Schuster N H 1962 Early days of Roentgen photography in Britain. British Medical Journal ii: 1164–1166

Skalpe I O 1977 Adverse effects of water soluble contrast media in myelography cisternopgraphy and ventriculography. Acta Radiologica Supplementum 355: 359–370

Skalpe I O 1980a Metrizamide myelography. In: Post M J D (ed) Radiographic evaluation of the spine. Masson Publishing USA, New York, ch 22, p 514–532

Smith F W 1981 Whole body nuclear magnetic resonance. Radiography 564 (XLVII): 297–300

Smith M S, Laguna J F 1980 Confusion, dysphasia and asterixis following metrizamide myelography. Canadian Journal of Neurological Sciences 7 (4): 309–311

Smith P 1976 Lumbar epidural myelography. Radiography 498 (XLII): 125–129

Sortland O, Magnaes B, Hauge T 1979 Functional myelography with metrizamide in the diagnosis of lumbar spinal stenosis. Acta Radiologica Supplementum 355: 42–54

Wall B F, Green D A C, Veerappan R 1979 The radiation dose to patients from EMI brain and body scanners. British Journal of Radiology 52: 189–196

Young I R et al 1982 Initial clinical evaluation of a whole body nuclear magnetic resonance (NMR) tomograph. Journal of Computer Assisted Tomography 6 (1): 1–18

# 6

# Urodynamic studies in the patient with severe cord injury

The application of physiological measurements to the act of micturition has shown a steady increase in urological centres throughout the world during the past 10 years. As the investigation cannot be truly physiological, controversy still exists upon the value of and the reproducibility of its findings. The introduction of catheters or supra-pubic needles into the bladder, an unphysiological filling-rate with a non-urine-equivalent fluid, and a request to micturate in a situation far removed from the privacy of the lavatory, all have an inhibitory effect on the act of micturition in the normal patient. The reverse may apply to those patients with neurogenic bladders and the result may be a voiding pattern far removed from the patient's usual responses.

There is also some disagreement on the role of radiology combined with physiological measurements in those patients with cough/stress incontinence. There is general agreement however that in patients with neurogenic bladders and especially those patients with cord injuries or spina bifida or other neuropathic conditions such as multiple sclerosis a combined approach is essential.

Urodymoradiology is impossible unless there is a concerted effort to develop a team approach to these studies and to interface effectively the X-ray apparatus and the urodynamic equipment. To achieve maximum benefit both the clinician and the radiologist need to be present and to achieve optimum results the work of the radiographer, physicist, nurse and physiological measurement technician or medical physics technician needs to be co-ordinated within the fluoroscopy room. Except for the investigation of ureteric reflux during both the filling phase and during the act of micturition urodynamic studies relate almost

entirely to the changes which occur from within the bladder to the external meatus. There are two different investigations. The first is used to show changes which take place within the urethra and is called the urethral profile without the need for fluoroscopy. The other — generally termed urodynamic investigation — consists of measurements of changes in pressure of and the rates of flow during the act of micturition, and may include electromyographic studies of the detrusor muscle by inference.

## THE ACT OF NORMAL MICTURITION

To maintain the well-being and survival of the human being poisonous waste products need to be systematically disposed of. The kidneys, ureters bladder and urethra represent a filtering system, internal connectors, a store with a variable volume and an external connector which includes a controllable valve. One can say that the act of micturition starts with the tubules of the kidney and is only complete on the satisfactory expulsion of urine via the urethra.

There are no anatomical differences in the bladder, trigone or their innervation in either the male or female. The bladder neck and urethra are different. Pelvic floor support does differ and in the male there is a relationship between the prostatic bed and the initial part of the urethra and the distal segment. Throughout the text the term 'external sphincter' relates only to that muscle associated with the distal part of the urethra.

### The bladder wall

The whole of the bladder wall with the exception

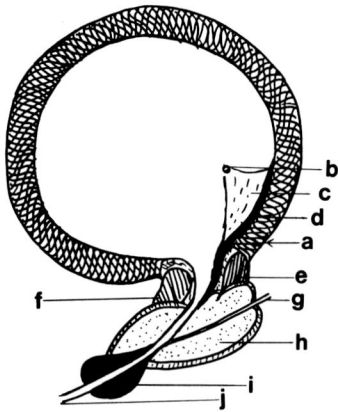

**Fig. 6.1A** Lower urinary tract in the male. This represents a mid-line section of the bladder with the patient in the posterior-oblique position. The bladder neck and the sphincters are wide open with the urethra dilated as if the patient were micturating with a good stream. Key:
(a) detrusor muscle is virtually continuous around the whole of the bladder including that area known as the trigone;
(b) ureteric orifice; (c) superficial trigone muscle triangular in shape which extends inferiorly into the bladder neck and into the posterior urethra; (d) smooth muscle of the bladder neck;
(e) internal urethral sphincter — posterior part; (f) internal urethral sphincter — anterior part; (g) seminal canal;
(h) prostate gland; (i) external sphincter consisting of striated muscle; (j) proximal urethra.

**Fig. 6.1B** Lower urinary tract in the female. This represents a mid-line section with the patient in the posterior oblique position. Key: (a) detrusor muscle; (b) ureteric orifice; (c) trigone; (d) posterior connective tissues; (e) anterior connective tissues; (f) external sphincter comprising of striated muscle. Note its elongated oval shape; (h) vaginal urethral orifice — The dark area represents the peri-urethral striated muscle. A thin layer of smooth muscle extends along the internal surface of the urethra.

of trigone and bladder neck is comprised of groups of smooth muscle cross-meshed and interrelated. In the male and female these muscles exhibit the same histological and histochemical characteristics. If the bladder wall is transected there is no clear demarcation between these bundles of muscle although there are more longitudinal muscles on its internal and external surfaces. As the bladder fills the series of muscles relax allowing the bladder to stretch and assume a spherical shape.

During micturition as the bladder empties, the detrusor of the bladder wall contracts in all directions and continues to do so as the total bladder volume diminishes. (Under cystoscopy unco-ordinate muscle contractures in the bladder of a paraplegic patient appear as if the fingers of the hand were entwined with each finger moving without reference to the others.) The whole of the detrusor muscle is involved and its action is not peristaltic in nature (Figs 6.1A and B).

The most recent research has re-inforced the view that detrusor smooth muscle is predomi-

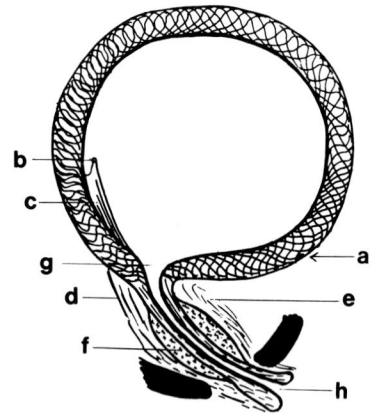

nantly innervated by the parasympathetic nervous system via large numbers of cholinergic terminals. In this respect it is markedly different from smooth muscle elsewhere within the urinary system for these muscles are rich in the enzyme acetylcholinesterase which relates to the degradation of acetylcholine.

On the other hand sympathetic noradrenergic nerve supply is almost entirely associated with the blood supply of the bladder. Because of this sympathetic nerves can have no direct influence on detrusor muscle activity. Inhibition of this activity is achieved somewhat remotely via ganglionic transmission from the pelvic plexus to the parasympathetic cell bodies supplying cholinergic fibres to the detrusor muscle within the bladder wall. There is still some disagreement of the real function and its implication of a third component in autonomic innervation. One theory is that the cholinergic nerves in this area release another substance which has a functional-transmit code. Another suggestion is that there are highly specialised nerves which are specific for bladder innervation. The significance of this substance is in the future development of special drug regimes to combat intractable bladder dysfunction.

## Trigone

Anatomically the trigone is that area of the bladder on the postero-inferior wall bounded by the ureteric orifices with its lateral walls both angled medially until they merge with the internal urethral meatus. The deep trigonal muscles are directly related to the rest of the bladder wall. The superficial muscle has a very different structure from the rest of the trigone being of quite tiny bundles of muscles. There is little parasympathetic innervation of this muscle, confirmed by the absence of acetylcholinesterase in its cells. In the male the superficial muscle extends much further through the urethral crest and ceases at the verumontanum whereas in the female it disappears in the proximal part of the urethra.

## Male bladder neck

Histologically and histochemically different smooth muscles form a complete ring around the bladder neck extending as far as the pre-prostatic part of the urethra. Here it becomes contiguous with the muscles within the prostatic capsule.

Autonomic innervation of the bladder neck is quite different from that of the detrusor muscle, in that cholinergic nerves are few and far between but there are abundant sympathetic non adrenergic nerve terminals similar to that of the prostatic nerve capsule. This close relationship leads to the closure of the bladder neck during seminal ejection preventing retrograde filling of the bladder. The converse applies and normally retrograde filling of the prostatic bed is prevented although this is often seen in paraplegic patients. The active role of the smooth muscle in preventing incontinence is now more clearly understood because of the continued research into the functions of the lower urinary tract.

### Intrinsic urethral sphincter

The pudendal nerve carries somatic fibres arising from nerve roots from the second, third and fourth sacral segments and it was originally thought to supply the striated muscle both within the external intrinsic urethral sphincter and the peri-urethral pelvic floor. Recently further research has shown that this sphincter is innervated via the pelvic

(splanchnic) nerve which means that this nerve has a dual role of carrying autonomic fibres to the detrusor muscle and of carrying somatic fibres to the external urethral sphincter.

The practical application of this finding is that the importance of EMG recording and the assumption that specific recordings from the levator ani do not have the same relevance as it was once thought. However, there is no doubt that in the paraplegic patient bipolar EMG readings directly from the anal sphincter do play a major role in the

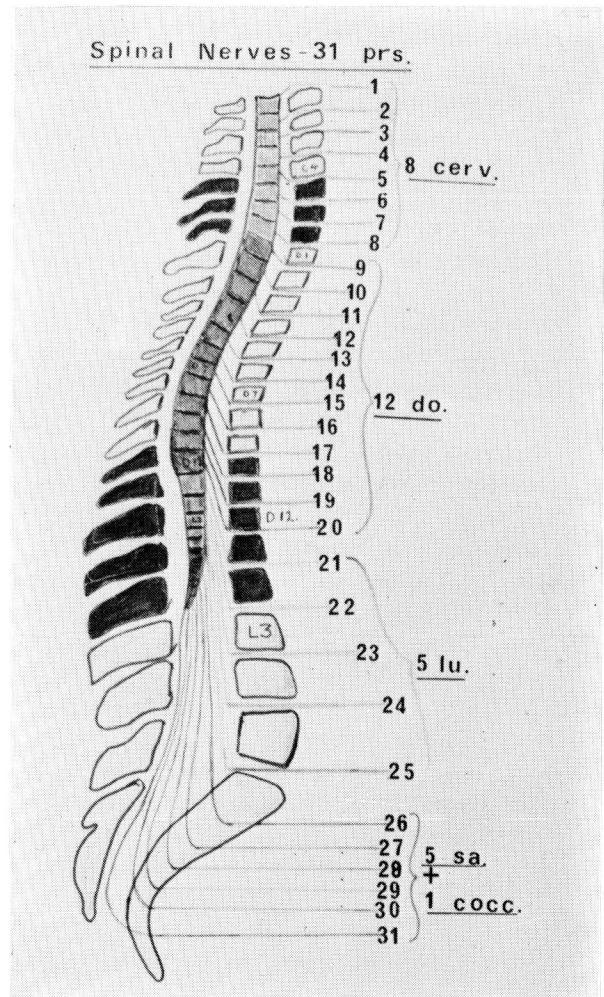

**Fig. 6.2** Schematic diagram of the spinal cord, the 31 pairs of spinal nerves and the spine. The two most common sites (in black) for fractures of the spine with cord involvement. Note the relationship between the dorsi-lumbar junction and the site of the sacral micturition centre.

management and the course of the urodynamic investigation in the paralysed patient.

This extended description of the anatomy the musculature and its innervation gives some indication of the complexity of the controls which prevent, allow or interrupt the act of micturition, and which initiate micturition when the bladder is only partially filled or prevent it when the bladder is full. This act of micturition involves instructions to the detrusor muscle to contract and coincidentally instructs the bladder neck, the internal or external sphincters, the urethral walls, as well as 'negative' instructions to the fast response muscles of the pelvic floor to stay relaxed. These local instructions emanate from the micturition centre in the sacral segments of the spinal cord which lie opposite the first lumbar vertebra. Most authorities agree that the second, third and fourth sacral nerves are intimately involved although others suggest that the first sacral nerve also plays a significant role (Fig. 6.2).

There is no doubt that the higher centres in the brain have an overriding control over the sacral micturition centre. The bladder response in the baby is a purely reflex one triggered off by a certain volume in the bladder. Bladder training, developing brain activity, amid a social awareness gradually allows the small child to exert overt control over the act of micturition until sufficient control is there to prevent mishaps well past the pain threshold engendered by an overfull bladder.

## Female bladder neck and urethra

The same histological and histochemical differences occur in the female between the detrusor muscle and the bladder neck as do those in the male. The infrastructure is quite different for there is no ring of smooth muscle and the fine bundles of smooth muscle continue longitudinally or at most diagonally within the wall of the urethra. In spite of the absence of an obvious valve the 'mid-stream arrest' mechanism in the normal female acts within a quarter of a second. It would appear from this that the efficiency of the mechanism is very dependent on the interplay of a number of nearby anatomical structures such as elastic tissue and striated muscle.

Another marked difference between the male and female bladder neck is the almost total absence of non-adrenergic nerves which confirm that the sympathetic nerves in the male in this region have a strong sexual bias for which there is no female equivalent. In the female the detrusor muscle, the bladder neck and the urethra all have numbers of similar para-sympathetic nerves. There is evidence that the female urethra widens and shortens during micturition. This also suggests why the relationship of the bladder to the pelvic floor and the position of the bladder neck, the very short urethra and the vaginal orifice are so important in continence and an adequate stream in the female.

### Female sphincter and pelvic floor muscle in the male and female

The pelvic floor in the male consists of mixed muscle divided between slow and fast response components innervated by the pudendal nerve. The peri-urethral levator ani consists of large diameter striated muscle some of which are of the fast response type but only capable of contraction for short periods. This may on demand reinforce the distal intrinsic external urethral slow-response striated muscle during coughing. These muscles play a part in stopping micturition in mid-stream by rapidly increasing mid-urethral pressure.

In the female although the anatomical relationships between the external urethral sphincter and the peri-urethral striated muscle are different, in configuration, the female is also capable of increasing mid-urethral pressure to cut-off in mid-stream. Both in the male and female this function is incapable of maintaining long-term continence. The intrinsic external sphincter in both the male and the female have circular muscle with a slow response time capable of sustained and prolonged contraction.

## THE PARAMETERS TO BE MEASURED

Only a brief description of the urethral profile will be given because there is seldom need to involve radiographic staff. The basic requirements are: a urethral catheter, a pressure line with a transducer, a constant speed withdrawal unit and a method of recording the changes in pressure as the catheter tip is withdrawn. However, if a radiographic image is required a low concentration of any of the water soluble contrast media, an image intensifier with television and visual display unit (VDU) recording facilities and a spot film device or some other method of hard copying is needed. Videotape recording and other dynamic methods of recording the image are not required since there are no rapid response sequences involved. The physics of fluid flowing through restricted tubes is made more complicated by the passage of this fluid through a flexible and distensible tube since the cross-sectional area is a function of the pressure and the elasticity of the walls.

An added complication relates to the variation in elasticity within the walls of the urethra from the bladder neck to the external meatus but calibration may be performed by taking measurements at the external meatus or in areas where there are known strictures.

Measurements may be taken with the urethra filled with liquid in steady flow or stationary. Because the urethra normally has very elastic walls pressure area measurements would appear in a recording as an almost straight line with a slight inclination until maximum distensibility of the urethral walls has been reached when a graph is produced showing measured pressure against cross-sectional area. Unavoidable errors occur when:

1. the diameter of the measuring probe is at variance to the true cross-section of the urethra
2. the transducers are at a different height to the urethra
3. if the small pressure balloon at the end of the catheter is not inflated to just slightly above atmospheric pressure
4. if the balloon becomes too spherical and distorts the lumen of the urethra
5. if the length of the balloon is too long relative to the cross-section it will even out pressure changes which may occur along its length. However, if all is well this method gives the true piezometric pressure.

A very simple method which suggests considerable theoretical difficulties but in practice does not produce them is to use a catheter tip transducer, this method being dependent on the urethra making an adequate seal around the catheter. The measurements do not produce direct peizometric measurement and need to be corrected.

The Brown-Wickham infusion method (1969) also works well if the urethra remains distensible and catheter diameter is not too great. It does present problems with paraplegic patients since there is sometimes quite high resistance to retrograde flow into the bladder.

Measurement of flow resistance within the tubing and catheter can be taken before insertion of the catheter and in this case a fairly viscous contrast medium can be used allowing 'spot film' which includes the bladder neck to be taken with ease. Some correction would need to be made to the pressure readings to give a true piezometric pressure since this requires a aqueous solution for direct readings.

## Urodynamic measurements

As has already been shown innervation of the lower urinary tract is extremely complex and interruption to the nerve supply will produce a wide range of sequelae. The bladder involvement from a particular spinal cord injury cannot be determined nor can its subsequent behaviour once spinal shock has subsided. It is therefore essential to make a full urological survey and the following regime should be instituted:

1. An accurate history and full neurological investigation
2. Blood urea estimation and urine culture
3. Estimation of amount of residual urine
4. Intravenous urography
5. Urethral profile
6. Urodynamic investigation with fluoroscopy
7. A continuous fluid charting of input and output of fluids.

Early investigations relied on micturating cystograms and the use of 'spot films'. The contrast medium used was 6% sodium iodide in sterile solution in quantities varying from 250–3000 ml. Unfortunately in a significant number of patients the investigation was followed by irritation of the bladder wall and flare up in the urinary tract. The opaque medium of choice is now Hypaque 15% made up into 1 litre sterile packs.

Another investigation which was undertaken was a metro-cysto-ureterogram. In this investigation the bladder catheter was also attached to an open-ended manometer. This allowed the variations in pressure to be indicated on a vertical scale as the detrusor contracted. Over-enthusiastic filling would produce massive changes in pressure which occasionally forced the fluid in the open-ended glass measuring column to hit the ceiling. When the number of pressure swings had settled down and there was a gradual increase in pressure, 'spot films' were taken as 50 ml, 100 ml, 300 ml and so on of contrast medium was introduced into the bladder. The latter films were taken with the patient resting and then straining in the supine position. The catheter was withdrawn after the patient had been placed in either oblique position with the knee nearest the X-ray table flexed and the other leg supported on pillows. As the catheter was withdrawn it was essential to have the film in position, the hand on the exposure switch with the X-ray tube rotor spinning since quite often the only time the patient voided was immediately on withdrawing the catheter. A subcutaneous injection of carbachol did induce a voiding sequence in a number of patients. Unfortunately the investigation was grossly unphysiological and the use of 'spot films' strictly limited information to a few seconds during the whole of the time spent on the investigation.

Some 8 years ago a research grant was made available and a Research Fellow was appointed. A Watnabe six channel recorder was purchased. Staff of the Physics Department and Renal Dialysis Department provided the physics back up as they still do, and also designed an EMG unit.

Initially 'spot films' were taken on an AOT changer so that there was a sufficient stock of film to take multiple repeat films with a reasonably rapid sequence using the 'single shot' mode (Fig. 6.3). Each specific exposure was electronically recorded on to the paper trace. Initial filling of the bladder was confirmed using the image intensifier and television viewing, the patient then being positioned over the AOT charger in the posterior oblique position as in the early micturating cystograms. The limiting factors using this method were:

1. slow needle response on the Watnabe Pen Recorder
2. need to reposition the patient from under the

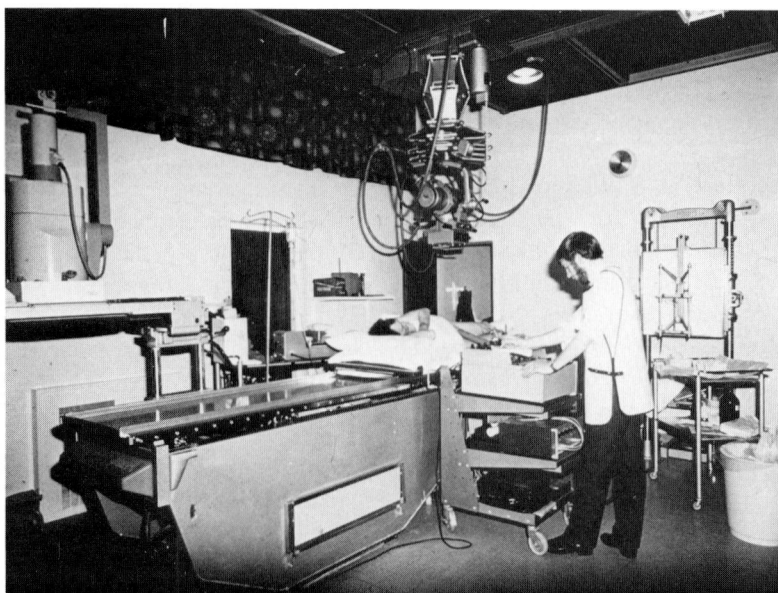

Fig. 6.3 The original arrangement with AOT changer and the six channel Watnabe pen-recorder in the fluoroscopy suite. The patient is positioned over the AOT changer.

image intensifier to over the AOT changer

3. problem of keeping one eye on the recorder and the other on the VDU of the image intensifier

4. delay in co-ordinating an event on the pressure recording instrument and in producing hard copy of that event in the patient.

## Requirements for the modern urodynamic investigation

For a successful investigation and, of equal importance, reproducibility of that successful investigation a number of criteria need to be fulfilled:

1. an integrated team
2. time to be made available in the X-ray department
3. the right apparatus for urodynamic measurements and the correct interfacing of that apparatus with the fluoroscopy unit
4. sufficient number of patients to allow the team to develop expertise
5. clinical use of the results of the investigation.

### 1. The integrated team

This should consist of the following staff:

a. a clinician with some responsibility for the case and/or a specialist with urological background who has some responsibility for running an investigating service
b. a radiologist
c. a radiographer
d. a nurse
e. a physicist and/or a physics technician; an alternative would be a physiological measurement technician.

This represents the complete team but in practice the numbers can vary considerably. An absolute minimum of four could make the necessary complement; this would be a clinician, a radiographer, a nurse and a technician.

At other times the ideal team may be increased by visiting clinicians, physicists and other members of the clinician's firm. In practice this means that adequate space needs to be reserved in the control cubicle area and adequate instructions

in radiation safety have to be given to casual visitors.

### 2. Time to be made available

It is absolutely essential to allocate specific sessions. A considerable quantity of ancillary equipment needs to be set up, the team to be gathered together (this may at times be rather difficult) and sufficient number of patients to make the allocation of the fluoroscopy room worthwhile. The time taken for a single urodynamic study varies considerably but there is a very obvious difference between those patients with spinal injuries and cord involvement and those other patients who have neurogenic bladders who may be classed as 'walking wounded'. The other difference is that the patients in the latter group will have part of their investigations conducted in the erect position.

Experience has shown that not more than four patients who are paraplegic or with spina bifida can be adequately investigated in any 3 hour session. Six patients in the second group may be investigated in the same period of time.

### 3. The equipment

As already stated, because of the difficulty of assessing the true urodynamic picture in the paraplegic patient it is essential to combine fluoroscopy and pressure/flow measurements. The X-ray apparatus should consist of a fluoroscopy table with under or over-couch image intensification, large spot film facilities, 100 mm camera with spot film or roll film and a video tape recorder (Fig. 6.4). The latter may either be an integral part of the urodynamic investigation unit or part of the imaging system of the fluoroscopy unit.

It is vital that a six channel recorder be used with a rapid response recording system with heat sensitive paper and a heated recording needle. It is possible to build and develop modules within an electronics laboratory and there are medical electronic firms such as Dantec in Europe and Life Tech Instruments in the USA. The purchase of state of the art equipment correctly interfaced with the X-ray television system requires a careful review of cost effectiveness of the use of and to the use the

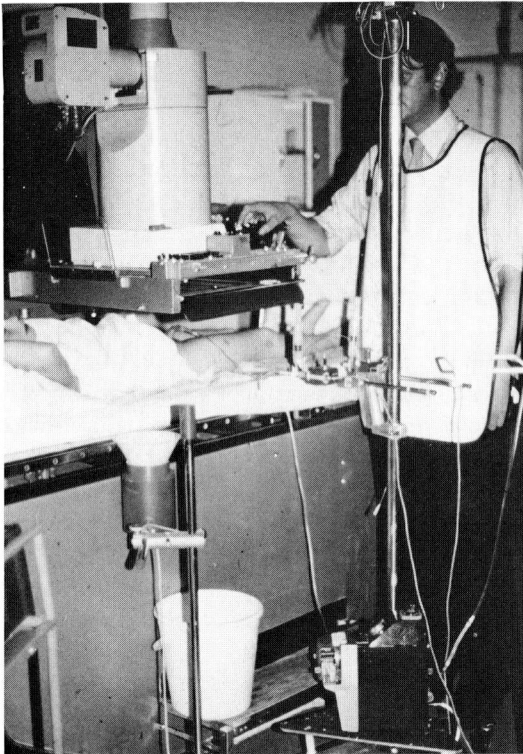

Fig. 6.4 The current situation with the infusion stand and urine collector. The best method of transferring urine from patient to collector is via a piece of 2 inch (50 mm) square PVC standard drainpipe fastened with skin-tape to the patient's thigh.

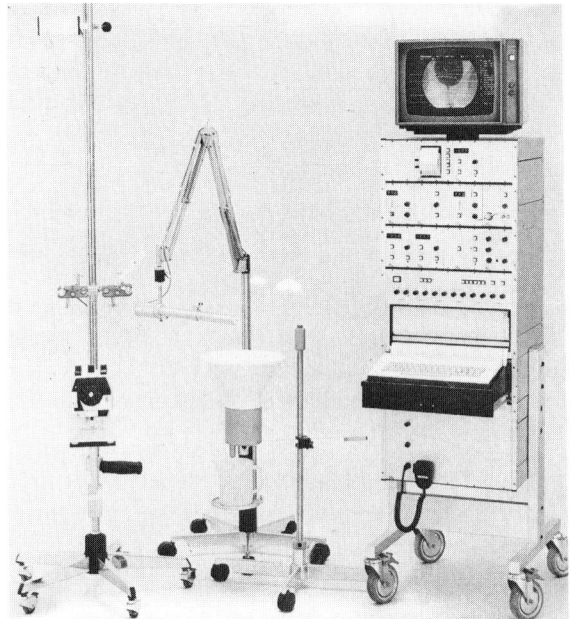

Fig. 6.5A The 2100 URO-Video colour system comprises a fully digitised six channel totally mounted units with integral monitor capable of showing pressure recordings and image intensifier/television recordings of the bladder simultaneously. The other instruments shown are a uroflow transducer, a transducer stand with pump and a patient remote control switch. The computer keyboard is situated in the lower half of the trolley.

information is put — when the equipment will cost in excess of £42 000 on top of the total cost of a fluoroscopy suite (Fig. 6.5).

There are two methods of co-relating the X-ray images and video-traces of the pressure studies. One is to make use of a screen-splitter so that X-ray images are related to one half only of the television screen with all the urodynamic recordings on the other half. The other method is to superimpose on to the whole of a time-based television picture the images from the output phosphor of the image intensifier. Attempts at superimposing the urodynamic recordings into the television chain have been only partly successful. This was because the automatic brightness control in the television chain corrected the inputs from both signals with inevitable degrading of the X-ray images. With current research into the digitisation of radiographic images, superior methods of inter-

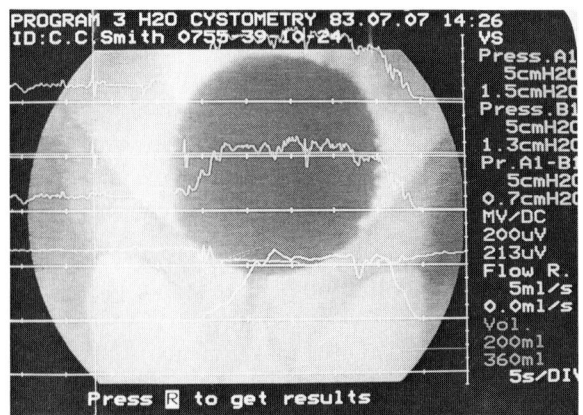

Fig. 6.5B A typical trace on the monitor display showing a radiographic image of the bladder, simultaneous tracings of the pressures and a read-out giving information. These facilities allow information to be compressed and there is available hard copy facilities to record the information shown on the monitor.

facing will become available. However there is little point in improving the imaging without being able to have clear simultaneous imaging so that a particular radiographic image is related to the appropriate moment in the dynamic recordings. The importance of simultaneous display on to a single visual display unit so that rapid sequences may be visualised cannot be over-emphasised.

### 4. Availability of patients

Some investigating techniques in medicine are capable of adequate management on an occasional or intermittent basis. There is little doubt that urodynamic studies do not fall into this category and they are best dealt with using a single centre with a large enough area to provide an adequate throughput. Some of the earlier research centres achieved this because they were the first in the field and so automatically became referral centres from gynaecological, urological and spinal injuries units.

In the centre in which the author worked the aim was to make maximum use of two half-day sessions with a throughput of ten patients per week, four of whom would be either patients with spinal injuries or spina bifida who were dealt with in one of the sessions, the other six either male or female being referred as outpatients from a number of centres either urological or gynaecological in nature covering the needs of a population of about $1\frac{1}{2}$ million. The catchment area for the spinal injuries unit was some $4\frac{1}{2}$ million. Outpatients referred were those patients whom it was thought would benefit particularly from combined X-ray and urodynamic studies. A number of centres in the area carried out studies on the group of patients who could be investigated on more basic apparatus; these patients chiefly women, belonged to the stress incontinence group.

### Parameters to be measured

*Intrinsic bladder pressure.* Measurement of the internal bladder pressure has two components. One from the pressure exerted by the detrusor muscle within the bladder wall and the other that is a product of internal abdominal pressure. Each pressure may be recorded separately with the difference between the total bladder pressure and the abdominal pressure being recorded on the third channel after being subtracted electronically. On the fourth the quantity of urine passed is recorded. The quantity thus voided is related to a time base and the rate at which urine is voided is shown. The sixth channel is reserved for electromyographic recordings. In this, electrophysiological measurements are taken from readings from the anal sphincter since there is a recordable response from the anal sphincter when the wall of the bladder or the urethra is stimulated.

Current research on innervation of the detrusor and sphincters throws doubt on the veracity of directly relating anal and bladder sphincters as closely as this. An alternative to the use of the first three channels is to produce the intrinsic bladder pressure on a single channel by combining the signals of the pressure changes within the bladder and abdomen integrating a positive and negative signal. This system works well when it works, but if the final signal is suspect it is extremely difficult to pinpoint the source of the trouble be it in the common or separate pressure lines.

### Methodology

The techniques described have two main aims the first being reproducibility of signals which have coherent physiological and anatomical relationships; the second the application of the data to the future well-being of the patient in a definable medical protocol.

The patient must be investigated in the supine position or the supine oblique position. There is considerable advantage in having trolleys with a flat top which are the same height as the table top of the fluoroscopy unit. The patient can then be transferred complete with mattress on to the X-ray table by simply sliding the mattress from trolley to table. One other advantage is that when, as occasionally happens, the patient and mattress become soiled, then both can be transferred back on to the trolley for return to the ward where the patient, mattress and trolley may be easily cleaned.

The two methods of introducing fluid or contrast medium into the bladder are by retrograde filling via a urethral catheter or by suprapubic needle for direct injection through the abdominal wall and the bladder wall. Both methods, have their supporters and opponents but both do work. The supra-pubic method leaves the bladder neck and urethra totally free of impediment but does not lend itself to multiple voiding investigations whereas the indwelling catheter may interfere in such a way as to make the results of the investigation suspect. The choice of catheter is critical to the success of the investigation. A number of bi-lumen catheters have been tried but it was found that if too much curvature of the catheter occurs, kinking of the inner wall all too often follows, interrupting either input of fluid or the pressure line depending upon the use of the central channel.

Over-flexibility of the walls of the catheter or over-rigidity of its walls create problems within the bladder. Two catheters have been used with success, each stuck to the other along most of its length and each not more than 1.2 mm in diameter giving a combined area of about 2.88 sq. mm. It would appear that catheters of this overall diameter do not interfere with the urethra, sphincters, bladder neck or bladder in either their resting, straining or voiding phases, compared with some of the larger catheters originally used.

Catheters of this size whether they be double or two single ones require a leader catheter. The two single or the double catheter is hooked into the side hole of the uretheral catheter and all of them are passed into the bladder using a sterile technique. Rotation of the urethral catheter will release the others and careful withdrawal usually leaves the smaller ones behind. The pressure lines are connected to the transducers. A suspended bottle of Hypaque 15% containing 1 litre of sterile solution is connected to one of the catheters via a pump (Fig. 6.6A). One of the transducers is connected to a fluid line around the end of which a finger-cot has been fastened. This is inserted into the rectum. This particular line will then measure the intra-abdominal pressure. While the patient is in the lateral position a bipolar electro-myographic needle is then connected to the urodynamic console for trace recordings and to an audible indicator which (when there is pelvic floor activity) makes a sound similar to that emitted from a geiger counter. An alternative method is to make use of an anal plug and a remote earth electrode.

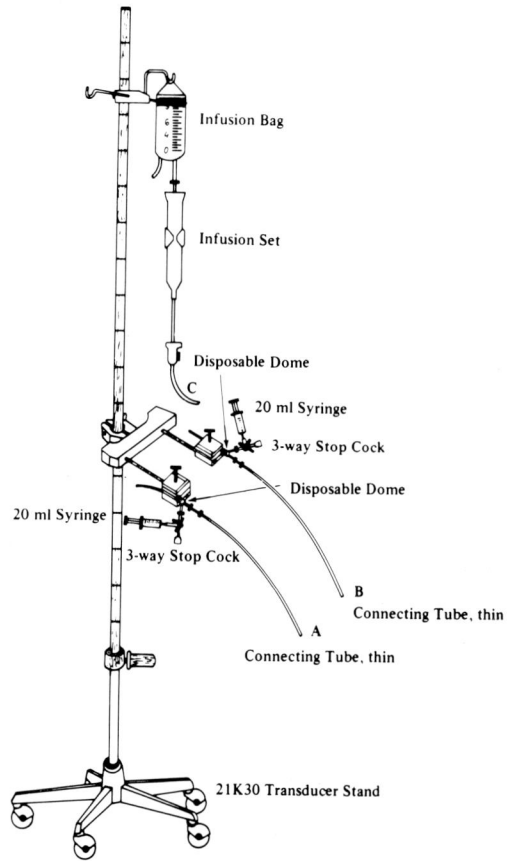

Fig. 6.6A A transducer stand complete with pressure lines and stop-cocks; provision is made for the introduction of a soluble contrast medium with a higher iodine content. The stand shown is by Dantec

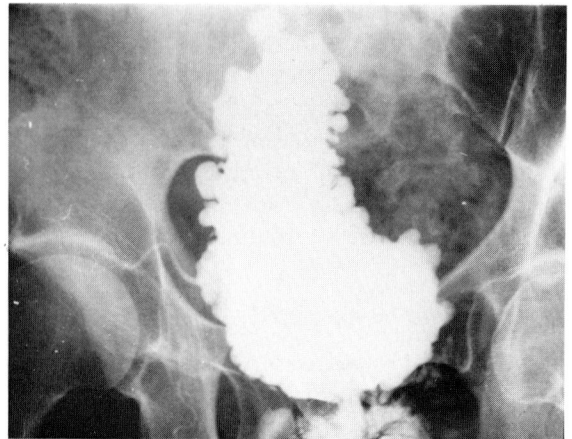

Fig. 6.6B The filled bladder with additional contrast medium. Note the gross trabeculation of the 'pine-tree' bladder. Just below the bladder there is prostatic bed filling.

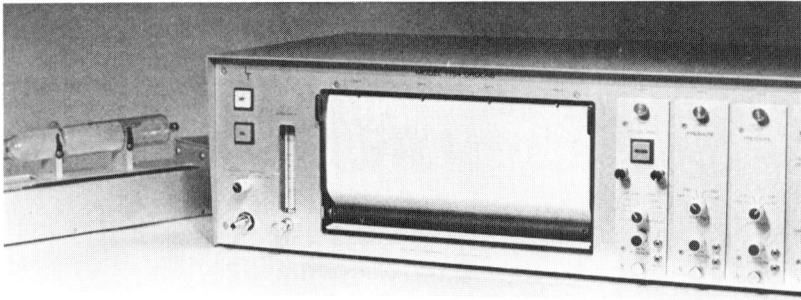

**Fig. 6.7** A small compact unit by Life-Tech USA with a four channel recorder, with a constant infusion pump on the left.

The anal plug is shaped like a bobbin and is made of acrylic and has two concentric electrodes. It is possible to dislodge this type of electrode quite easily especially where patients suffer from severe spasm.

Liberal use of skin tape anchors the plug in this position as the buttocks are pinched together. Carefully move the patient from the lateral position into the supine oblique position. The uppermost leg is made comfortable on pillows but the trunk is best supported upon an extra large positioning pad about 1 metre long and giving an angle of 35° to the horizontal. Another essential point is to ensure that this pad is well protected by plastic sheeting. The lowermost leg is flexed slightly. Another difference here between the paraplegic patient and other patients is that the bladder is not emptied of residual urine since complete emptying may produce a series of very unphysiological responses. For this reason it is convenient to have a three-way cock in the opaque medium input line so that if the urine/hypaque mixture proves to be too radio-translucent an opaque medium with higher iodine percentages may be introduced to improve the radiographic outline of the bladder (Fig. 6.6B).

Another difference of great importance is the rate at which the bladder is filled in the paraplegic patient. Some urodynamic studies can be quite clearly identified as a provocative investigation simply by choosing a high rate of input of fluid either by pump through a small diameter catheter or through a wide diameter catheter and a high position of the fluid reservoir. If this is attempted upon a paraplegic patient the bladder responds by reacting strongly thus producing very unco-ordinated detrusor activity ruining any chance of a meaningful investigation.

Initially a constant infusion pump (Fig. 6.7) was used giving an input rate of between 2 ml and 4 ml per minute retrogradely supplemented by an increased urine output from the kidneys by giving the patient plenty of fluids just before the investigation commences and bottles of beer during the investigation.

Peristaltic action within the walls of the ureters forces urine into the bladder through the ureteric orifices in small pulsed jets. To simulate this action a quite powerful peristaltic pump is used (Fig. 6.8). The action of this pump is exactly as in some replenisher pumps on automatic processing units. The feeder tube from the Hypaque reservoir passes through the pumping chamber where the tube is 'milked' by the action of three arms on a rotating spindle of the pump. This enables the input rate to be increased to 10–30 ml per minute thus reducing the length of the investigation to between 30 and 40 minutes, allowing the target number of four patients to be investigated per session. The significance of this type of pump action means that the investigation is that much closer to a true physiological investigation.

The other significant contribution that the use of indwelling catheters left in for the duration of the investigation makes is in that the voiding pattern may be recorded several times. It became apparent that the first recorded act of micturition does not indicate the actual state of the detrusor, the bladder neck and the external sphincter and these results are best discarded. By refilling the bladder several times and accepting the virtual invisibility in physiological terms of the very small

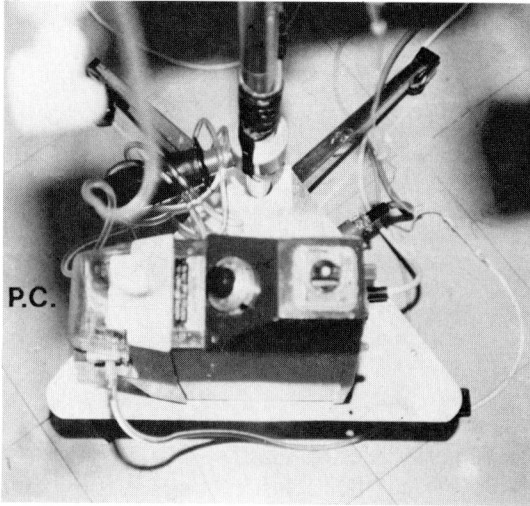

Fig. 6.8 The peristaltic pump fastened to a plate welded on to the base of the transducer stand. Plastic tubing containing the contrast medium passes through the pumping chamber (P.C.).

catheter, a number of voiding sequences can be recorded and compared.

Simultaneous recording of the fluoroscopic images and the pressure and flow measurement results are best recorded via a video tape recorder so long as independent physiological measurements are constantly recorded on to some form of paper recording medium. Alternatively the next best method is the use of either a roll or cut film in a 100 mm or 105 mm camera using a split imaging technique from the output phosphor of the image intensifier. The next would be the use of a large film format in a serial changer for here the delay of introducing the cassette into the exposure position is a problem. Finally the use of a film-changer may be tried but it is advisable to record electronically the timing sequence on the read-out of the multi-channel recorder (Fig. 6.7).

## CLINICAL CONSIDERATIONS

It was traditional to classify neurogenic bladder dysfunction on the site and degree of cord damage in paraplegics. However under the impact of modern investigation by urodynamic methods the importance of detrusor-sphincter dysfunction irrespective of the site has assumed great importance

but the end result cannot be divorced from the causative agent related to changes in understanding of the normal act of micturition and a greater knowledge of the complex innervation of the various muscles involved in the passage of urine. Due to incomplete understanding, definitions are contradictory, confusing and complicated.

The end result of bladder dysfunction may be similar in either neurological disease or to outflow tract obstruction and from this detrusor instability must not automatically relate to neurogenic origins. The close interrelation and interaction between continence and micturition means that if neural pathways are interrupted any, some, or nearly all types of dysfunction may occur. In anatomical and physiological order these will be:

1. Detrusor activity — weakened or totally absent
2. Functional
   a. unco-ordinated detrusor contractions whether they be spontaneous or instigated
   b. hypertonic filling phase
   c. inability to maintain a sustained bladder contraction.
3. Altered tone in the striated pelvic musculature
   a. detrusor-sphincter dyssynergia
   b. paresis
4. Lack of control of urethral smooth muscle activity
5. Lack of sensation in either bladder or urethra or both.

### Detrusor activity

Injuries to the sacrum and the cauda equina may radically alter bladder function yet give considerable sparing of neurological function in the lower limbs. There might well be total absence of detrusor contractions and the bladder may have to be emptied by catheter, abdominal straining or manually (compression of the abdominal wall just above the symphysis pubis).

If intermittent catheterisation is not carried out during the spinal shock phase and in the period before there is some return of detrusor activity,

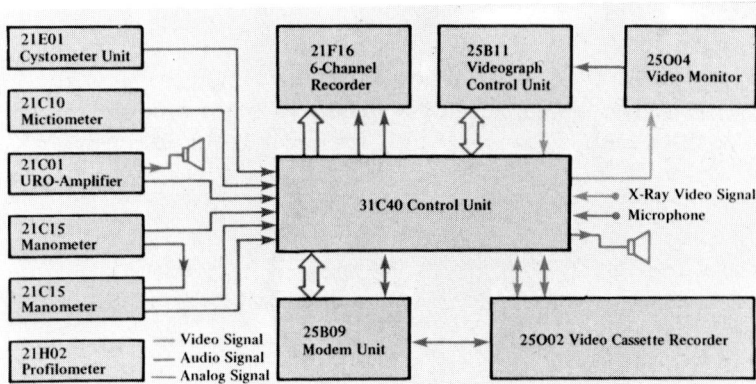

**Fig. 6.9** Flow diagram showing relationships between specific pieces of apparatus as suggested by Dantec, for a complete physiological measurement regime.

bladder distension itself may cause further dysfunction of the bladder because of the effect of bladder distension on innervation of detrusor activity giving nerve damage and long term paralysis to the muscles in the bladder wall.

If cord damage occurs above the spared uninvolved sacral cord segments, then a quite different bladder dysfunction occurs when either partial or total transection or irreversible damage within the cord takes place following trauma or disease. This manifests itself as inhibited detrusor contraction and total loss of voluntary control giving rise to frequency, urgency and incontinence, to complete emptying of whatever volume of urine happens to be in the bladder at the time.

## The detrusor/sphincter relationship

As shown earlier in the ideal detrusor contractile external sphincter, relaxant pairing is necessary to achieve a good stream and complete emptying of the bladder with no residual urine. Spinal lesions higher than the cauda equina and the sacral segments disorganises this interplay of these two functions, i.e. the external sphincter since it does not relax becomes obstructive and uncoordinated contractions of the detrusor with or without an open bladder neck markedly increases intrinsic bladder pressure irrespective of bladder volume and with varying amounts of residual urine. The term applied to this malfunction is detrusor sphincter dyssynergia.

*The rest of the urethra*

The importance of smooth muscle activity in the male and especially in the female urethra is an area in which further research is necessary but since the urethra in both sexes is not simply just a distensible tube, changes in the tone of the smooth muscle will have some effect on bladder emptying albeit somewhat overshadowed by the more easily recognised alterations to the act of micturition of the detrusor and external sphincter.

## INFORMATION GAINED FROM THIS TECHNIQUE

*1. Fluorography*

This clearly shows the shape of the bladder, an indication of bladder volume, trabeculation and presence of diverticulae since this can have a marked effect on internal bladder pressure as urine is milked from one diverticula to another. The importance of ureteric reflux, state of the kidneys and their future functions is related also to the timing of this reflux and the internal bladder pressure when this takes place.

*2. Urodynamic studies*

These will indicate the following:

In the detrusor — whether the patient has any control of bladder emptying or whether the bladder is totally hyperreflexic in action and if this is so, to what degree does this achieve complete emptying? It answers the question as to the cause of incomplete emptying that is either cessation of detrusor activity or of detrusor-sphincter dyssynergia. Do weak detrusor contractions continue after closure of the bladder neck?

Do coughing, abdominal straining, pubic tapping or hand pressure initiate or change any of the previously recorded responses?

How the intrinsic bladder pressure relates to detrusor, bladder neck and external sphincter.

The very important relationship of symptoms of autonomic dysreflexia in those patients with cord lesions above dorsal 4 vertebra. If the symptoms of autonomic dysreflexia occur during the investigation, then this must stop and steps are taken to empty the bladder. These symptoms of abnormal sympathetic activity start with sweating, then bradycardia, a rapid increase of blood pressure and sudden onset of severe headaches. A small group of patients have low residual urine or complete emptying but exhibit very high bladder pressures as the detrusor contracts with a wide open bladder neck and total closure of the external sphincter. It is not until the external sphincter relaxes and allows voiding to take place, that the pressure drops.

The whole spectrum of bladder-urethral dysfunction may take place from the onset of injury to eventual kidney failure: at first the bladder is flaccid with very little detrusor activity and if the bladder is filled with any of the water-soluble media the bladder neck remains closed and its site is identified by a small dimple in the outline on the inner wall on the inferior aspect of the base of the bladder. Next there is a slow return of weak detrusor activity and only limited bladder neck opening a tight sphincter with no, or at best, a poor stream. Reflex detrusor contractions continue to increase in strength improving voiding performance although there is no change in detrusor sphincter dyssynergia. The time it takes before the external sphincter opens and the length of a detrusor contraction will have a considerable bearing on total urine voided and the amount of urine retained whilst the maximum recorded pressure relates to detrusor tone.

The combination of high residual urine and high internal pressure will inevitably lead to upper tract dilatation, hydronephrosis and nephritis. With increasing blood urea concentrations and bladder infections it only requires the development of skin sores to begin a downward spiral of debilitation leading to death, a sequence which applied to great numbers of patients who died in the years after the First World War.

## APPLICATION OF THE INFORMATION

The previous description relates purely to those patients who suffer spinal cord injury but comparable complicated symptoms from children with spina bifida (spinal dysraphism) and those patients with demyelinating disease make them an essential category for the type of investigation with the results having a bearing on the subsequent treatment and control of bladder dysfunction.

The ideal solution is to have a patient who does not require intermittent catheterisation, maintains an infection-free urine, achieves good emptying, and achieves a socially acceptable level of continence. Initially one of the most important symptoms to alleviate is to ensure adequate emptying of the bladder by division of the external sphincter by surgery. The sort of patient to benefit from this type of interventional surgery can best be assessed by combined urodynamic-radiological studies.

Other patients to benefit from this type of surgery are those whose urological investigations indicate that long standing upper tract complications are taking their toll of cortical substance of the kidneys, have grossly trabeculated bladders, and a history of repeated urinary infections. It has not been clear why some patients with the same neurological deficit avoid later upper tract complications but by long-term assessment across a broad spectrum of patients, the causative agent may be isolated.

Good recovery from incomplete cord lesions above the sacral micturition centre allows the patient to walk out of hospital and may give the patient a sense of well-being. He may depend upon uninhibited contractions for voiding but because he has control over voluntary sphincter contraction maintains continence. In spite of this the pattern has been set for high intrinsic bladder pressures leading to upper tract problems.

The other major problem is presented by a patient with a root lesion or sacral cord lesion below the sacral micturition centre. His bladder is atonal with no detrusor activity. He has to rely on external pressure either from manual compression or abdominal muscle contraction increasing overall abdominal pressure. This induced pressure can be very high because it may have to overcome partial bladder neck opening and urethral resistance contributed to by smooth and/or striated muscle. Under these conditions high residual urine levels and ureteric reflux are an all too frequent sequelae. As in assessing previously outlined problems combined studies will help to pinpoint the specific restrictions to flow which added together make a poor prognosis.

The problems so far discussed have particular relevance to male anatomy mainly because the majority of patients are male paraplegics. The female paraplegic has an additional problem compared with her male counterpart. This is the purely mechanical difficulty of remaining dry since in her case there is no convenient 'pipe' onto which external collecting systems may be attached to maintain dryness.

The importance of combined studies relates to accurate assessment of urethral competence against detrusor instability and the need to develop a bladder training regime socially acceptable but physiologically suitable to avoid upper tract dilatation and urological problems.

## SOME POST-SURGICAL PROBLEMS

Unselective operations without complete urological assessments and combined urodynamic-radiological studies are bound to present to the clinician on follow-up patients for whom trans-urethral resection will merely decrease bladder neck-urethral sphincteric function since the wrong parameters were used in early selection. Two possible causes would be: (a) operation failure by insufficient resection of the sphincter or postoperative complications which add to the restriction by fibrotic stenosis, (b) diagnosing outlet obstruction when the cause was weak detrusor contractions.

The choice of operating technique on an early series of patients was a transurethral resection of

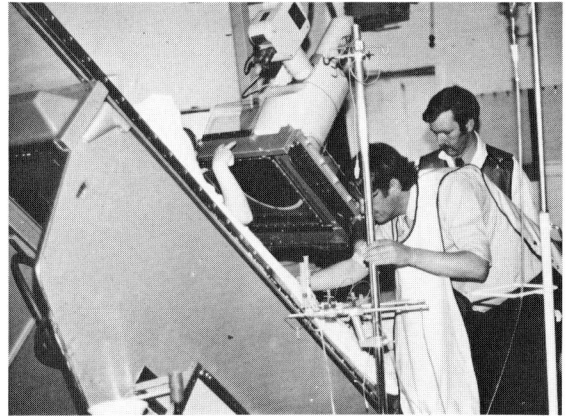

**Fig. 6.10** Re-adjusting the height of the pressure transducers upon changing the height of the fluoroscopy table.

the external sphincter with a V-shaped cut at 10 minutes to 2 o'clock. Unfortunately some males were made impotent by this technique and in all the later series of patients a single cut at 12 o'clock has been used.

### Comparison of urodynamic studies with the radiographic studies

To show the comparisons between the recordings and the radiographs sketch drawings in the positive mode rather than radiographic reproductions have been used for greater clarity. The patient lies firstly in the supine position and then is rotated through 30–35°. Usually the side raised is that away from the operating side of the table especially so if the image intensifier pedestal and tracks are likely to interfere with any of the pressure lines.

### Patient with a cervical spine injury

Figure 6.11 shows the first investigation when the patient is out of spinal shock but there has been no return of detrusor activity and the bladder neck is closed. If abdominal pressure is manually applied or if the patient is capable of straining the only change which takes place is a flattening of the bladder so that it becomes ovoid in shape and the bladder neck may appear as a small dimple (Fig. 6.11 and Fig. 6.12). Some 2 months later detrusor activity is returning although it is reflexic in nature. Electromyographic activity is now

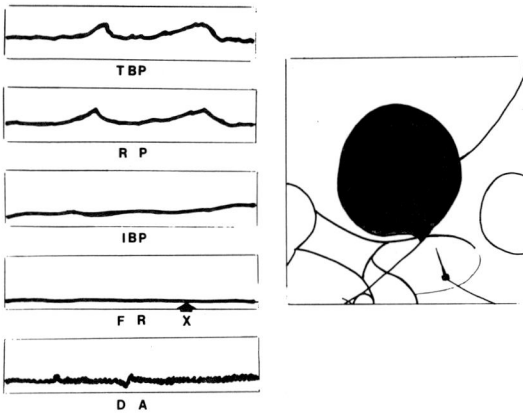

**Fig. 6.11** The urodynamic study of a patient with a cervical spine injury soon after the accident. No detrusor activity no flow. X marks the point at which the radiograph was taken (as in all the illustrations of the urodynamic studies from Fig. 6.10 to Fig. 6.15). Bladder emptying would be via urethral catheter.

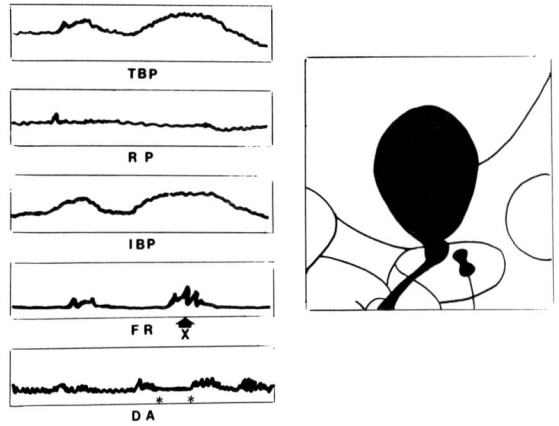

**Fig. 6.12** The same patient 2 months later. EMG activity is now continuous; detrusor activity is poor and unco-ordinated; the bladder has a purely reflexive response.

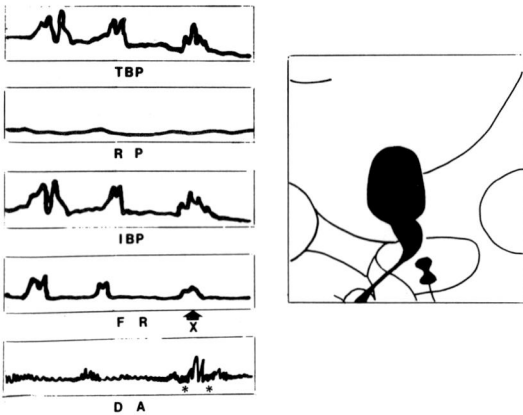

**Fig. 6.13** Some months later. Improved detrusor activity with the best flow rate being achieved when the EMG is silent, i.e. when the external sphincter is relaxed.

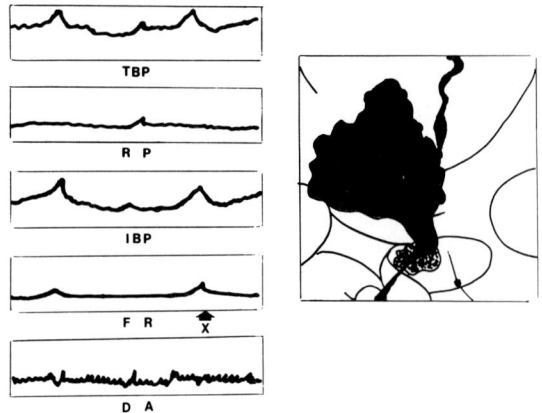

**Fig. 6.14** The end result if the patient is not treated. Trabeculated bladder with ureteric reflex. Prostatic bed filling. High residual urine. Poor flow with high bladder pressures. Marked detrusor sphincteric dyssynergia.

apparent, i.e. it is no longer 'silent'. The distal urethral sphincter remains tight showing no relaxation. The stream is poor.

A few months later there is improved detrusor activity — improved in the sense that the bladder would be capable of emptying completely but because of detrusor-sphincteric dyssynergia it fails. It is clear that the flow rate is at its best when the sphincter is relaxed (where the EMG recording is silent) (Fig. 6.13). If the patient is not treated, the long-term consequences are as follows: the bladder becomes grossly trabeculated; the intrinsic bladder pressure is high; there is ureteric reflux up one ureter which is tortuous and widened (a mega-ureter); the bladder neck is wide open, the small section of the posterior urethra is widened; the urethral sphincter is closed and there is evidence of prostatic bed filling (Fig. 6.14). The patient could have had a transurethral resection. In Figure 6.15 can be seen the improvement that this operation achieves in these patients with cord involvement above the level of dorsal 4.

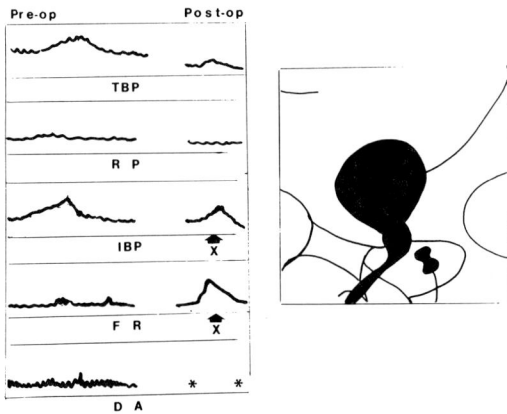

**Fig. 6.15** Patient with cervical spine injuries. Comparison of the urodynamic studies before and after a trans-urethral sphincterotomy. The radiograph was taken postoperatively. Note the marked improvement in flow rates.

The first shows a typical radiographic appearance and urodynamic recording with detrusor-sphincter dyssynergia. Postoperatively there is no sphincteric resistance and there is a good flow rate. The patient is now incontinent requiring the wearing of one of the permanent urine collecting devices but the bladder now exhibits a far lower overall intrinsic pressure. So far most patients have maintained their improvement over 5 years and the exceptions have already been discussed.

### Acknowledgement

The methods for the investigation of bladder problems in paraplegic patients have been developed over a period of some 8 years. Although it is true that the results achieved were due to the contribution of 'the team' comprising radiologists radiographers, physicist and physics technician, and nursing staff; the co-ordination, the drive and the farsightedness of David Thomas, Consultant Urologist at the Spinal Injuries Unit, Lodge Moor Hospital, Sheffield, made a vital contribution (Thomas et al 1975, Thomas 1979).

## APPENDIX

If the bladder is considered as a sphere the contained fluid acts as a circular recipient of detrusor muscle contractions. The following formula applies:

$$T = 2\pi R T^1$$

where T = total tension

R = radius of spherical bladder

$T^1$ = tangential surface tension per unit area.

Equilibrium of forces between this total tension and the detrusor muscle gives an equation:

$$T = \pi R^2 \, p \, det \qquad therefore$$
$$2\pi R T^1 = \pi R^2 \, p \, det$$
$$2T^1 = R \, p \, det$$
$$p \, det = \frac{2T^1}{R}$$

where p det is the pressure related to detrusor activity.

### Measurements of bladder pressure

The practical measurement of pressure in urodynamics is in centimetres of water (cm $H_2O$). Urine in the bladder when it is in equilibrium has a straightforward vertical pressure gradient.

The intravesicular pressure is that pressure in the bladder measured just above the symphysis pubis with respect to atmospheric pressure. Since micturition pressures have parameters of 20–150 cm ($H_2O$) very small gravitational pressure differences can be discounted.

It has become standard practice to use a fluid filled line attached to an indwelling catheter and externally attached to a pressure transducer. So long as the fluid (the contrast medium) is not too dissimilar to urine in its density and the external transducers are zeroed at the supero-anterior aspect of the symphysis with the patient, in the supine oblique position the final placing of the catheter tip has no significance and correct intravesicular pressures are measured. This measured pressure is the summation of abdominal and intrinsic bladder pressures.

p1 = measured pressure within bladder

p2 = measured pressure within rectum

p3 = true bladder pressure

p3 = p1 − p2

The close anatomical relationship between bladder and rectum gives the above equation a more meaningful result. Any activity which increases abdominal pressure, coughing, sneezing, straining for example would balance the equation

$$p1 = p2 + p3$$

## REFERENCES

Brown M, Wicklam F R A 1969 The urethral pressure profile. British Journal of Radiology 41:211

Thomas D G 1979 Clinical urodynamics in neurogenic bladder dysfunction. Symposium on Clinical Urogenics. The Urologic Clinics of North America 6(1) Saunders, Philadelphia

Thomas D G, Smallwood R, Graham D 1975 Urodynamic observations following spiral trauma. British Journal of Urology 47:161


## BIBLIOGRAPHY

Abrams P H, Martin S, Griffiths D J 1978 The measurement and interpretation of urethral pressures obtained by the method of Brown and Wicklam. British Journal of Urology 50:33

Bates C P, Corney C E 1971 Synchronous cine-pressure-flow cysto-urethrography; a method of routine urodynamic investigation. British Journal of Radiology 44:44

Bates C P, Whiteside C G, Turner Warwick R 1970 Synchronous video pressure flow cysto-urethrography. British Journal of Urology 42:714

Caine M, Edwards D 1958 The peripheral control of micturition; a cine-radiographic study. British Journal of Urology 40:747

Diokno A C, Koff S A, Bender L F 1974 Periurethral striated muscular activity in neurogenic bladder dysfunction. Journal of Urology 112:743

Enhorning G, Miller E R, Hinman F Jr 1964 Urethral closure studied with cine-roentgenography and simultaneous bladder-urethral pressure recording, Surg. Gynecol. Obs. 118:507

Gosling J A, Dixon J S, Lendon R G 1977 The autonomic innervation of the human male and female bladder neck and urethra. Journal of Urology 118:302

Griffiths D J 1980 Urodynamics. The mechanics and hydromechanics of the lower urinary tract. Adam Hilger, Bristol

Hinman F Jr, Miller G M, Nickel E, Miller E R 1954 Vesical physiology demonstrated by cine-radiography and serial radiography. Radiology 62:713

Martin S, Griffiths D J 1976 Model of the female urethra Parts I and II. Medical Biological Engineering 14: 512, 519

McGuire E J 1977 Combined radiographic and manometric assessment of urethral sphincter function. Journal of Urology 118:632

McGuire E J, Wagner F C, Weiss R M 1977 Urethral closing pressure after spinal cord injury and its relationship to autonomic dysreflexia. Urologia Internationalis 32:97

Meirowsky A M, Scheibert C P, Hinchey T R 1950 Studies on the sacral reflex arc in paraplegia. Response of the bladder to surgical elimination of sacral impulses by rhizotomy. Journal of Neurosurgery 7:33

Rockswold G L, Chov S N, Bradley W E 1978 Re-evaluation of differential sacral rhizotomy for neurological disease. Journal of Neurosurgery 48:773

Sunder G S, Parsons K F, Gibbon N O K 1978 Outflow obstruction in neuropathic bladder dysfunction. The neuropathic urethra. British Journal of Urology 50:190

Tanagho E A 1971 Interpretation of the physiology of micturition. In: Hinman F Jr (ed) Hydrodynamics of micturition. Thomas, Springfield, Illinois

Thomas D C 1979 Clinical urodynamics in neurogenic bladder dysfunction. Symposium on clinical urogenics. The Urologic Clinics of North America 6(1). Saunders, Philadelphia

Turner Warwick R T, Whiteside C G 1977 A urodynamic view of clinical urology. Recent advances in urology 2:44

Turner Warwick R T, Whiteside C G, Milroy E G et al 1979 The intervenous urodynamogram. British Journal of Urology 51:15

Vereecken R L, Verduyn H 1970 The electrical activity of the para-urethral and perineal muscles in normal and pathological conditions. British Journal of Urology 47:161

Whiteside C G, Turner Warwick R T 1976 Urodynamic studies: The unstable bladder. Scientific Foundation of Urology, vol. II. Heinemann, London

Yalla S V, Rossier A B, Fam B 1975 Synchronous cysto-sphincterometry in patients with spinal cord injury studies with continuous bladder and urethral infusions and physical factors influencing interpretation Urology 6:777

# 7

# Aspects of skeletal radiography

The term skeletal survey implies an examination of the skeleton by an imaging process. 20 years ago this would have been almost certainly a radiographic survey, but these days other means of surveying the skeleton are also employed. These include radionuclide imaging, ultrasonics and whole body computed tomographic scanning.

The common groups of diseases for which skeletal surveys are employed include:

1. Malignant disease and secondary deposits
2. Metabolic disorders including acromegaly
3. Assessment of skeletal maturity
4. Paget's disease
5. Osteodystrophies and skeletal syndromes
6. Surveys for non-accidental injury (battered baby syndrome).

In this chapter the skeletal imaging involved will be discussed including any aspects of image quality appropriate.

## RADIOGRAPHY

It is essential if using the standard radiographic procedures, that good quality radiographs are produced. This term refers, amongst other things, to clarity or distinctiveness with which the radiographic image detail is viewed.

Production of a satisfactory radiographic image depends upon the radiographer's skill in manipulating the various equipment and photographic facilities at his disposal. Some examinations of the skeleton employ densitometric analysis, and objective contrast is measured as a means by which alteration in calcification of the skeleton can be detected.

Whenever viewing of radiographs takes places it is essential that the conditions are satisfactory, for it is by these means that density differences are detected visually. Subject contrast will depend upon the patient and the patient part under radiographic examination. The composition of the body material whether it be bone, muscle or soft tissue, determines subject contrast by physical density and atomic number of tissue under investigation. Additionally, the quality of radiation will also determine subject contrast and this is dependent upon the kVp used to penetrate the part and also upon the scattered radiation produced.

As it is important to obtain good radiographic detail, attention must be paid to reduction of the unsharpness factors. This will involve the use of smaller X-ray tube focal spot sizes, smallest object film distances, largest focal film distances, immobilisation of the part under investigation, use of short exposure time and use of 'fast' film/screen combinations. To reduce photographic unsharpness it may on occasions be necessary to use non-screen film techniques. Finally, careful film processing is essential to produce the optimum image contrast and detail.

### The bony skeleton

The bony skeleton is a tissue which is supportive and protective in function and in addition it acts as a major depository for inorganic salts required for daily metabolic processes. It is constantly refashioning to meet the requirement of these functions and is dependent upon the activity of osteoclastic bone resorption and osteoblastic new bone formation.

These processes are dependent upon:

1. endocrine interplay
2. chemical and vitamin content of blood and tissue fluid
3. availability of all basic substances necessary for bone formation
4. other stimuli such as pressure and traction.

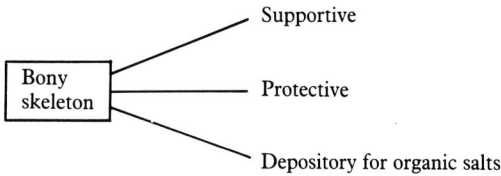

Fig. 7.1 Functions of the bony skeleton

*1. Endocrine interplay*

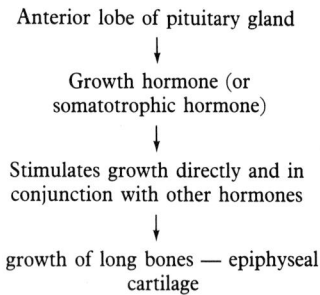

Fig. 7.2A Effects of pituitary gland upon growth

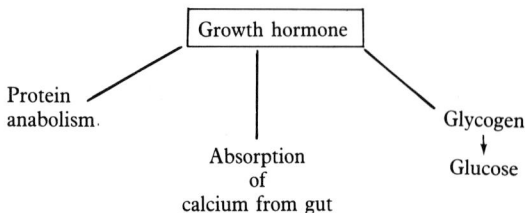

Fig. 7.2B Effects of growth hormone

*2. Chemical and vitamin content of blood and tissue fluid*

Calcium and phosphorus in diet

Vitamin D required for absorption of calcium and phosphorus from gut.
Vitamin C required for formation of bone matrix.

*3. Basic substances necessary for bone formation*

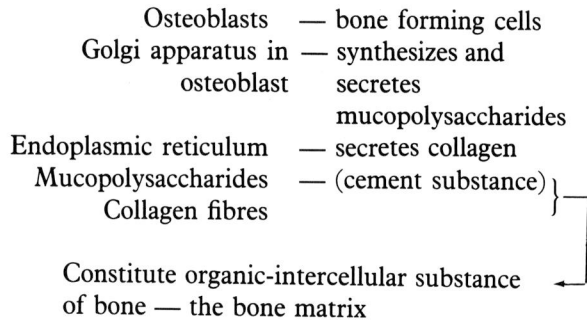

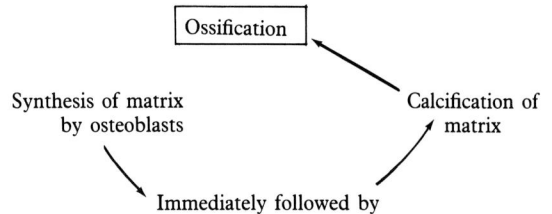

Osteoblasts — bone forming cells
Golgi apparatus in — synthesizes and osteoblast    secretes mucopolysaccharides
Endoplasmic reticulum — secretes collagen
Mucopolysaccharides — (cement substance)
Collagen fibres

Constitute organic-intercellular substance of bone — the bone matrix

Fig. 7.3 Ossification

The following processes go on together throughout life:
Bone formation — Osteogenesis
Bone destruction — Resorption
Imbalance causes bone disease.

*4. Other stimuli*

Pressure
Traction

## MALIGNANT DISEASE AND SECONDARY DEPOSITS

### Radiographic views required include:

Lateral skull
Antero-posterior and lateral of dorsal and lumbar spine
Postero-anterior view of chest (ribs can also be viewed on this projection)

Antero-posterior view of pelvis (hips and upper femora also seen on this projection)

Additionally, any other area where the patient complains of pain may be examined. The radiographic appearances will be taken in conjunction with the patient's case history and results of any biochemical or pathological tests which have been carried out.

A most common occurrence in primary neoplastic lesions is inflammation of the periosteum (periostitis) and bony erosion. Periostitis will result in local reaction such as new bone growth and periosteal elevation, both of which are visible on a radiograph. This is not seen however in secondary deposits. Bony erosion will appear as 'nibbled' margins of otherwise straight bone edges. The effects are most often seen in the long bones so that the antero-posterior view of the femur is important in this respect.

## Secondary deposits

Secondary deposits in the vertebral bodies may result in a weakening of the structure with subsequent wedging and collapse of vertebra itself (Fig. 7.4A). It may affect one, or more than one, vertebra, but the vertebral discs are extremely resistant to destruction by metastases. Other causes of vertebral collapse are known and the differential diagnosis may depend upon the quality of the radiographs and upon the full clinical history and biochemistry.

Secondary deposits in bone may occur from various primary tumours and are sometimes few and sometimes numerous. For example, the incidence of bone metastases from Wilms' tumour is very low compared with the incidence of pulmonary metastases (Bond & Martin 1975). Similarly the distribution of these lesions has no particular pattern.

Bone metastases from carcinoma of the nasopharynx principally affect the trunk bones and bone pain is the main clinical indicator of skeletal metastases (Tan & Oon 1976). The pain often precedes the radiographic manifestations by a long interval. Numerous adenocarcinomas are prone to produce osteoblastic bone lesions.

Skeletal metastases are often the first sign of malignant spread in patients with breast cancer

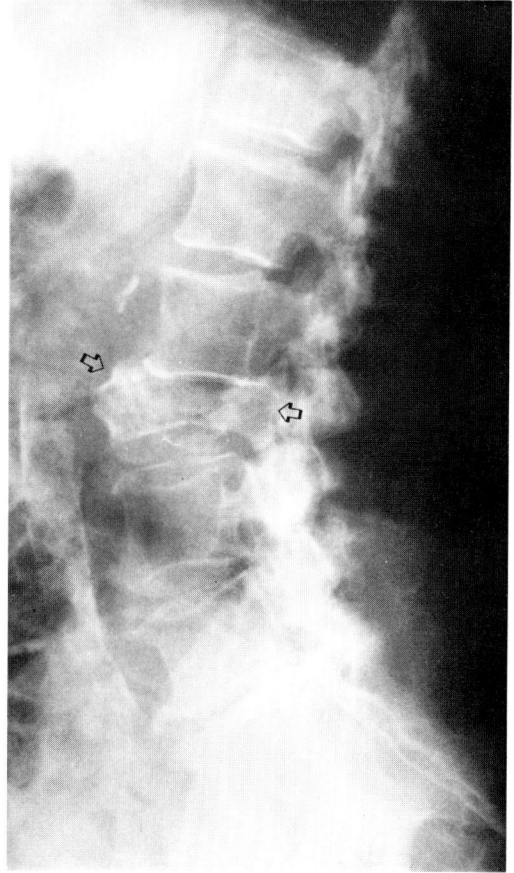

Fig. 7.4A Collapse of vertebral body as a result of metastases indicated by arrows.

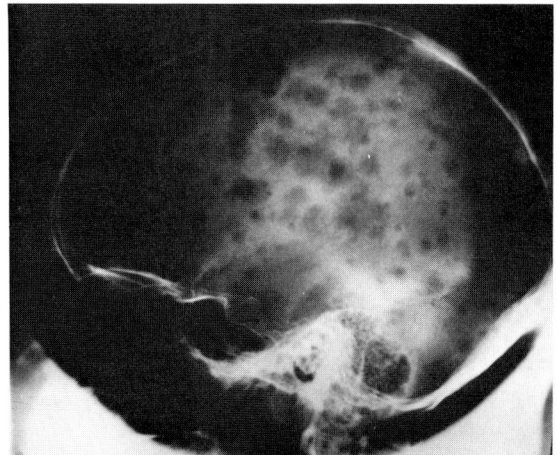

Fig. 7.4B Lateral view of skull showing multiple secondary deposits from carcinoma of the breast.

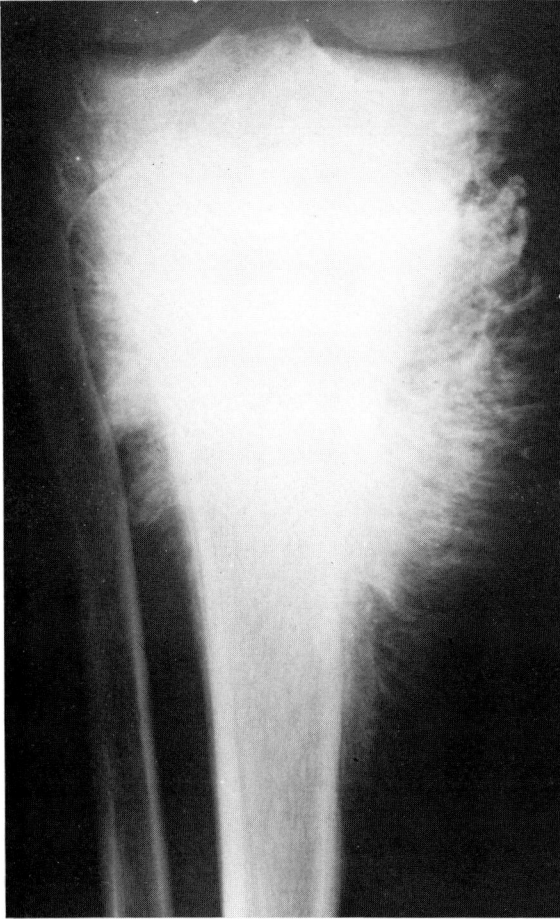

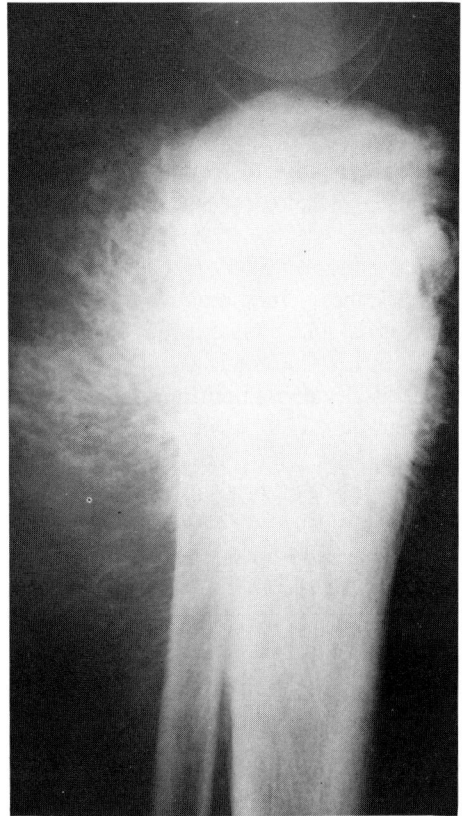

**Fig. 7.5B** Lateral view of tibia and fibula.

**Fig. 7.5A** Antero-posterior view of tibia and fibula. Note new bone growth — 'sun-ray' appearance, and periosteal elevation.

(Fig. 7.4B). The early radiographic changes produced by metastases include osteoporosis and, following radiotherapy treatment, there may be signs of radiation osteitis (Ratzkowski et al 1967).

## Primary bone tumours

These may be benign or malignant but both are extremely rare. Pain often accompanies malignant tumours and there is often an associated soft tissue swelling.

Age is important in deciding the type of tumour, for specific tumours occur at specific ages.

### Osteogenic sarcoma

Osteogenic sarcoma is probably the most malignant bone tumour and occurs in the 10–25 year age group, being more common in males than females. It often affects the distal part of the femur and knee. It is a tumour which characteristically presents as a dense non-homogeneous cortical lesion, usually situated on the metaphysis of a long bone (Figs 7.5A & B). There is variable thickening at the base from where it originates but not infrequently beyond this point as well. Because of cross sectional display, computed tomography offers a unique image of osteosarcoma and other bone tumours (Destouet et al 1979).

Routine radiography however does not always allow differentiation of cortical and medullary involvement whereas computed tomography shows the exact site and extent of macroscopic bone destruction. Computed tomography can also determine accurately the extent of soft-tissue involvement.

## Osteoid osteoma

This is an example of a benign bone tumour of oestoblastic origin. It affects the 20–30 year age group and is more common in males than females. It usually affects the long bones of the leg and the bones of the foot. It is a tumour with well-established radiographic features. Omojola et al (1981) advocate standard radiography as the first course of action, and reserve radionuclide studies for use in patients with obscure bone pain whose standard radiographs appear normal. They further suggest that computed axial tomography is of particular value in locating foci in the axial skelton.

## Multiple myeloma

This is another frequent bone disease which occurs mainly in males over the age of 50 years. The vertebrae, ribs, skull, shoulder, pelvis and long bones may be affected. The radiographs show 'punched-out' lytic lesions (Fig. 7.6 A–D).

In the last few years there has been much consideration of the appropriate imaging process for each bone lesion. Sometimes plain radiography is employed, sometimes radionuclide imaging, on other occasions a combination of both. McCormick et al (1975) recommend a combination of radiography and radionuclide imaging in the search for bone metastases arising from primary breast cancer. They used Strontium $87^m$ as the isotope. In their survey, in general, more lesions were detected using isotope scanning than by straight radiography, except in lesions of the pelvis. The use of $99^mTc$ has also been reported in the detection of secondary deposits from primary bone tumours. Gildray et al (1977) compared the sensitivity and accuracy of $99^mTc$ phosphate bone scans with conventional radiographic bone surveys in the diagnosis of bone metastases. They confirmed that greater overall sensitivity occurs from use of radioisotope bone scans, but widespread bone metastases can occasionally give rise to a uniform distribution of

$99^mTc$ methylene diphosphonate resulting in a superficially normal appearance on bone scan. This condition is thought to be more frequently associated with prostatic carcinoma than with other conditions. However, according to Condon et al (1981), increased uptake of radioisotope $99^mTc$ methylene diphosphonate is not necessarily proportional to the progression of disease when following bone lesions in patients suffering from cancer of the breast and prostate.

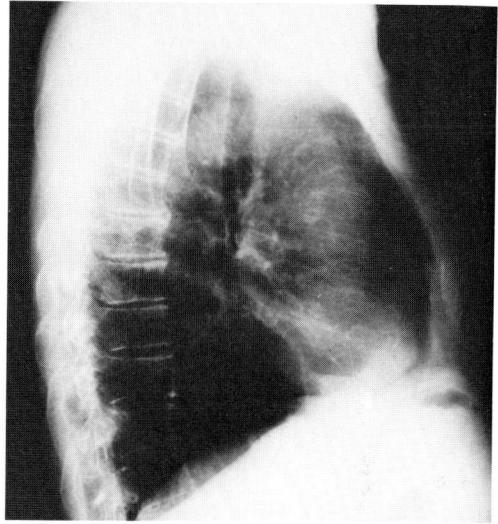

**Fig. 7.6A** Lateral view of chest. There is erosion at lower end of sternum.

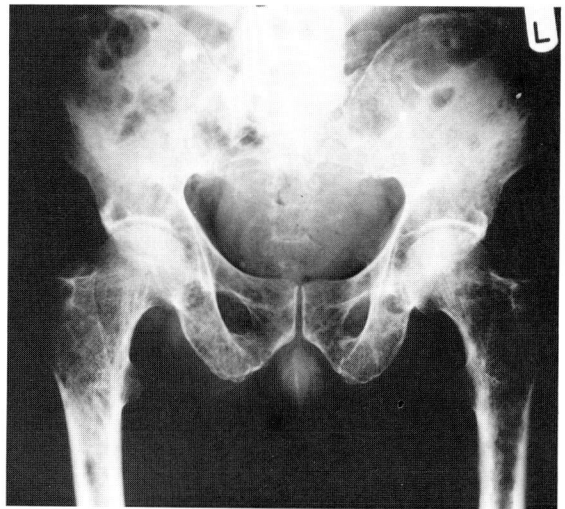

**Fig. 7.6B** Antero-posterior view of pelvis showing lesions in ilium, symphysis and femora.

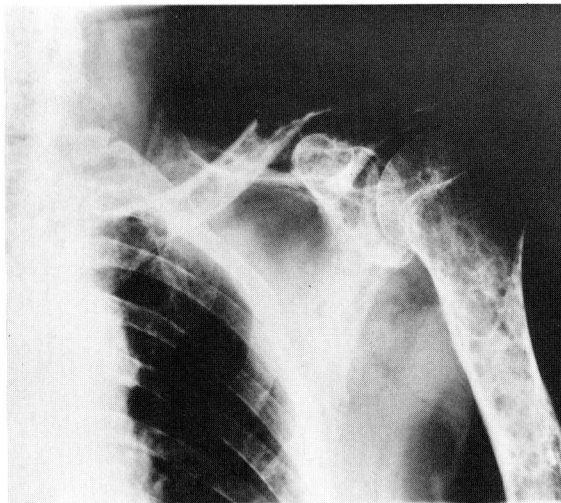

**Fig. 7.6C** Left shoulder — note involvement of scapula and humerus and widespread destruction of distal end of clavicle.

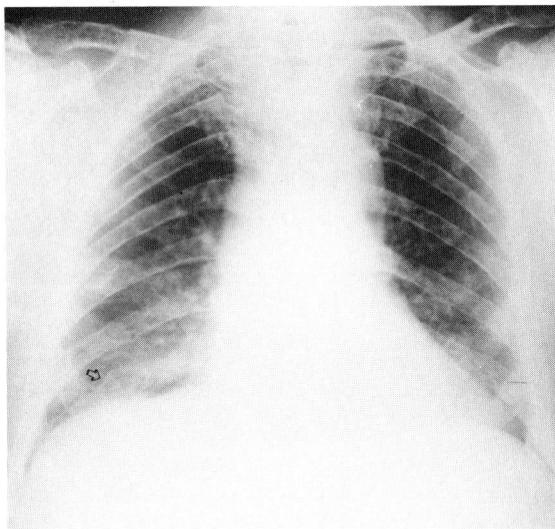

**Fig. 7.6D** Chest — postero-anterior view — note expanded lesion of right lower rib (indicated by arrow); again note involvement of clavicles.

Radionuclide skeletal surveys may be employed for initial screening procedure for bony metastases. If a lesion is seen, then complete radiographic evaluation of involved bone is performed (Gildray et al 1977). Radionuclide skeletal surveys are far more sensitive in children than radiographic skeletal surveys.

Charkes et al (1966) confirm the complementary nature of bone scanning and conventional radiographic bone surveys. The extent of disease as judged from scan appearances corresponds closely to that seen on the radiograph according to Harmer et al (1969). Furthermore, in their opinion, only occasionally do scans of primary tumours give more information than good quality radiographs. In cases where standard radiography is undiagnostic, and if there is doubt as to whether a tumour is benign or malignant, arteriography may be useful. It may also define the extent of the lesion and help in selecting the best site for biopsy (Voegeli & Fuch 1976).

## METABOLIC DISORDERS

Some of the more common metabolic disorders lead to a reduction in bone density which can be detected radiographically, i.e. osteoporosis. This category of disorders includes Cushing's syndrome, hyperthyroidism, hypogonadism and hyperparathyroidism.

## OSTEOPOROSIS

All the above conditions result in osteoporosis, which is characterised by a reduced volume of bone tissue relative to the volume of anatomical bone (Nordin 1973).

There is a tendency for the skeleton to lose bone with age, particularly in women after the menopause. There is a tendency for fasting plasma and urinary calcium to rise after the menopause (Young & Nordin 1967) due to increase in bone resorption.

### Radiographic views required include

| | |
|---|---|
| Femora | — particularly upper two-thirds |
| Radius and ulna | — at the wrist and in some cases the whole forearm |
| Dorsal and lumbar spine | — lateral views |
| Humerus | — whole length. |

Osteoporosis is accompanied by crush fractures and biconcavity of the vertebrae, loss of cortical

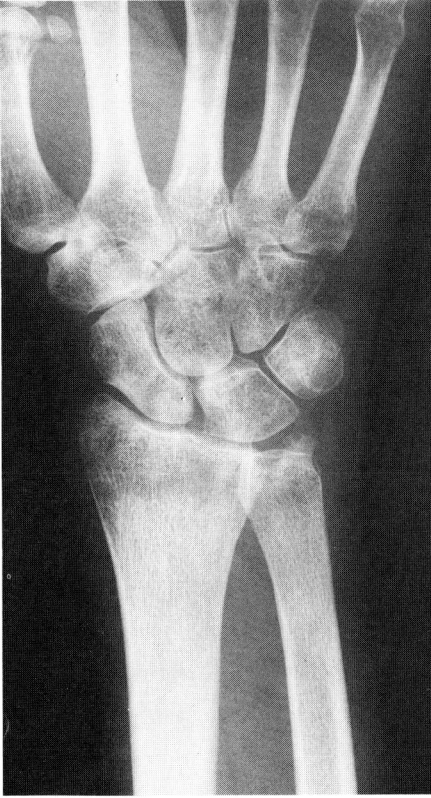

Fig. 7.7A Hand and wrist — note thinning of lower end of radius and ulna and carpal bones.

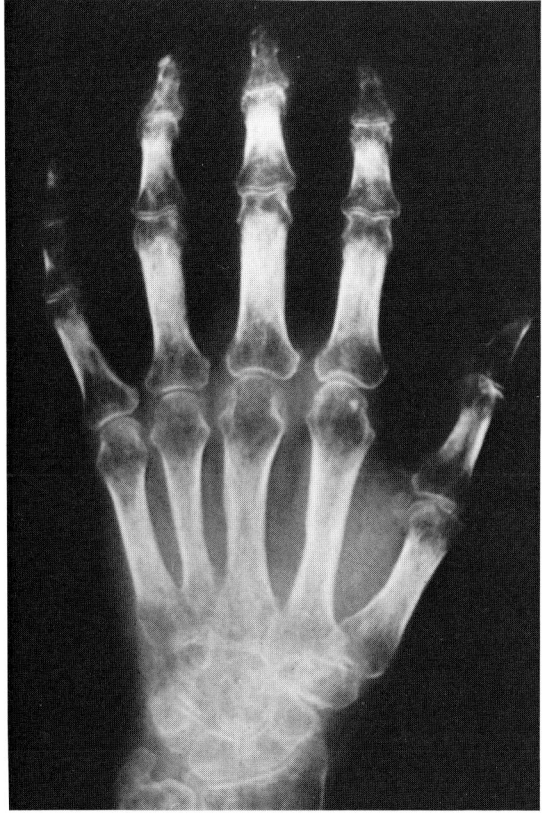

Fig. 7.7B Disuse osteoporosis as a result of herpes zoster.

bone in the long bones, e.g. humerus, radius, ulna and femur.

A *single* crush fracture may however be due to trauma or malignant disease, so that osteoporosis is assessed by viewing the skeleton as a whole.

Bone loss occurs for several reasons. Cortical bone is the strong supportive material which surrounds the spongy cancellous bone. During life there is a constant remodelling of bone substance and cancellous bone remodels faster than cortical bone. The remodelling process involves two types of bone cells, the osteoblasts which are bone forming and osteoclasts which destroy bone.

Calcium is required for healthy bone formation and the calcium pool is the source. This pool includes the kidneys, the gut and bones themselves. Calcium metabolism is controlled by the parathyroid glands which secrete parathormone. If the serum calcium level falls, the glands secrete more hormone and if the serum calcium level rises,

the glands will secrete less hormone. The parathormone acts on the organs of the calcium pool. The serum calcium level also affects the secretion of calcitonin from the 'C' cells of the thyroid, parathyroid and thymus glands. Calcitonin inhibits osteoclastic bone resorption thus controlling normal bone remodelling.

Vitamin D is also required for healthy bone processes. Vitamin D in the form of cholecalciferol is synthesised by the skin or absorbed by the gut and is converted by the liver to a 25-hydroxycholecalciferol and is further hydroxylated by the kidney to 1,25 dihydroxycholecalciferol. Again the secretion of this metabolite is controlled by the calcium level in the blood. When, as a result of physiological interplay, calcium leaves the bone faster than it goes in, there will be a consequent reduction in bone density and this will present osteoporotic appearances on the radiograph.

The appearances of osteoporosis may occur in

association with disuse of a part, and bear a relation to the age of the patient, and to the period of immobilisation (Fig. 7.7 A & B). Jones (1969) described a 'scalloping' or 'lamellation' and this affects predominantly the outer third of the cortex. The cortical indices, however, are not affected. The cortical indices are described by Barnett & Nordin (1961) as the ratio of cortical dimensions to total width of the second metacarpal. It is postulated that the local increase in the vascularity of immobilised bone plays a part in the distribution of the radiological changes.

## SECONDARY OSTEOPOROSIS

This is a feature of some endocrine disorders and a few of these are described.

### Cushing's syndrome

This was described by Harvey Cushing (1932) who reported that in this connection there is a tendency for spontaneous fractures. Primary Cushing's disease occurs as a result of hyperactivity of the supra-renal glands and is rare.

An appearance similar to Cushing's syndrome occurs as a result of long courses of steroid therapy.

Radiographic appearances in this disease include osteoporosis of the spine and axial skeleton. There may be crush fractures of the vertebrae, fractures of the ribs, pelvic bones and long bones. Albright (1942) also noted the tendency to fracture as a form of osteoporosis and Murray (1960) noted that excessive callus formation occurs after such fractures. The urinary calcium level is often high in patients with Cushing's syndrome which may, according to Ross et al (1966), be a reason why they develop renal stones.

### Hyperthyroidism

Patients with this condition may have severe osteoporosis involving the whole of the skeleton. Crush fractures of the spine may also occur. Fraser et al (1971) found a reduction in metacarpal density in patients suffering from hyperthyroidism. It is probable that thyroid hormone can cause bone

resorption which would explain the increased bone destruction and secondary parathyroid supression (Nordin 1960).

### Hypogonadism

In the male the testes have two functions: production of sperm and production of the sex hormone testosterone. The female ovaries produce the female sex hormone oestrogen and its effects on bone metabolism will be discussed later. The sex hormones in each case are responsible for the appearances of the secondary sex characteristics and sexual development from puberty onwards. In hypogonadism regression of this process takes place and this is accompanied by reduction in bone matrix synthesis and an increase in bone resorption. This may produce osteoporotic effects in the axial skeleton and in the spine.

### Hyperparathyroidism

This condition may be either 'primary' or 'secondary'. The former results from a benign tumour (adenoma) of the parathyroid or generalised parathyroid hyperplasia. The latter is produced when excess parathormone secretion is stimulated by low blood calcium resulting from, e.g. a leak of calcium in the urine in renal failure or inadequate calcium absorption in vitamin D deficiency.

### Primary hyperparathyroidism

This condition results in hypercalcaemia, hypophosphataemia, osteoporosis, hypercalcuria and, therefore, formation of renal stones. Hyperparathyroidism may also result in the bone having multiple cystic lesions (osteitis fibrosa cystica). The loss of cortical bone can be seen in the hands, in particular, and can be detected by measurement of the metacarpal cortical area ratio. Fractures of the vertebrae are seldom seen in this condition.

Thinning of the upper cortex of the clavicle is reported in adults under the age of 45 as a sign of osteomalacia and hyperparathyroidism (Anton 1979). Measurement can be made to the nearest 0.5 mm using a simple rule at the mid-point of the clavicle. One of the earliest signs of this condition is seen in the terminal phalanges of the hands, while bone cysts, which may be solitary or multiple, are usually seen in the shafts of the long bones. Fractures sometimes occur through the cysts because of the generally weakened bone structure.

### Secondary hyperparathyroidism

This condition may arise as a result of chronic renal disease when the plasma calcium level is low. In this case there is stimulation of the parathyroid glands to secrete parathormone. Vitamin D deficiency may also cause secondary hyperparathyroidism. There may be sclerosis of the vertebral bodies, the epiphyses and the skull. The sclerotic bands in the vertebrae result in 'rugger jersey' appearances of the spine. A 'cottonwool' appearance of the skull is seen in severe cases of secondary hyperparathyroidism.

Radiographically there are sub-periosteal erosions particularly of the terminal phalanges of the hand and irregular bone cysts, as well as coarsening of the trabecular pattern in the area of the femoral neck. Xeroradiographic and macroradiographic techniques have been employed in examination of the hand to demonstrate the changes in the terminal phalanges.

### Osteomalacia and rickets

Another condition which results in a reduction of calcium in bones is osteomalacia. By definition this condition is one in which there is a delayed mineralisation of new bone as and when it is laid down (Nordin 1973). Osteomalacia presents in adults in whom the epiphyses have fused, whereas rickets occurs during childhood during the periods of growth. In this condition there is a reduction in the ash content of the bone and excess of uncalcified osteoid or cartilage.

Osteomalacia may be present due to several reasons, but the most common of these is dietary vitamin D deficiency (Fig. 7.8A). Rickets may be associated with hypocalcaemic tetany in which the patient may present with carpopedal spasm, laryngismus which causes respiratory obstruction, or grand mal fits.

The characteristic features of this condition are

manifested in the epiphyses of long bones, the ribs and scapulae (Fig. 7.8B). As the cartilage cells at the metaphysis of the long bones do not calcify,

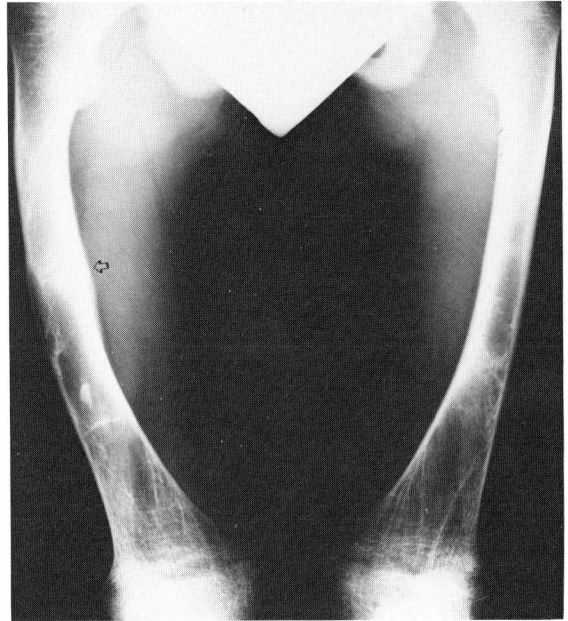

Fig. 7.8A Femora showing Looser zones of the right femur — indicated by arrow — due to vitamin D deficiency.

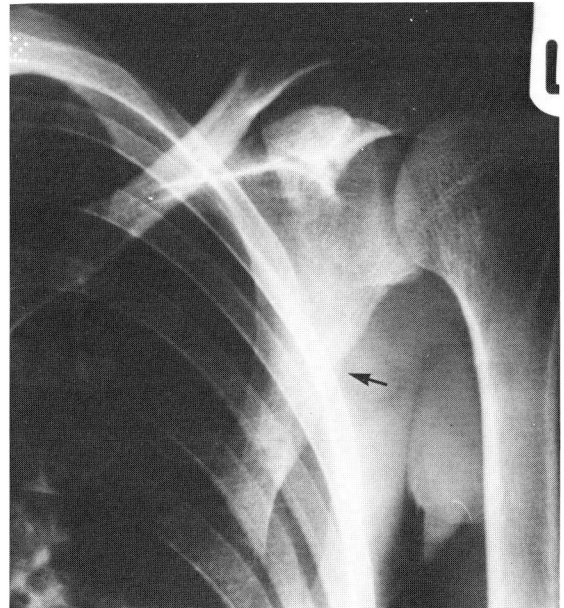

Fig. 7.8B Left shoulder. Note Looser zone in lateral border of scapula indicated by arrow.

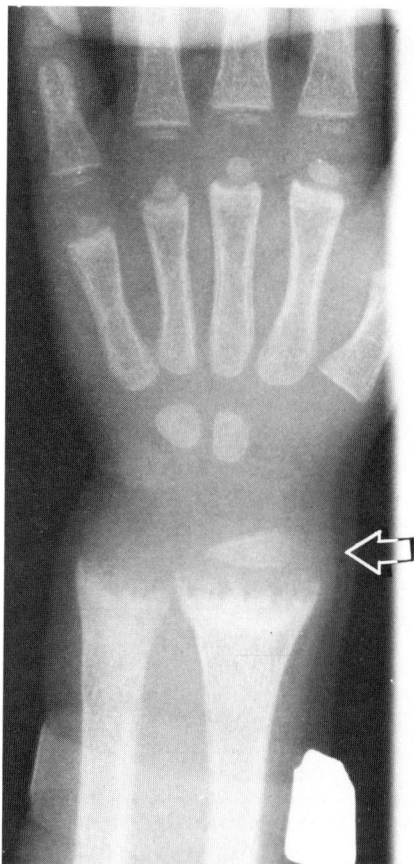

**Fig. 7.8C** Hand and wrist. Note 'saucer-shaped' deformity lower end radius and ulna indicated by arrow.

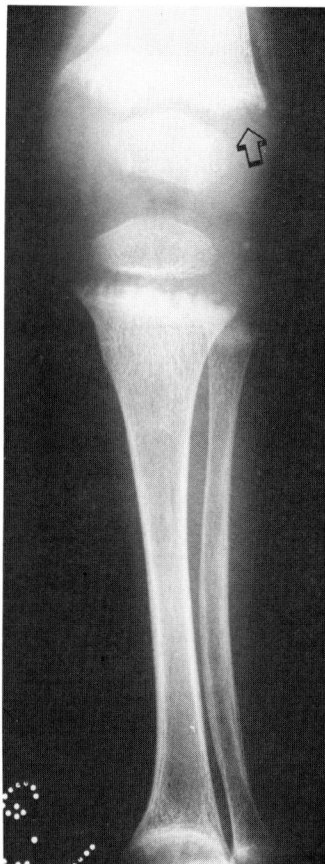

**Fig. 7.8D** Lower limb. Note deformity of lower femur indicated by arrow.

the cartilage becomes widened and the distal ends of the long bones become 'saucer-shaped' and irregular. This is sometimes known as the 'saucer-shaped' deformity of rickets (Fig. 7.8 C & D). In severe cases it may affect the ends of all long bones. There is a reduction in the thickness of the cortex of the long bones.

One important feature associated with rickets and osteomalacia is pseudo-fracture or Looser zones (Looser 1920). These are characterised by short zones of resorption which are transradiant. They are fractures without displacement in which callus is deposited but it fails to calcify adequately. Looser zones are often bilateral and symmetrical and are usually found in the lateral borders of the scapulae, medial borders of the upper two-thirds of the femur (around the neck) and in the ischio-pubic rami. If radiographs of the hands, femora

and shoulders are examined, there is loss of cortical bone in the metacarpals, femoral shaft and clavicles. The spine is also affected and manifestations occur such as biconcavity of the vertebral bodies. Although biconcavity may occur in osteoporosis the two conditions are different and in osteomalacia the radiographic density of the vertebra is usually normal and may even be increased.

In severe cases of osteomalacia there may be involvement of humeri, tibiae and fibulae, ribs and occasionally the ulnae. Radiographically the long bones may be bowed and the ribs may have nodular appearances at the costochondral junctions, this manifestation being known as the 'rickety rosary'.

Dietary deficiency of vitamin D or malabsorption syndromes in which vitamin D is not absorbed from the gut may give rise to rickets or osteomalacia. Malabsorption is associated with steatorrhoea (Nordin & Smith 1967), and following gastrectomy (Morgan et al 1966). Subsequently the lack of vitamin D causes the malabsorption of calcium which may lead to osteoporosis or severe lack of vitamin D may cause osteomalacia. Other causes of osteomalacia include hypophosphataemic rickets due to vitamin D resistance.

Osteomalacia may occasionally occur in association with a neoplasm (Hosking et al 1975), and Salassa et al (1970) reported instances of removal of the neoplasm being followed by regression of the osteomalacia without additional treatment being given. There is sometimes redistribution of skeletal calcium observed in association between osteomalacia and prostatic carcinoma.

Comparisons have been made between bone scanning and routine radiography in patients with metabolic bone disease (Fogelman et al 1980). A bone scan was carried out 4 hours after the intravenous injection of 15 m Ci $^{99}$Tc$^m$ hydroxyethylene — diphosphonate (HEDP). A radiographic skeletal survey carried out simultaneously included antero-posterior and lateral views of the cervical, dorsal and lumbar spine, postero-anterior view of the chest, antero-posterior of the pelvis and views of the long bones and hands. The following data is recorded from their survey:

(i) Osteoporosis — in 81% (22 cases) the radiographs showed osteoporosis. The bone scan was not suggestive of metabolic bone disease.

(ii) Primary hyperparathyroidism — radiographs showed evidence of the condition in 21% of cases whilst the bone scan was suggestive of metabolic bone disorder in 50% of cases (7 out of 14).

(iii) Osteomalacia — radiographs showed evidence of osteomalacia in 60% (9 cases) whilst the bone scan suggested metabolic bone disorder in all patients with osteomalacia. Ten of the patients had pseudofractures in a total of 24 sites, of which 58% were evident on radiographs and 78% were detected by bone scan. It is stated that when bones become abnormally brittle with pathological fracture and

vertebral collapse then this is demonstrated on the bone scan as a focus of increased tracer uptake.

*Renal osteodystrophy*

For convenience it would be useful to consider here some skeletal disorders which arise in patients having long-term haemodialysis. Patients on regular haemodialysis often suffer from bone disease attributed to osteomalacia, hyperparathyroidism and osteoporosis occurring either alone or in combination. Simpson et al (1976) described a series of skeletal surveys undertaken on 70 patients receiving long-term haemodialysis for chronic renal failure and these findings were correlated with histological findings in a specimen obtained by iliac crest biopsy.

Skeletal surveys were carried out every 6 months or annually. The radiological findings included: subperiosteal erosions in the phalanges (Fig. 7.9),

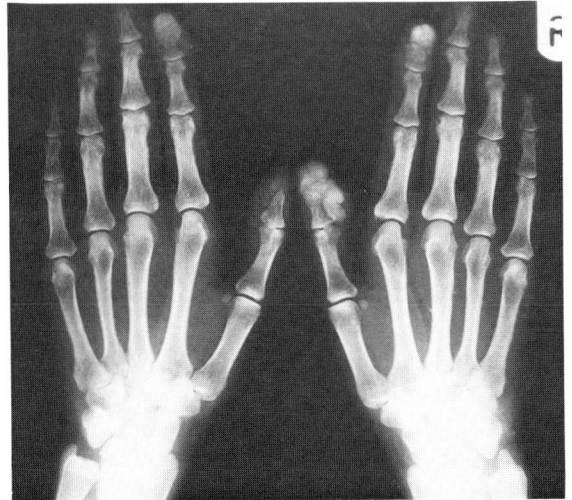

**Fig. 7.9** Note subperiosteal erosions of the phalanges of both hands, also ectopic soft tissue calcification.

erosions at other sites, abnormal mottling in skull vault, fractures or Looser zones, vertebral collapse, sclerosis of spine, hands, femoral heads and whole skeleton, juxta-articular rarefaction in hands and feet. The main findings of these authors were that fractures and severe medullary rarefaction appear to be most commonly the result of osteomalacia; subperiosteal erosions are associated with the more severe grades of osteitis fibrosa cystica. Cortical striations and sclerosis are associated with an increased amount of osteoid and sclerosis is diagnosed more frequently by radiological means than by iliac crest biopsy.

Steinbach & Noetzli (1964) described the radiological features of osteomalacia associated with haemodialysis. It is generally assumed that subperiosteal erosions in the phalanges indicate hyperparathyroidism and Looser zones indicate osteomalacia. Certain radiological features have been known to occur in more than one primary disease, e.g. decreased radiodensity, pathological fracture and osteosclerosis. Iliac crest biopsy is useful in determining the accuracy of radiological signs in diagnosing the nature of bone changes.

Another series of skeletal surveys reported by Ritchie et al (1975) is of interest. In this series the surveys were carried out at 6-monthly intervals in patients on renal dialysis and included radiographs of skull, pelvis, chest, lumbar spine and hands. The hands were radiographed on industrial X-ray film to provide better bone detail. Periosteal new bone was noted along the inner borders of the pelvic inlet and this appeared as a thin layer of calcification lying parallel to the inlet and sometimes extending down the superior surfaces of the pubic rami.

The average duration of dialyses before recognition of periosteal new bone was 47 months and this was usually preceded by the appearances of osteosclerosis and subperiosteal bone erosions. This new bone formation may be a feature of hyperparathyroidism and may be due to increased mass of osteoid tissue in patients with chronic renal failure (Garner & Ball 1966, Simpson et al 1976). The level of parathyroid hormone is raised in chronic renal failure and there is a possibility that increased osteoid mass is due to high parathyroid concentrations.

## Acromegaly

This condition results from an overactivity of the anterior lobe of the pituitary gland which provides growth hormone. Once the epiphyses are closed, increase in stature cannot occur as a result of overactivity of the anterior pituitary so that growth occurs 'laterally' rather than longitudinally. The patient will notice that the hands, feet, head and jaw become larger. This overactivity may be due to an eosinophilic pituitary adenoma.

There will be pronounced radiographic features and a skeletal survey should include:

lateral view of the skull to include the facial bones and mandible
antero-posterior and lateral views of dorsal and lumbar spine
postero-anterior view of hands
postero-anterior view of feet
lateral view of the os calces.

There is a view held in some quarters that osteoporosis is present in acromegaly but in a series carried out by Doyle (1967) where ulnar densitometric measurements were recorded, very few patients (out of 60 cases) had osteoporosis.

Radiographically the hands show an increased thickening of the tufts and there is soft tissue thickening in the fingers causing typical 'spade-like hands' (Fig. 7.10A). There is also enlargement of the frontal sinuses ('cauliflower'

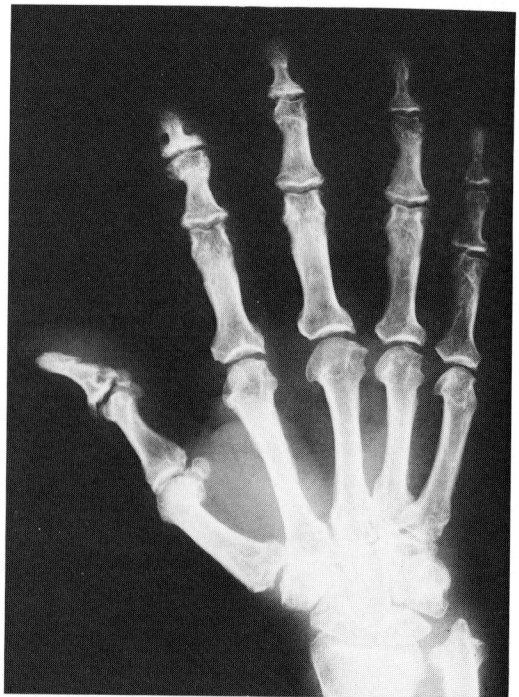

**Fig. 7.10A** Hand and wrist. Note 'beaking' of heads of metacarpals and tufting of phalanges.

sinuses) and elongation of the mandible with increased prominence. Thickening of the skull tables occurs and there may be enlargement of the

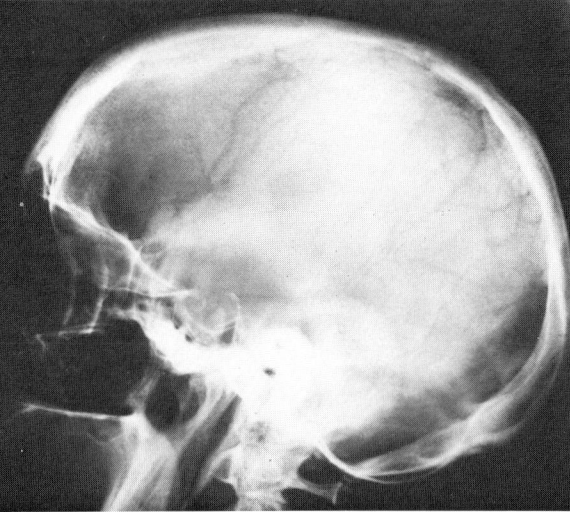

**Fig. 7.10B** Lateral view of skull. Note erosion of clinoid processes and enlargement of sella turcica.

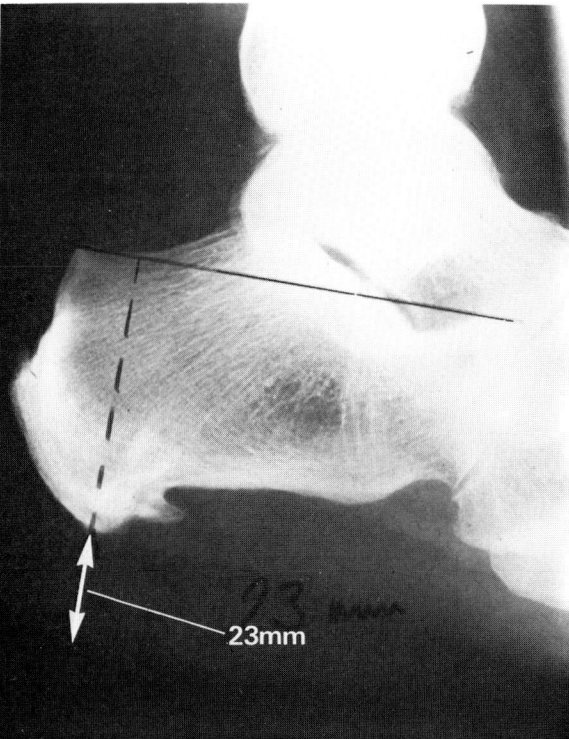

**Fig. 7.10C** Lateral view of heel showing area of measurement — 23 mm — just within normal limits. Note calcaneal spur.

sella turcica if a tumour is present. This may be accompanied by erosion of the clinoid processes (Fig. 7.10B). Tomographic examination of the sella turcica may be undertaken in the lateral projection. Radiographs of the feet show enlargement, mainly due to increase in soft tissue thickness. Additionally the heel pad is increased in thickness. This measurement can be made from the lateral radiograph of the os calcis (Lusted & Keats 1972). The measurements are different between the sexes.

## ASSESSMENT OF SKELETAL MATURITY

The developmental status of children may be assessed from postero-anterior radiographs of the

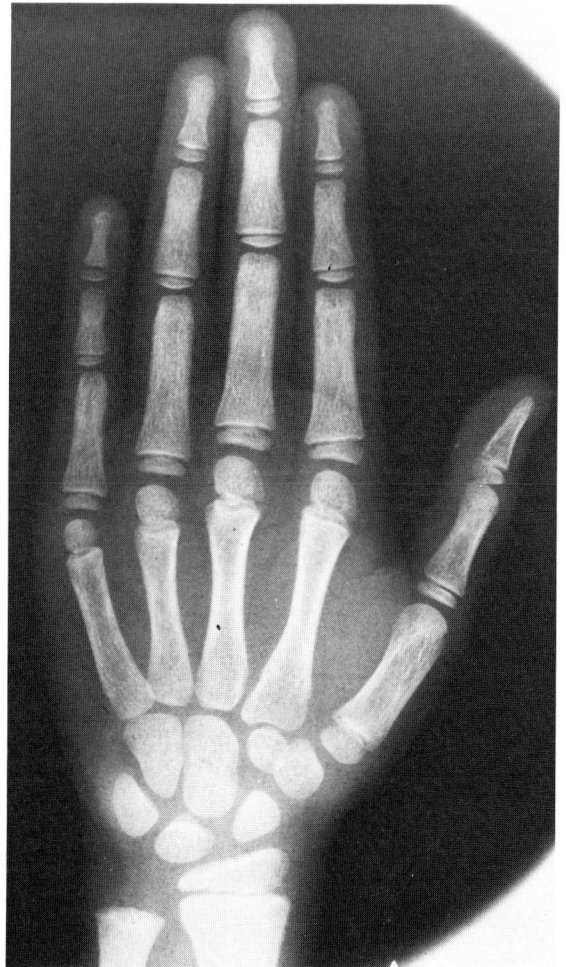

**Fig. 7.11** Left hand and wrist — assessment is made of carpal bone sizes and sizes of epiphyses of long bones.

hand and wrist. Growth and development proceed constantly in the normal child. The skeleton of an adequately nourished child develops as a unit and because of this the general development can be assessed from radiographs of the hand, but occasionally the skeleton as a whole needs to be assessed (Fig. 7.11).

**Radiographic views required include:**

Postero-anterior hand and wrist
Antero-posterior elbow joints
Antero-posterior pelvis.

Advantages of assessing maturation of the skeleton from radiographs of the hand and wrist include:

a. the size, shape and dimensions of the carpals and metacarpals can be measured, as can the status of the epiphyses.
b. it is possible to compile a record of progress or regress from birth to adulthood.
c. maturative changes in the skeleton are intimately related to those of the reproductive system so that one reflects changes in the other.

There are various atlases for measuring maturity: Tanner et al (1962), Greulich & Pyle (1959), Pyle, Waterhouse & Greulich (1971). There are other books which deal with measurement from radiographs, e.g. Lusted & Keats (1972). Atlases also exist for measurement of knee joint to assess maturation.

*Paget's disease*

The other name for this condition is osteitis deformans which perhaps helps to describe the results of the condition in the skeleton. It was first described by Sir James Paget in 1877. Earlier incidences of the condition have been reported by Deuxchaisnes & Krane (1964) and are said to have occurred in the Egyptian and Neolithic eras. Incidence of the condition increases with age and is highest in both sexes in the 90–100-year-old age range and is rare below the age of 40 years.

In 1957 Pygott (1957) found an incidence of the disease in about 2.2% of patients in a series who

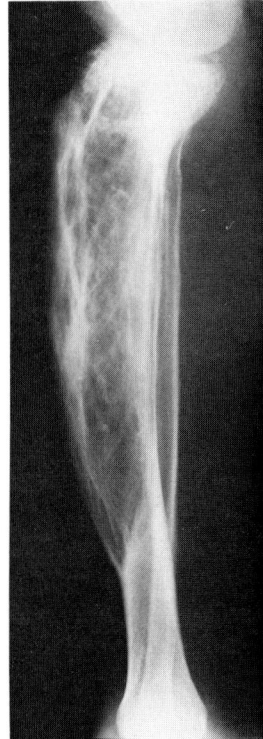

**Fig. 7.12A** Lateral view of tibia. Note coarse trabecular pattern.

had been X-rayed for spine and pelvis conditions. The diseased bone demonstrates an increased vascularity and increased uptake of isotope $^{99m}Tc$ stannous polyphosphate. The bone scintiscan is a more sensitive indicator of the extent of polyostotic Paget's disease than conventional radiographs, demonstrating 34% more lesions (Lentle et al 1976). This condition may affect any of the bones of the skeleton but it is most prevalent in the spine, skull and pelvis. It may affect one bone or more than one and may be found in adjacent bones (Collins 1956).

Deformity often occurs and this sometimes leads to enlargement of the skull or bowing of long bones. The deformities are due to remodelling rather than bending of the bones. There may be increased skin temperature over the bones affected and this is thought to be due to increased blood supply. A complex condition exists in the bone and there are lytic as well as sclerotic phases. All the stages may be present in one patient or indeed in one bone. Osteoclastic resorption occurs at the

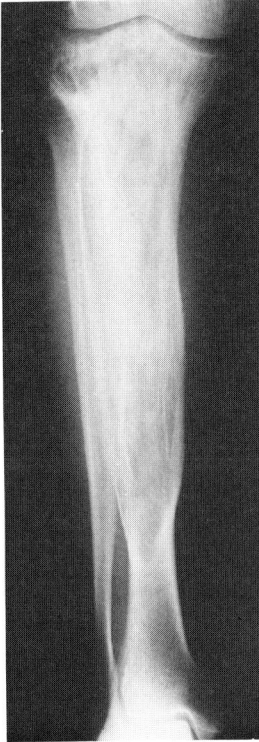

Fig. 7.12B Antero-posterior view of tibia.

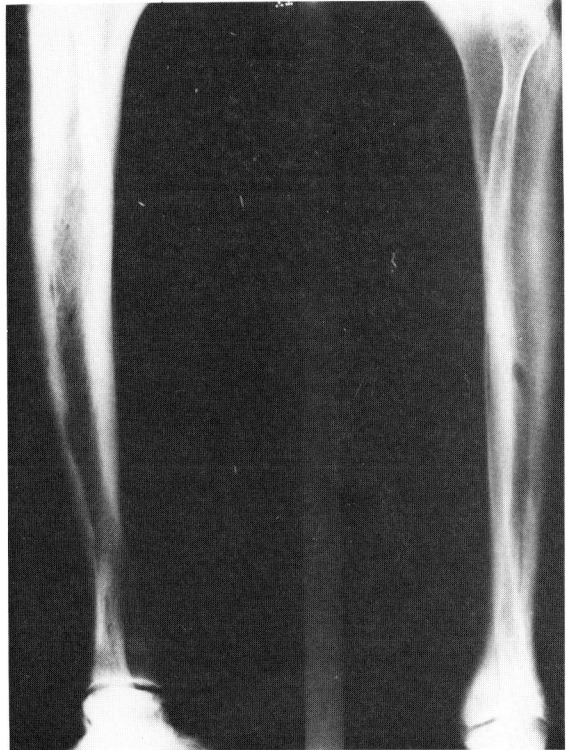

Fig. 7.12D Lateral views of tibia showing increased density of right one.

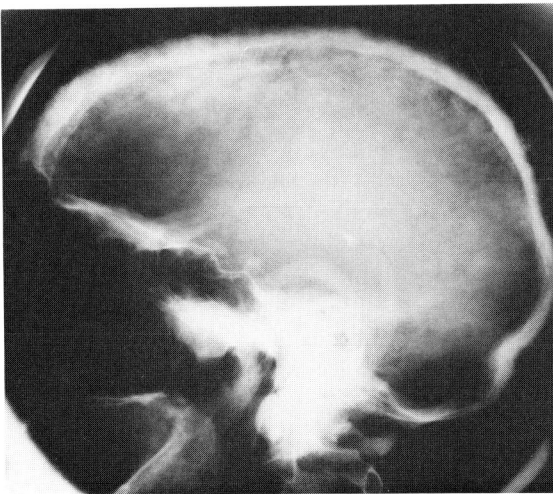

Fig. 7.12C Lateral view of skull showing extensive involvement of vault and base. Note 'woolly cap'.

advancing edge of the lesion, and osteoblastic regeneration behind it. The bone is gradually destroyed and is replaced by disorganised trabeculae of coarse-fibred bone. In osteoporosis circumscripta, which occurs only in the skull, osteolysis is not followed by new bone formation (Collins & Winn 1955).

In the final stages of the disease, collagen is laid down in a disorganised manner which creates a mosaic pattern giving the deformity described in the name of the disease (Fig. 7.12A & B). In this condition there is an increase in the plasma alkaline phosphatase. The rate of bone turnover can be measured by radionuclide methods using radiocalcium or strontium.

Bones affected by the disease have a high uptake of these radioactive materials. Qualitative bone scanning can show a response to treatment in Paget's disease whereas radiographs rarely change (Lavender et al 1977). A survey carried out by Guyer & Clough (1978) revealed that in 100 patients who received full skeletal surveys, the order of frequency of site was pelvis, lumbar spine, sacrum, femur, skull and dorsal spine. These authors also recorded an increase in the frequency of involvement in lumbar spine, pelvis and femora with increasing age. They also conclude that Paget's disease of bone is an environmental disease which is related to stress on bone.

**Radiographic technique**

Views required include:

lateral skull to include facial bones
lateral cervical spine
postero-anterior chest
antero-posterior shoulders
antero-posterior pelvis
antero-posterior femora and knees
antero-posterior tibiae.

The radiographic appearances of the condition include vertebral bodies which have a more dense appearance and coarse trabecular pattern. In the long bones the cortex is thicker with coarsening of the trabecular pattern. The three most common characteristics are: expansion, thick cortex and coarse trabecular pattern. Micro-fracture together with bowing of the long bones may also appear. A mottled appearance is characteristic in the skull and there may be increased thickening of the table. In osteoporosis circumscripta the appearances in the skull are of a patch of reduced density. The pelvis is more dense with coarse trabeculations and protrusio acetabulae of the hips may be present.

Fractures are sometimes seen in bones in which osteitis deformans is present but tumours may also develop in these bones, osteosarcoma being the most common In rare cases sarcoma may arise in Paget's disease but this is not usually a pure osteosarcoma.

During the sclerotic stage of the disease when the bone may be enlarged and heavy there will be a necessary increase in the radiographic exposure factors, particularly the kVp, to penetrate the increased dense bone (Fig. 7.12C & D).

## OSTEOCHONDRODYSTROPHIES AND SKELETAL SYNDROMES

There are many complicated congenital and other conditions which can affect the skeleton in the

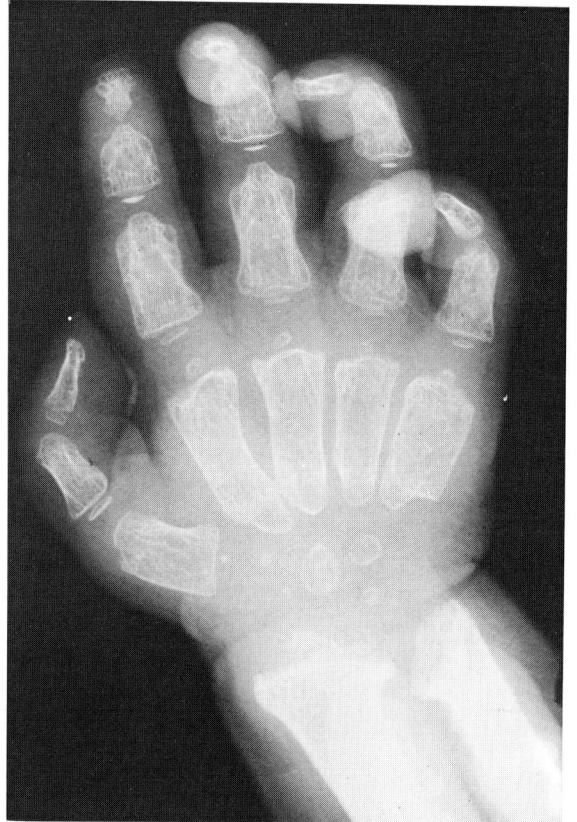

**Fig. 7.13** Hand and wrist of child. Note proximal tapering of metacarpals.

region of the epiphyses. Some of these conditions may be familial and may produce developmental defects in the same regions.

There are three main types of dysplasia due to:

## 1. Disturbances of chondroid production

These include Morquio-Brailsford syndrome and Hurler's syndrome among the mucopolysaccharidoses.

*Morquio-Brailsford syndrome*

This condition is due to a gene defect which leads to an enzyme defect and excess of mucopolysaccharides is found in urine. In the Morquio-Brailsford disease there is excess of keratosulphate. The condition is seen radiographically as flattening of vertebral bodies with a tongue-like

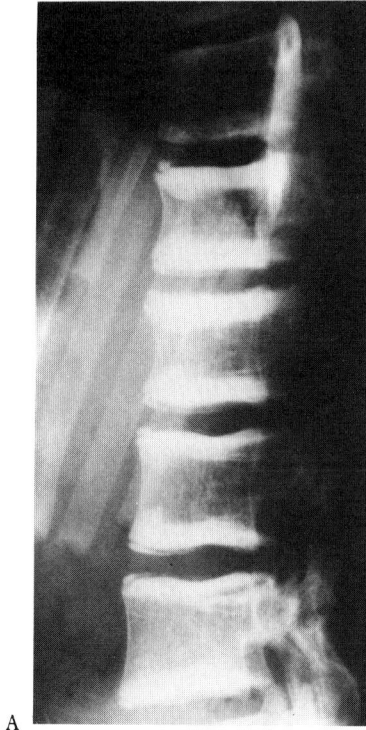

A

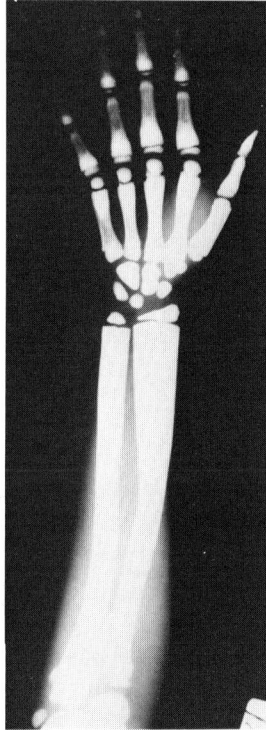

B

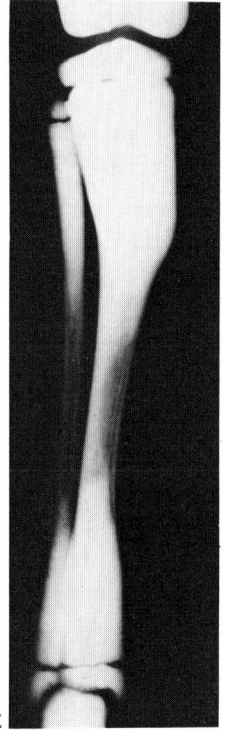

C

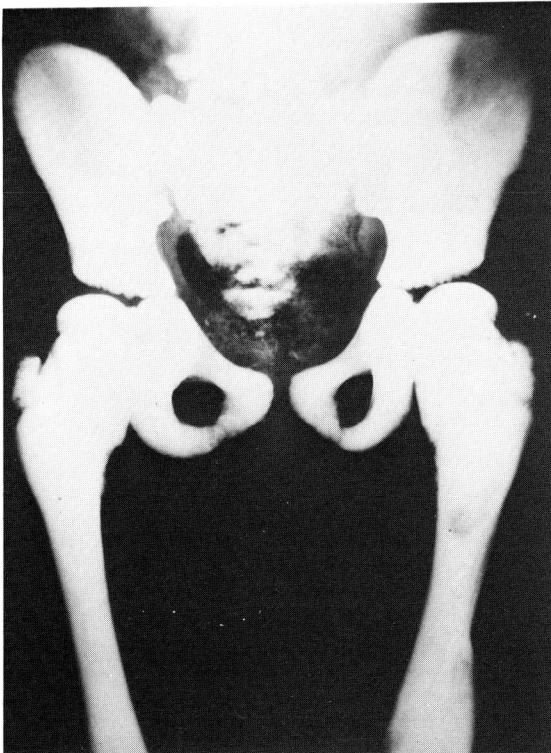

D

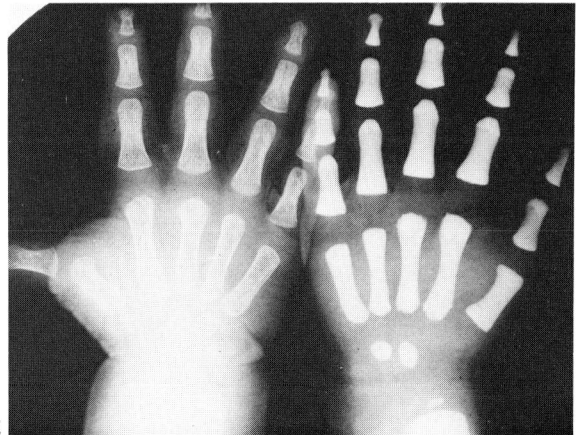

E

**Fig. 7.14A** Lateral view of spine showing 'rugger-jersey' appearance.

**Fig. 7.14B** Hand, wrist and forearm. Note increased density of bones.

**Fig. 7.14C** Antero-posterior view of tibia and fibula.

**Fig. 7.14D** Antero-posterior view of pelvis and femora. Note femoral deformity due to healed fracture in the dense but brittle bone.

**Fig. 7.14E** Both hands and wrists of baby, right hand normal and left hand osteopetrotic.

projection on the anterior part. The metaphyses of the long bones are widened and have irregular and fragmented epiphyses.

### Hurler's syndrome

There are significant radiographic differences in appearance which include backward displacement of a single wedge-shaped vertebra in the lumbar spine region together with a j-shaped appearance of the sella turcica, and characteristic proximal tapering of the metacarpals (Fig. 7.13).

## 2. Disturbances of osteoid production

Amongst these dysplasias are those involving abnormal epiphyseal ossification such as multiple epiphyseal dysplasia and those involving a deficiency in osteoid production of which osteogenesis imperfecta is an example.

The group of conditions due to excessive osteoid production includes osteopetrosis and osteopoikilosis.

A third group related to abnormal osteoid production includes fibrous dysplasia and neurofribromatosis.

### Multiple epiphyseal dysplasia

This condition is often familial and may affect many of the growing epiphyses. The radiographic appearances of the affected epiphyses include normal shape and size but fragmentation and flattening of the bony nucleus.

### Osteopetrosis (Albers-Schönberg disease (1904) — marble bones)

This condition involves sclerosis of bone which results in an increased density radiographically. Due to the increased density an increase in kVp is required by 5–10 over that required for the normal bone examination.

The cause of the disease is unknown but it is thought that there is failure of bone resorption and remodelling. Areas generally affected include the skull, pelvis and long bones. This may extend to include the metacarpals and metatarsals. Occasionally there is an increase in density in the

ribs. Sometimes a 'rugger-jersey' type appearance may be seen in the vertebrae (Fig. 7.14A–E).

### Osteopoikilosis

This is another condition first described by Albers-Schönberg (1915) and is a rare familial disorder. Radiographically the appearances are of small areas of increased density scattered throughout the bone structure. These areas may be circular or oval in shape and appear in any area of the body. Isolated ones are sometimes referred to as bone islands (Fig. 7.15A & B). This is insignificant to the patient.

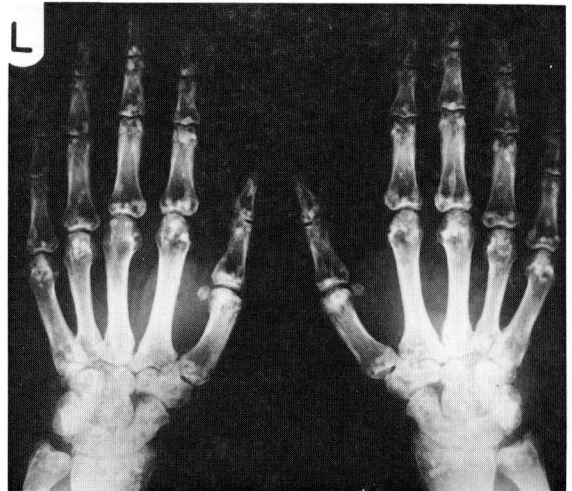

**Fig. 7.15A** Left and right hand showing numerous circular and oval dense areas.

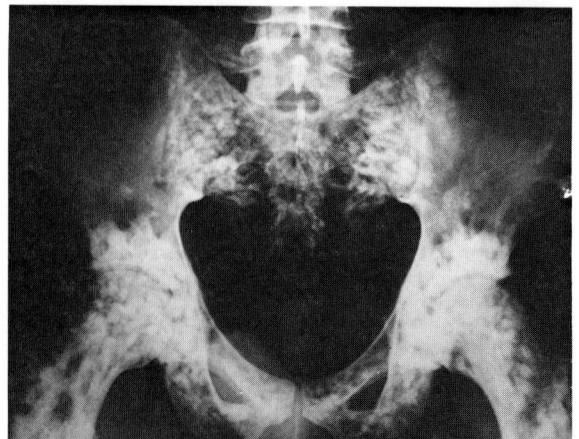

**Fig. 7.15B** Antero-posterior view of pelvis and femoral necks again showing areas of increased density.

*Fibrous dysplasia*

This condition may affect a small area of one bone or it may affect a whole bone or even a whole limb. It may result in a thinning of the cortex of a bone or the cortex may be scalloped out and there is loss of trabecular pattern ('ground glass' appearance). When only small areas of bone are affected the upper end of humerus and lower end of tibia as well as femoral neck are diseased. In cases of whole limb involvement it is often the middle-third of the shaft of the long bone which is diseased.

*Neurofibromatosis* (Von Recklinghausen disease)

This condition is a familial disorder involving neurological tissue. In approximately 50% of all cases there may be bone disease. These manifestations include areas of bone erosion in the cortices of long bones, spinal deformities such as kyphosis, scoliosis and spina bifida and pseudarthrosis in long bones.

## Miscellaneous group of dysplasias

Amongst the miscellaneous group of dysplasias the one of most interest is Marfan's syndrome.

*Marfan's syndrome* (arachnodactyly)

The long bones of the fingers are longer than usual which sometimes gives the name 'piano-fingers' to the condition. There is a metacarpal index for estimating the presence of the condition and this involves measurements from the radiographs of the lengths and widths of each metacarpal — from 2 to 5 (excluding thumb) down the centre of the long axis and across the mid-point (Fig. 7.16). The length is divided by the width in each case. The resulting figures from each of the four metacarpals are added together and divided by 4. The rating, in normality, should be less than 8. As accurate measurements are required it is obviously important to have good radiographs on which the bone boundaries can be clearly defined. A pair of needle-point, direct reading calipers would be useful for recording measurements.

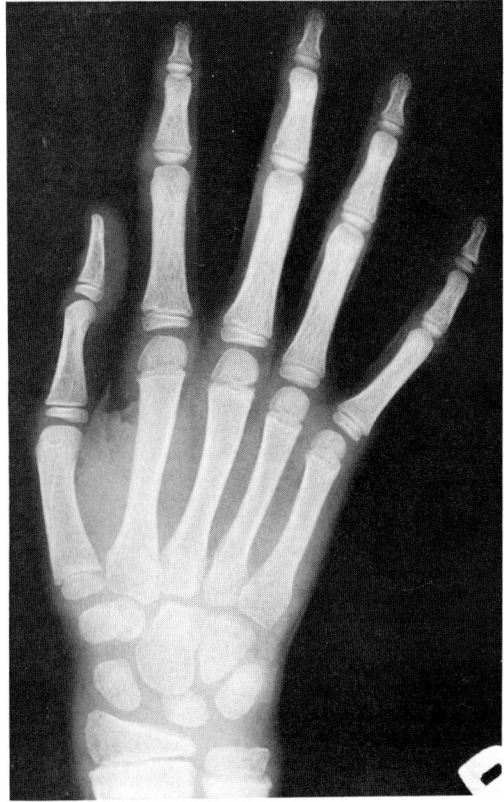

Fig. 7.16 Postero-anterior hand and wrist. Measurements are made of lengths and widths of metacarpals 2 to 5, from the radiograph.

## NON-ACCIDENTAL INJURY (battered baby syndrome)

In the last two or three decades attention has been drawn to cases of children suffering from injuries which have arisen in suspicious circumstances. Radiographers, along with other groups of medical and paramedical staff, are being asked to look out for injuries seen radiographically which are either not associated with the current history, or injuries which appear to be more severe than could be accounted for by the history provided.

A child may appear with multiple bruises, scars, burn marks, black eyes and when X-rayed old fractures or healing fractures may be detected. Injuries to epiphyses also occur as a result of violence from parents and others. Children may be picked up and swung round by the arms or legs.

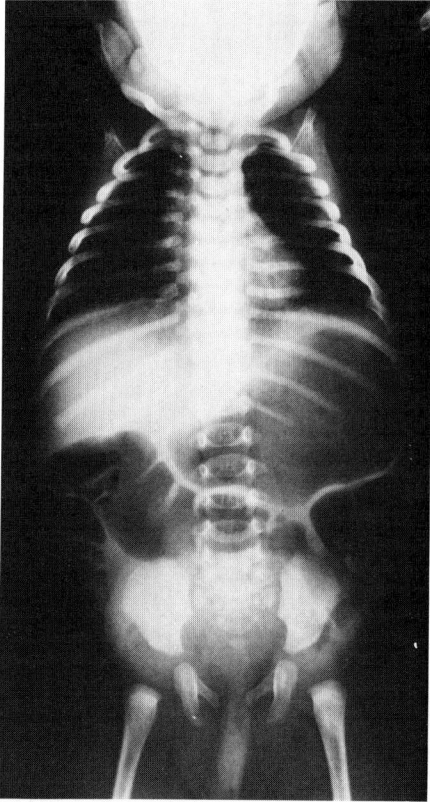

Fig. 7.17A Antero-posterior body survey showing most of chest, spine, pelvis and upper femora. This is often the first film taken; in this case no injuries are seen.

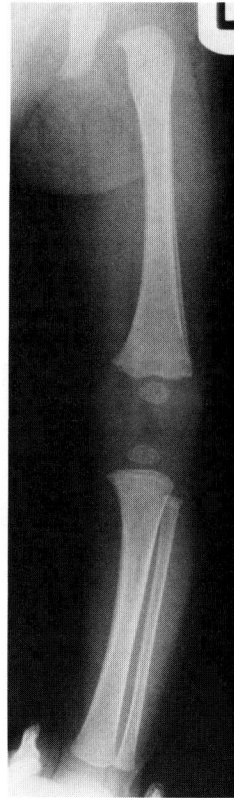

Fig. 7.17B View of left leg. Note periosteal elevation of femur. This is indicative of trauma and associated haemorrhage under periosteum.

Child violence occurs in middle class and upper class homes as well as in poor homes and from couples with low IQs or psychological problems.

Skeletal trauma due to violence needs to be separated from other skeletal abnormalities (Fig. 7.17A–C). A child suffering from osteogenesis imperfecta (fragilitas ossium) may suffer skeletal injury during normal loving care and handling or if the child stumbles during play. Fractures which occur in the long bones of children suffering from this condition are usually diaphyseal whereas metaphyseal fractures are more common in child battering. The child who has suffered from a long history of battering may have radiographic evidence of old healed fractures or healing fractures. These may involve long bones, ribs and skull fractures without signs of osteoporosis which normally accompany osteogenesis imperfecta.

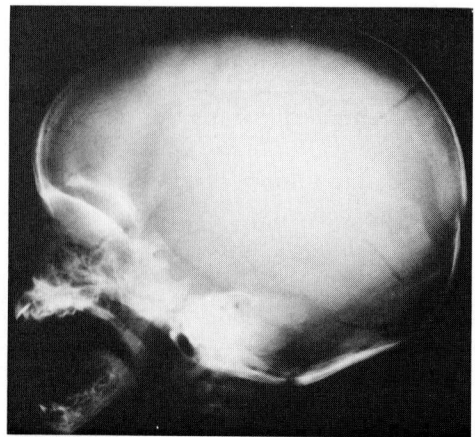

Fig. 7.17C Lateral view of skull. There are multiple fractures with bone displacement posteriorly.

Another condition which may mimic osteogenesis imperfecta or the battered baby syndrome is hyperphosphatasia. This condition is also associated with fragile thickened bones which fracture and there may be dwarfism deformities and abnormal dental findings. Yet another condition in which the radiographic findings are similar to or may mimic battered baby syndrome is vitamin C deficiency (scurvy). This results in a reduction in bone density with thinning of bone cortex and trabeculae. The epiphyses are also regular and small and there may be fractures through the metaphyses. There is also bleeding into the tissue and beneath the periosteum causing this to become elevated. Eventually calcification under the periosteum may occur at the site of the haemorrhage.

## Acknowledgement

The author is grateful to Dr G.J.S. Parkin, consultant Radiologist at The General Infirmary at Leeds and Wharfedale General Hospital, Otley, for kindly reading the script and offering constructive criticism.

## REFERENCES

Albers-Schonberg H E 1904 Röntgenbilder einer seltenen Knockenerkrankung. Munchener Medizinische Wochenschrift 51: 365–366

Albers-Schonberg H E 1915 Eine seltene, bisher nicht bekante Strukluranomalie des skellettes. Fortschritte Röntgenstrat 23: 174–177

Albright F 1942–43 Cushing's syndrome. Harvey Lectures 38: 123–186

Anton H C 1979 Thinning of the Clavicular cortex in adults under the age of 45 in osteomalacia and hyperparathyroidism. Clinical Radiology 30: 307–310

Barnett E, Nordin B E C 1960 The radiological diagnosis of osteoporosis: a new approach. Clinical Radiology 11:166

Bond J V, Martin E C 1975 Bone metastases in Wilm's tumour. Clinical Radiology 26: 103–106

Charkes N D, Sklaroff D M, Young I 1966 A critical analysis of strontium bone scanning for detection of metastatic cancer. American Journal of Roentgenology, Radium Therapy and Nuclear Medicine 96: 647–656

Citrin D L, Bessent R G, Grieg W R 1977 A comparison of the sensitivity and accuracy of the $^{99}Tc^{cm}$ — phosphate bone scan and skeletal radiographs in the diagnosis of bone metastases. Clinical Radiology 28: 107–117

Collins D H 1956 Paget's disease of bone. Incidence and subclinical forms. Lancet 2: 51–57

Collins D H, Winn J M 1955 Focal Paget's disease of the skull (osteoporosis circumscripta). Journal of Pathology and Bacteriology 69: 1–9

Condon B R, Buchanan R, Garvie N W, Ackery D M, Flemming J, Taylor D, Goddard B A 1981 Assessment of progression of secondary bone lesions following cancer of the breast or prostate using serial radionuclide imaging. British Journal of Radiology 54: 18–23

Cushing H 1932 Basophil adenomas of the pituitary body and their clinical manifestations (pituitary basophilism). Bulletin of the Johns Hopkins Hospital 50: 137–195

Destouet J M, Gilula L A, Murray W A 1979 Computed tomography of long-bone osteosarcoma Radiology 131: 439–445

Deuxchaisnes C N De, Krane S M 1964 Paget's disease of bone: clinical and metabolic observations. Medicine 43: 233–266

Doyle F H 1967 Radiologic assessment of endocrine effects on bone. Radiologic Clinics of North America 5: 289–302

Fogelman J, Carr D 1980 A comparison of bone scanning and radiology in the evaluation of patients with metabolic bone disease Clinical Radiology 31: 321–326

Fraser S A, Smith D A, Anderson J B, Wilson G M 1971 Osteoporosis and fractures following thyrotoxicosisis Lancet i: 981–983

Garner A, Ball J 1966 Quantitative observations on mineralised and non-mineralised bone in chronic renal azotaemia and intestinal malabsorption syndromes. Journal of Pathology and Bacteriology 91: 545–561

Gildray D L, Ash J M, Reilly B J 1977 Radionuclide skeletal survey for pediatric neoplasms. Radiology 123: 399–406

Greulich W W, Pyle S I 1959 Radiographic atlas of skeletal development of the hand and wrist, 2nd edn.

Guyer P B, Clough P W L 1978 Paget's disease of bone: some observations on the relation of the skeletal distribution to pathogenesis. Clinical Radiology 29: 421–426

Harmer C L, Burns J E, Sams A, Spittle M 1969 The value of Fluorine-18 for scanning bone tumours. Clinical Radiology 20: 204–212

Hosking D J, Chamberlain M J, Shortland-Webb W R 1975 Osteomalacia and carcinoma of prostate with major distribution of skeletal calcium. British Journal of Radiology 48: 451–456

Jones G 1969 Radiological appearances of disuse osteoporosis. Clinical Radiology 20: 345–353

Lavender J P, Evans I M A, Arnott R, Bowring S, Doyle F H, Joplin G F, MacIntyre I 1977 A comparison of radiography and radioisotope scanning in the detection of Paget's disease and in the assessment of response to human calcitonin. British Journal of Radiology 50: 243–250

Lentle B C, Russell A S, Heslip P G, Percy J S 1976 The scintigraphic findings in Paget's disease of bone. Clinical Radiology 27: 129–135

Looser E 1920 Über spatrachitis und Osteomalacie. Klinische, rontgenitogische und pathologischanatomische. Untersuchunger Deutsche Zeitschrift fur Chirurgie 152: 210–357

Lusted L B, Keats T E 1972 Atlas of roentgenographic measurements. 3rd edn. Year Book Medical Publishers, Chicago, p 171–172

McCormick J St C, Summerling M D, Aldrich J E, Langlands A O 1975 A review of strontium 87 m scintigraphy in the detection of skeletal metastases from mammary cancer. Clinical Radiology 26: 185–189

Morgan D B, Pulvertaft C N, Fourman P 1966 Effects of age on the loss of bone after gastric surgery. Lancet ii: 772–773

Murray R O 1960 Radiological bone changes in Cushing's syndrome and steroid therapy. British Journal of Radiology 33: 1–19

Nordin B E C 1960 Osteoporosis, osteomalacia and calcium deficiency. Clinical Orthopaedics 17: 253–258

Nordin B E C, Smith D A 1967 Pathogenesis and treatment of osteomalacia. In: Hioco D J ed L'Ostéomalacie. Masson, Paris

Omojola M F, Cockshott W P, Batty E G 1981 Osteoid Osteoma: An evaluation of diagnostic modalities. Clinical Radiology 32: 199–204

Paget Sir J 1877 On a form of chronic inflammation of bones (osteitis deformans). Medico-Chirurgical Transactions 60: 37–63

Pygot F 1957 Paget's disease of bone. The radiological incidence. Lancet i: 1170–1171

Pyle S I, Waterhouse A M, Greulich W W 1971 A radiographic standard of reference for the growing hand and wrist. Case Western Reserve University Press, Cleveland, Ohio

Ratzkowski I E, Frankel L M, Hockman A 1967 Bone metastases osteoporosis and radiation necrosis in breast cancer. Clinical Radiology 18: 146–153

Ritchie W G M, Whiney R J, Davison A M, Robson J S

1975 Periosteal new bone formation developing during haemodialysis for chronic renal failure. British Journal of Radiology 48: 656–661

Ross E J, Marshall-Jones P, Friedman M 1966 Cushing's syndrome: diagnostic criteria. Quarterly Journal of Medicine 35: 149–192

Salassa R M, Jowsey J, Arnaud C 1970 Hypophosphataemic osteomalacia associated with nonendocrine tumours. New England Journal of Medicine 283: 65–70

Simpson W, Ellis H A, Kerr D N S, McElroy M, McNay R A, Peart K N 1976 Bone disease in long-term haemodialysis: the association of radiological with histological abnormalities. British Journal of Radiology 49: 105–110

Steinback H L, Noetzli M 1964 Roentgen appearance of the skeleton in osteomalacia and rickets. American Journal of Roentgenology 91: 955–971

Tan B C, Oon C L 1967 Bone metastases in carcinoma of the nasopharynx. Clinical Radiology 18: 69–73

Tanner J M, Whitehouse R H, Healy J B 1962 Standards of skeletal ages. (International Children's Centre, Paris)

Voegel E, Fuch, W A 1976 Arteriography in bone tumours. British Journal of Radiology 49: 407–415

Young M M, Nordin B E C 1967 Effects of natural and artificial menopause on plasma and urinary calcium and phosphorus. Lancet ii: 118–120

**8**

# Radiography of the heart and great vessels

Radiographic examinations of the heart necessitate a good knowledge of anatomy by the cardiac radiographer who is required to produce appropriate projections to illustrate various abnormalities. Figure 8.1 has proved to be an invaluable source of information both to the inexperienced radiographer and cardiologist alike.

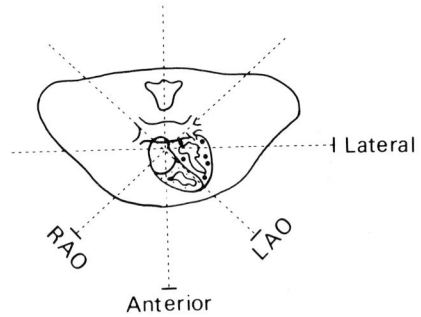

A

B

**Fig. 8.1** The four cavities of the heart.

Figure 8.1A illustrates the appropriate projection to demonstrate a ventricular or atrial septal defect. Figure 8.1B illustrates the position of the valves and chambers of the heart. These illustrations are used to optimise projections to demonstrate valve incompetence or abnormalities such as persistent ductus arteriosus (PDA).

A diagram depicting the position of the coronary arteries in the left and right oblique projections is also useful especially for illustrating the individual branches of the left coronary artery (see Fig. 8.2A).

The equipment which produces the extremely complex oblique angulations necessary for coronary

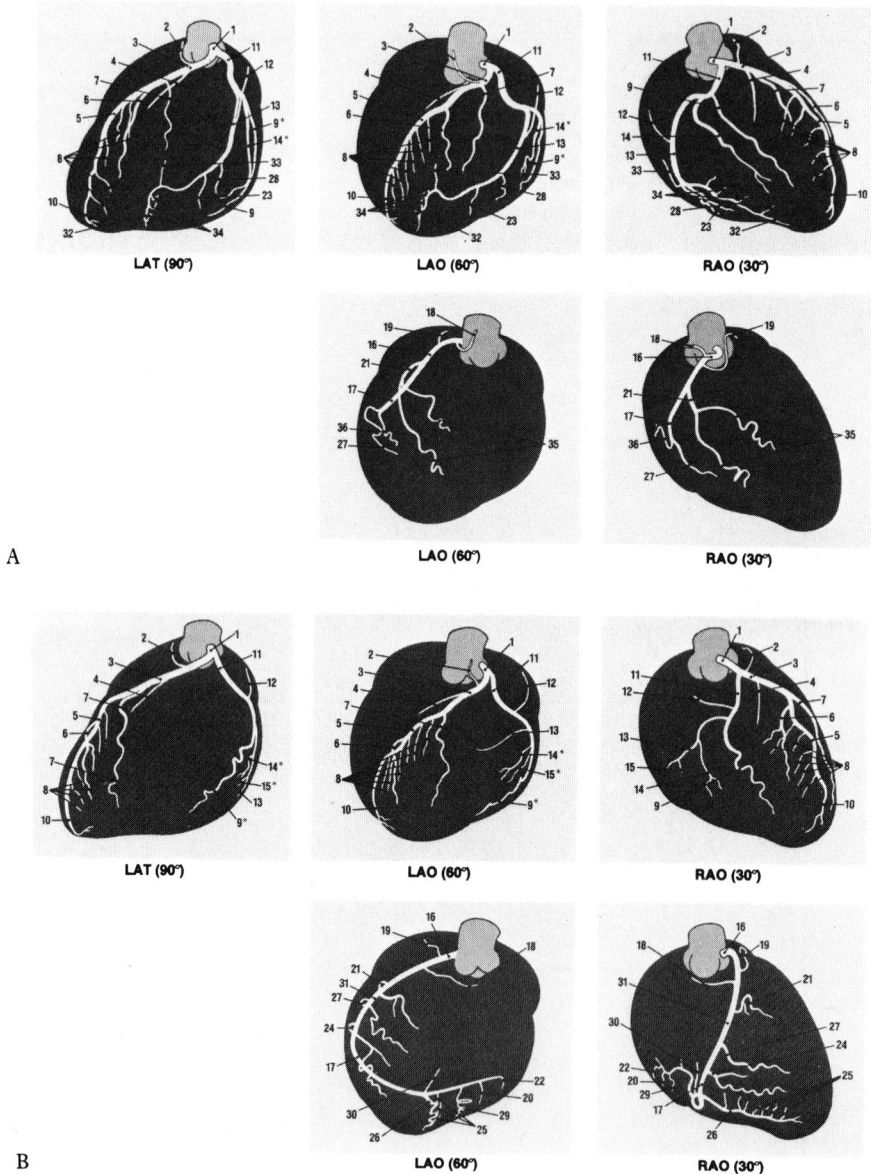

**Fig. 8.2** Position of coronary arteries (A) left dominance (B) right dominance. Key: 1 — left main stem; 3, 10 — left anterior descending branch; 9 — obtuse marginal branch; 11, 13 — circumflex branch; 16, 17 — right coronary. (Reproduced by courtesy of Kodak Ltd.)

angiography has been developed over the last few years but, before this, simple fixed angiographic equipment was used and is still used to a limited degree is some cardiac departments.

## EQUIPMENT FOR CARDIAC ANGIOGRAPHY

### A.  Large film techniques

For many years Elema-Schonander (AOT) cut film changers and then roll film changers were the mainstay of cardiac angiography. Their equal disadvantage is that they are fixed antero-posterior and lateral changers and should oblique projections be necessary the patient must either be rotated on radiolucent foam pads or possibly in a cradle device which may be fitted to the X-ray table.

The generators and X-ray tubes for these changers need not be as sophisticated or high powered as is necessary for fast ciné angiography.

### Generator and X-ray tube requirements

#### 1. Cut Film Changer

Generator:  750 mA, 75–80 kW and switching capability matched to speed of changer.
X-ray tube: High Speed, maximum — 50 kW broad focus, 30 kW fine focus.

#### 2. Roll film changer

Generator: As for cut film changer with faster switching capability.
X-ray tube: High Speed, maximum 100 kW broad focus, 50 kW fine focus.

It is preferable to have a cross-hatch grid for the antero-posterior changer to reduce the scattered radiation caused by the greater exposure used on the lateral changer. These fast speed changers have now, in general, been superseded by the use of ciné radiography or possibly photofluorography using a 70 mm or 100 mm camera. The large film techniques now involve the slower, 2 or 3 films per second, Puck Changer. These are proving very successful for the examination of such conditions as dissection of the aorta and pulmonary embolus, where detail of the anatomy of the structure is of

**Table 8.1** The advantages and disadvantages of the two changers

| Cut film | Roll film |
|---|---|
| 1. Radiographs produced are easy to view especially away from the X-ray department | Requires special viewing facilities |
| 2. Maximum speed 6 frames/second | Maximum speed 12 frames/second (This can be advantageous for paediatrics.) |
| 3. Radiographs are easy to process especially when the processor is used by other X-ray rooms | Roll film takes over the processor for approximately 20 minutes (can be developed at specified times of the day) |
| 4. Control films can easily be taken | Only *comparative* control films on grid cassettes can be taken, ensuring that the types of scattered radiation grids are matched with those in the changer |
| 5. Loading of film magazines can easily be done in a darkroom. The receiving box containing the exposed radiographs is easily transported to the processing darkroom | Loading and unloading of roll film can only be done in darkroom facilities or by blacking out the X-ray room itself. |

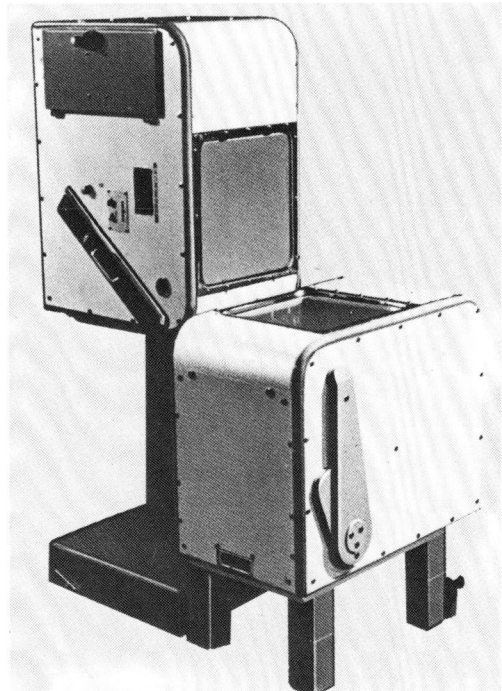

**Fig. 8.3** Cut film changer.

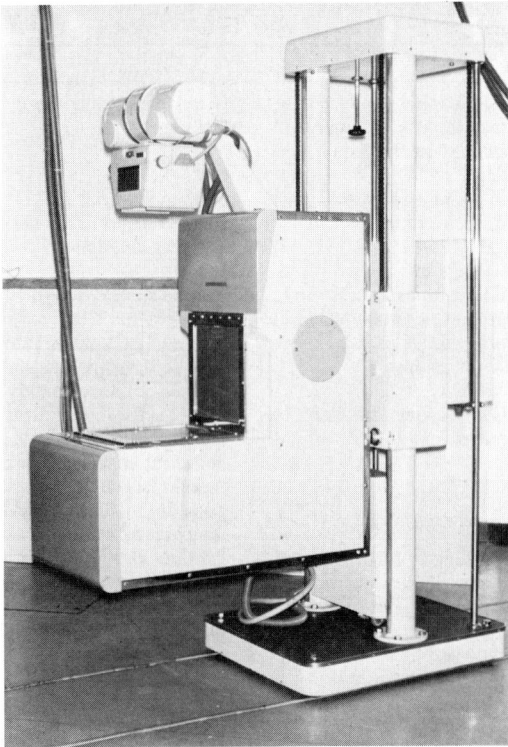

**Fig. 8.4** Roll film changer.

prime importance, rather than speed to show anatomical movement. Although most radiography of the heart can now be devoted to ciné angiography it is necessary to provide facilities for large film techniques.

At the Leeds General Infirmary, en suite to the bi-plane ciné angiography room, a single plane room is sited. This includes a 'U'-arm apparatus on which is mounted an image intensifier which can be motor driven downwards to be replaced by a 35 cm square Puck-'U' film changer. This provides the opportunity to have 'in-line' fluoroscopy and radiography, together with the ease of producing oblique projections without having to move the patient. The equipment is particularly invaluable in the examination of the critically ill who may require such examinations as thoracic aortography to demonstrate a dissection of the aorta.

Puck film changers can also be mounted as separate units similar to that of the cut or roll film changers. They are either wheeled into position, or the table is appropriately positioned when angiography is required.

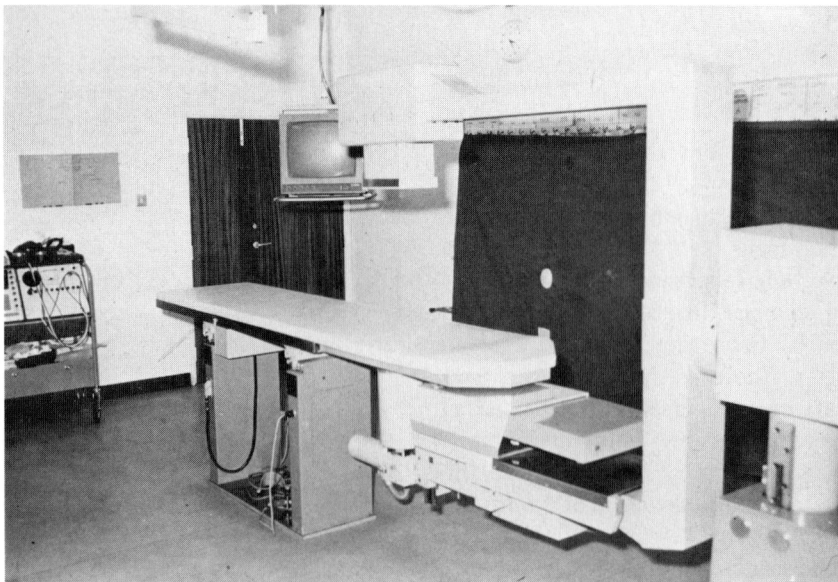

**Fig. 8.5** Antero-posterior fluoroscopy.

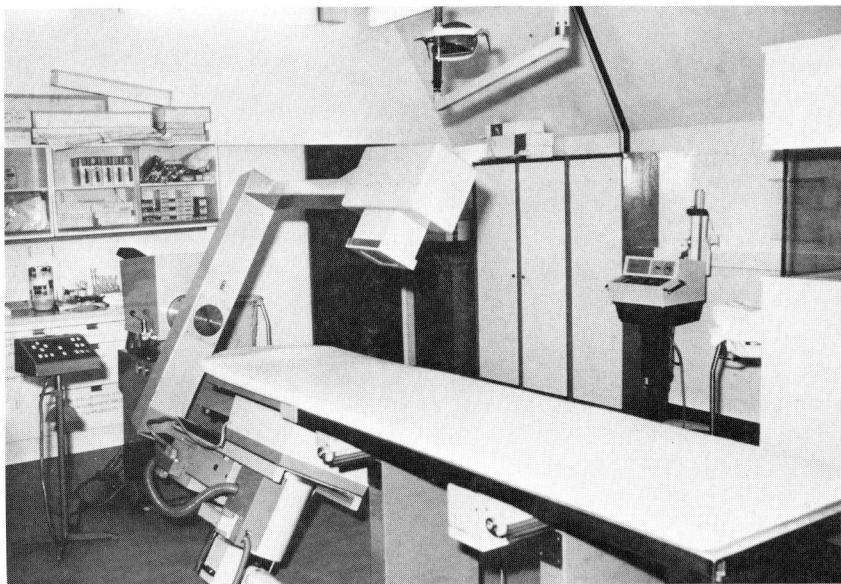

**Fig. 8.6** Oblique radiography.

## B. Ciné angiography

This is by far the most commonly used method of documenting cardiac angiography at the present time using frame speeds in the order of 30–50 per second in adults and 75–100 per second in children — the faster speed provides better detail in the film at the increased heart rate of children. The ability to produce these high speed recordings together with the output necessary to produce magnification techniques in coronary angiography necessitates a high powered ciné generator and sophisticated electronic switching.

## Generator and X-ray tube requirements

### 1. Single-plane ciné

Generator: 125 kV, 400 mA.
X-ray tube: High Speed, maximum 100 kW broad focus, 50 kW fine focus.

### 2. Bi-plane ciné

Generator: Second system as above
X-ray tube: Second system as above.

### 3. Dedicated coronary angiography installation

Generator: 150 kV, switching in excess of 1000 mA.
X-ray tube: High speed and capability of high heat dissipation.

## C. Coronary angiography

The equipment that is needed to give adequate demonstration of the coronary arteries is necessarily more complicated than the fixed anteroposterior and lateral techniques of the heart described previously. Oblique projections are vital in coronary angiography and these views cannot be ideally produced by the patient constantly moving, since the catheter may be dislodged from the artery. This then necessitates either a 'U'-arm or a 'C'-arm piece of equipment, the latter being extremely flexible but is mechanically less stable. The 'U'-arm is more robust and may be less likely to produce movement blur during high speed cine angiography.

The 'U'-arm on its own, however, will only produce oblique projections across the chest wall but it is thought that lesions could be missed or not easily shown in a tortuous coronary artery,

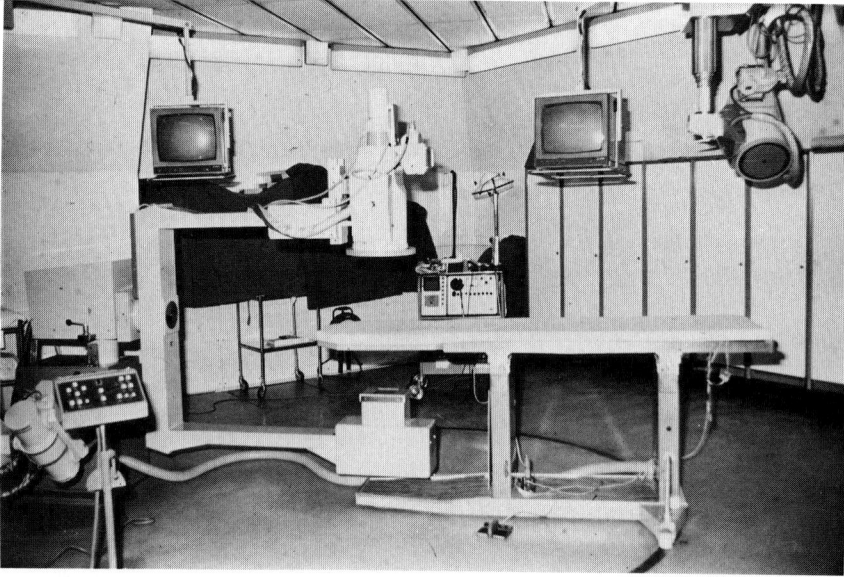

**Fig. 8.7** Philips U-arm.

**Figs 8.7–8.9** Typical coronary angiography installations.

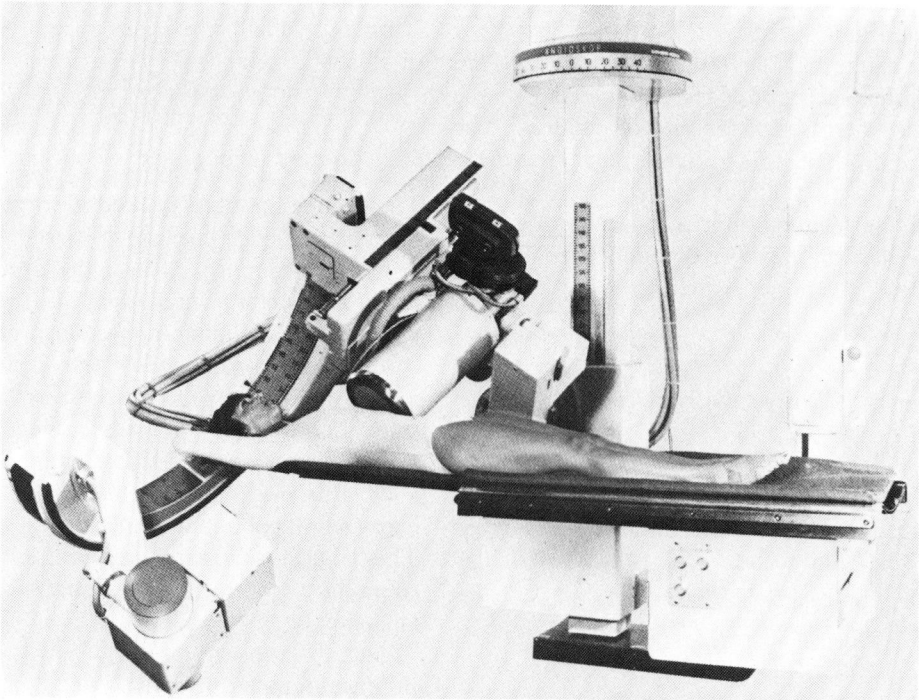

**Fig. 8.8** C-arm model (by courtesy of Siemens).

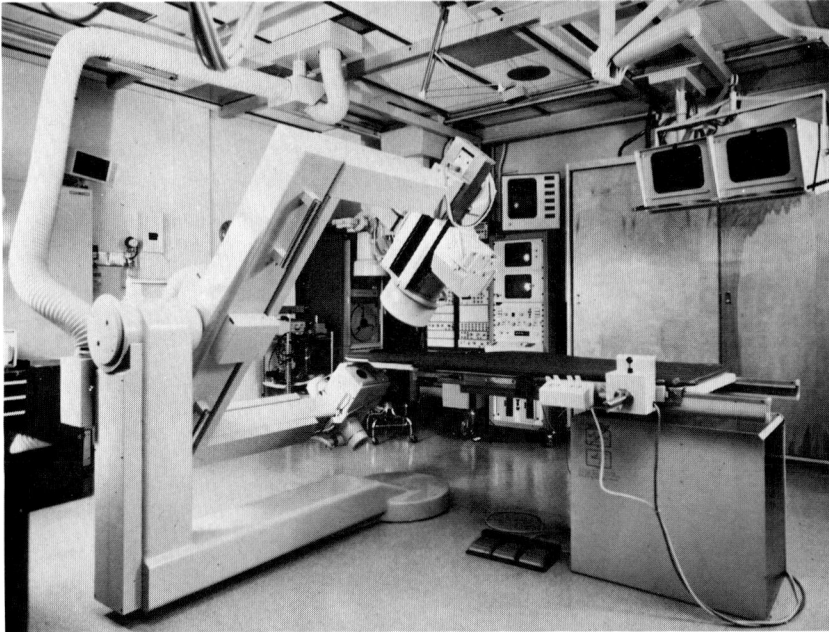

Fig. 8.9 U-arm model (by courtesy of IGE(NY)).

therefore it is necessary to try to view obliquely above and below the heart in addition.

The parallelogram allows these projections to be carried out with ease. If a parallelogram facility is not available, then similar facilities can be obtained by using a 'U'-arm and rotating the arm (note IGE equipment — Fig. 8.9) or rotating the table, on its base, around in a clockwise and anti-clockwise direction, from the neutral position.

These views, at first, may be difficult to interpret and so set angulations may be established as a routine. Additional views, however, may be obtained if a particular area is not clearly demonstrated in the routine projections. It is most important with equipment of this versatility, especially the Poly DIAGNOST C, that it is used to its fullest capabilities and projections altered according to the anatomy of the patient (see Fig. 8.10).

A bi-plane facility is of great value especially in congenital heart disease or the examinations of left ventricular aneurysms, the extent of which could be obscured in only one projection. Ideally, these planes should be at right angles to each other but to achieve this the 'U'-arm can give the right anterior oblique (RAO) projection and the second

channel if possible should be angled to produce a left anterior oblique (LAO) projection. If this is not possible, then a right anterior oblique together with a fixed lateral view will give extra detail. The ultimate versatility for bi-plane angiography has recently been marketed by Philips Medical Systems for coronary angiography incorporating the Poly DIAGNOST C and Lateral ARC, and called the 'LARC' (Figs 8.10–8.16).

Bi-plane angiography reduces the number of injections of contrast medium necessary, if one has been left in any doubt with the interpretation of only one projection. It does, however, necessitate either a second generator or a very sophisticated generator capable of coping with a bi-plane facility itself. The dangers of multiple high dose injections of contrast medium to the patient does warrant the extra expense of a bi-plane facility in a busy cardiac centre, more especially in the examination of children where the number of injections of contrast medium should be kept to a minimum. The speed of any cardiac catheterization is of paramount importance and if bi-plane angiography can significantly shorten the examination time this is all to the patient's benefit.

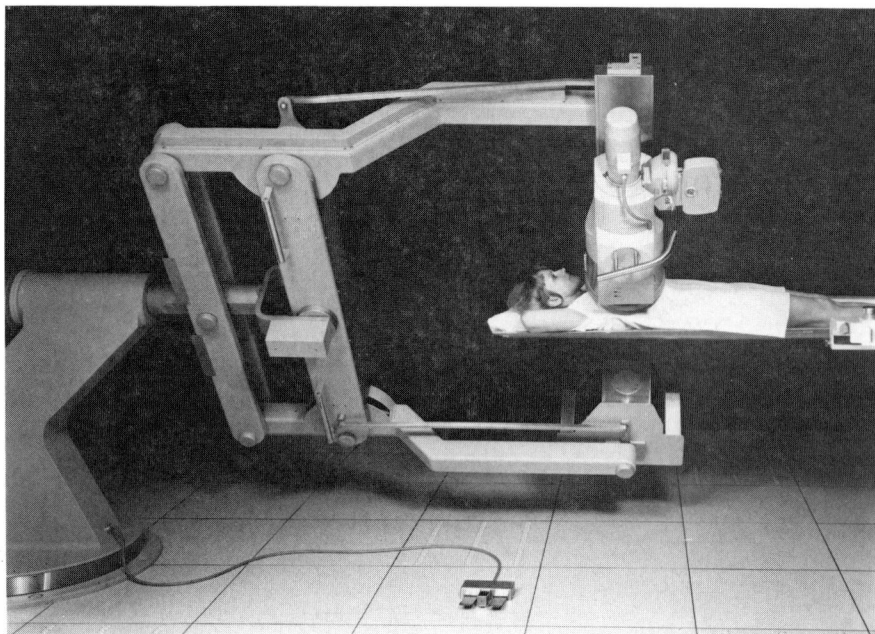

**Fig. 8.10** Poly DIAGNOST C (by courtesy of Philips Medical Systems).

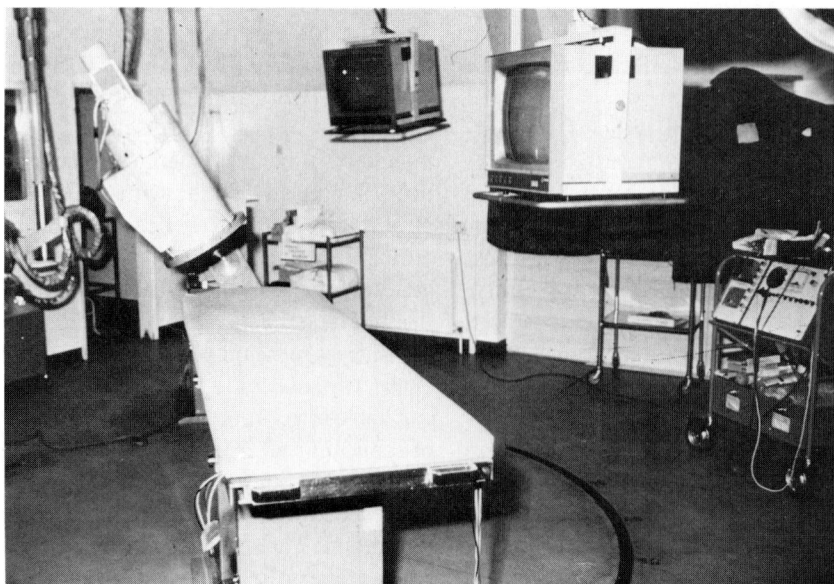

**Fig. 8.11** Table neutral, U-arm 45° right anterior oblique (RAO).

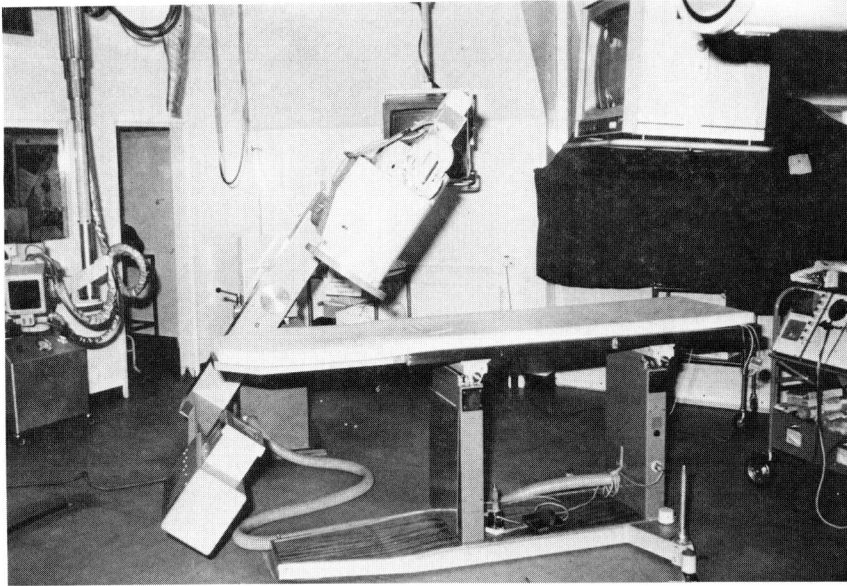

**Fig. 8.12** Table 90°, 30° caudal view.

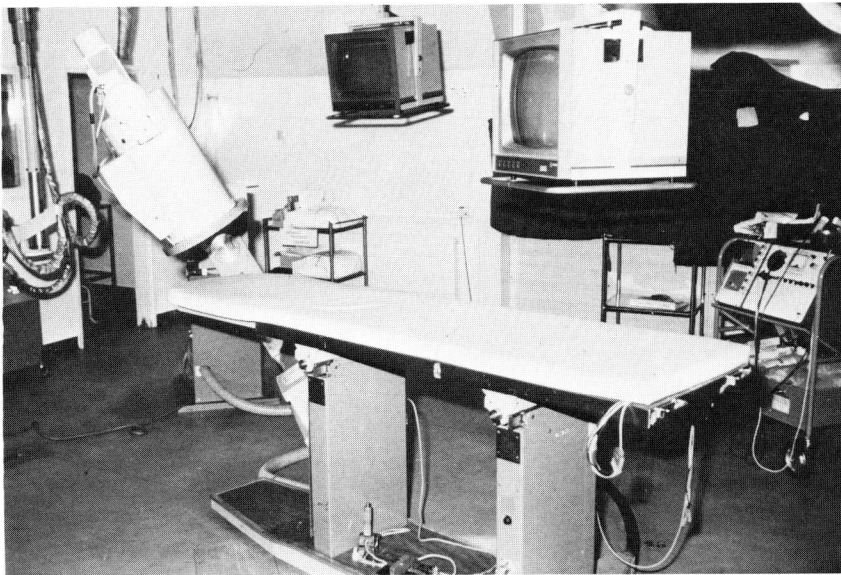

**Fig. 8.13** Table 45° anti-clockwise, 30° RAO cranial view.

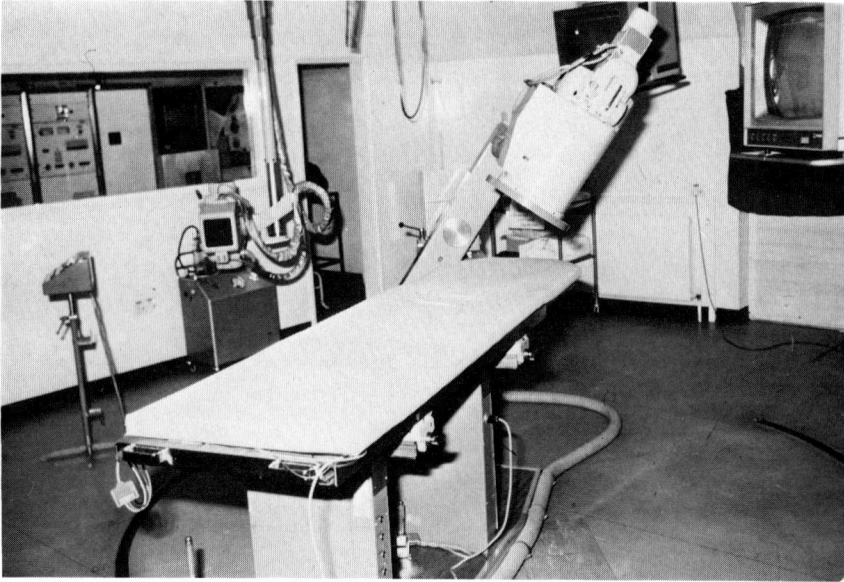

**Fig. 8.14** Table 45° clockwise, 30° left anterior oblique (LAO) cranial view.

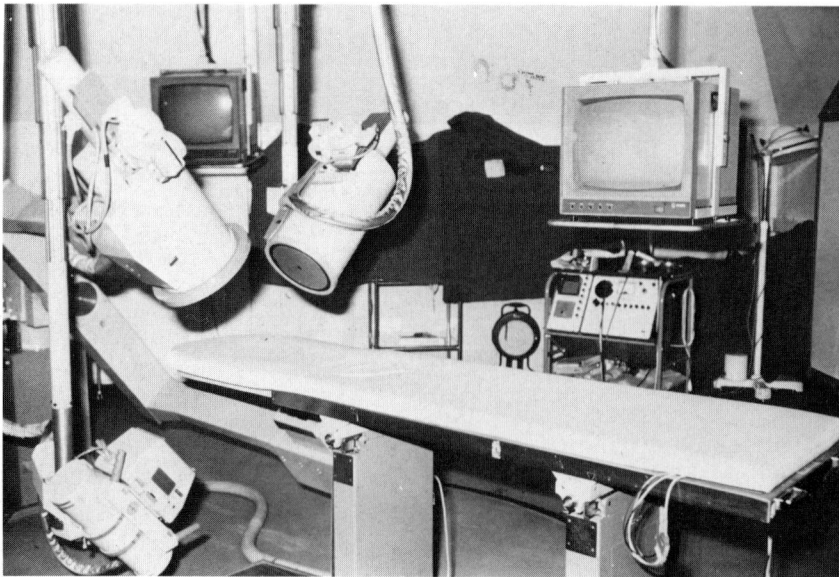

**Fig. 8.15A** Right anterior oblique and left anterior oblique.

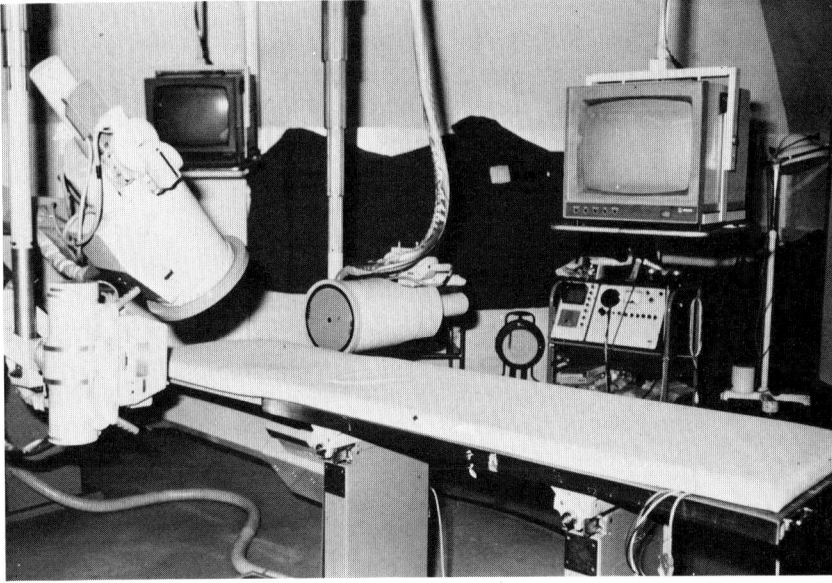

**Fig. 8.15B** Right anterior oblique and left lateral.

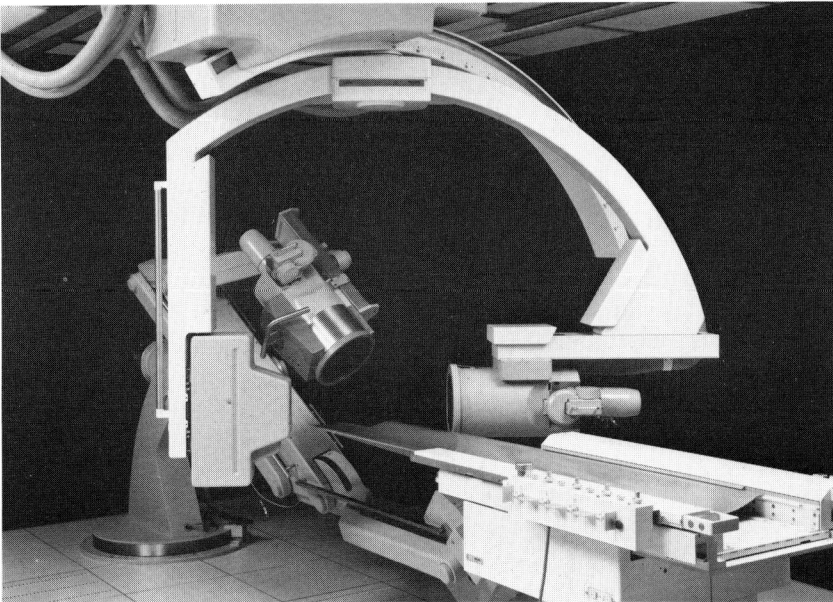

**Fig. 8.16** 'LARC' (by courtesy of Philips Medical Systems).

To complement all these facilities the equipment must include a good quality image intensifier, usually providing an enlargement facility, such as 10"/6" or 9"/5". The larger image makes catheterization of the chambers of the heart much easier, for a full cardiac outline can be seen. In addition, the enlarged image gives good visualization of the coronary arteries. The imaging system must be correctly 'set-up' both for the visualization on the television monitor and for the ciné film.

The television picture quality must be adequate for performing such procedures as an atrial transseptal puncture and for demonstrating small amounts of calcification in the heart or its valves. As the heart is a fast moving object it is best to have a plumbicon tube insert for the television camera as this has faster resolution, whereas a videcon tube is more sluggish. The videcon tube can have an advantage in other areas of the body as the noise level is lower and possibly more acceptable to some who find 'noisy' pictures difficult to view.

A useful piece of ancillary imaging equipment is a video-tape recorder. This is generally used as a back-up facility, for instant recall during the examination. Rarely would it be used solely as a diagnostic tool as the quality is inferior to that of the ciné film. Two video-tape machines are useful where there is a bi-plane facility which is in constant use.

## Ancillary equipment essential in a cardiac catheterization room

All cardiac catheterization rooms must have equipment to monitor the patient's electrocardiogram (ECG) and pressures within the cardiovascular system. Ideally, the tracings should be presented on a monitor alongside the X-ray monitor and some are colour coded to make each tracing easily and quickly recognisable throughout the catheter procedure. In Leeds the ECG and pressure tracings are displayed on the monitor together with the radiographic image. The bright white trace is constantly updated in a 'slipping' trace so that at no time is there an absence of a complex as found in many forward running traces of an oscilloscope. It is also under constant observation by the operator performing the catheterization so that any instability in either cardiac rhythm or pressure value is quickly noticed by the operator as well as by the physiological measurement technician. This same facility also provides an ECG trace on the ciné film, which has been useful if doubt exists as to the presence of ectopic heart beats at a particular moment during an injection of contrast medium.

An important machine, that one hopes will never or hardly ever be used, but is nevertheless very necessary is a defibrillator. The heart can react in a very adverse way when catheters are being manipulated or doses of contrast medium are being injected under pressure. Some reactions may be severe bradycardia or even asystole or ventricular fibrillation.

Ventricular fibrillation may revert to sinus rhythm by simply withdrawing the catheter from the heart, if not it may be reverted by shock using external cardiac massage or the defibrillator. Bradycardia and asystole will need temporary pacemaking facilities to be on hand to induce the heart to beat electrically. A pacemaker inserted under these conditions can generally be removed at the end of the examination if the patient's heart rate has returned to normal sinus rhythm. In some patients a temporary pacemaker may be inserted at the beginning of the procedure if it is felt clinically advisable to do so.

Finally, a piece of equipment capable of measuring the patient's cardiac output is useful where valvular heart disease patients are examined. This is now generally carried out using special catheters passed, via the ante-cubital vein or the femoral vein through the right heart chambers, to the pulmonary artery and then connected to a cardiac output computer. Attempts are being made in Leeds to produce a machine to calculate a patient's cardiac output without necessitating cardiac catheterization; this would be especially valuable in the critically ill patient where catheterization is clinically inadvisable.

*Summary of equipment for coronary angiography*

1. A high powered generator capable of producing high kV and very short exposure times.

2. A mains synchronised ciné-pulse unit.
3. A high speed rotating anode X-ray tube with as small a focal spot as is compatible with the work to be undertaken and the generator capabilities.
4. A dual mode image intensifier e.g. 9"/5" with a 35 mm ciné camera.
5. A 'U'-arm, 'C'-arm or Parallelogram to achieve oblique projections, and a table (or 'U'-arm) on a rotating base to achieve cranio-caudal projections.
6. Bi-plane film exposure facilities if possible.
7. Large film technique facilities depending on requirements.
8. Possibly a 70 mm or 100 mm camera facility but this is not commonly used in cardiac work.
9. Good image quality both on television monitor and photographic film.
10. A processor specially designed for ciné film; to allow flexibility in development time and temperature, for use with different types of film or alterations in X-ray equipment.
11. A film of medium contrast should be used for coronary angiography as detail in fine vessels is lost at high contrasts.
12. Essential ancillary equipment:
    ECG and pressure monitoring equipment
    Video-tape recorder
    Cardiac output measurement facility
    Defibrillator and emergency resuscitation equipment and drugs
    Pacemaker equipment
    High Pressure Pump for contrast medium injection.

### Preparation of the patient

Most cardiac catheterizations are performed as inpatient procedures. In the case of neonates cardiac catheterization is usually performed under general anaesthetic as an emergency procedure prior to immediate surgery to correct congenital heart defects. Older children may be sedated for an examination carried out to assess a known or recently diagnosed congenital heart defect. Adults presenting for cardiac catheterization have generally undergone exhaustive treatment with drugs, but in some patients the point is reached where surgery becomes necessary. In these patients the extent of the patient's condition is generally examined angiographically and possibly haemodynamically.

Prior to an invasive examination of this nature non-invasive techniques may be employed to provide clinical information. These may be echocardiography and electrocardiography at rest and possibly after exercise. Postero-anterior and left lateral chest radiographs should also be taken. If valve calcification is suspected, fluoroscopy of the heart may be performed before catheterization if this is clinically desirable. The information from these examinations and tests should accompany the patient to the catheterization room.

A mild premedication may be given to the patient, if desired, but nothing so strong as to make the patient unresponsive, since the patient's co-operation is required during angiography for such exercises as holding the breath or coughing in order to clear contrast medium from a coronary artery. A test dose of the contrast medium to be used should be administered, preferably on the ward the day before the examination to make sure there are no adverse reactions to the medium. If there are adverse reactions then an alternative contrast agent test dose should be tried 24 hours later. If the patient is receiving anti-coagulation therapy it is advisable to check the blood clotting (prothrombin) time and an adjustment made in the dosage of this drug, if necessary, to avoid haemostasis problems in the femoral artery after catheterization. Also if the patient is on a high dose of beta-blockade drugs the dosage of these may need to be reduced prior to coronary angiography, but this may depend upon clinical assessment.

### Contrast medium

The most commonly used contrast media are Cardio-conray, Urografin 370 or Triosil. If very adverse reactions occur to these contrast agents then a non-ionic contrast medium (e.g. Hexabrix or Iopamidol) can be used, as it is non-irritant. It is costly for routine use but can be used in patients who suffer severe bradycardia during coronary angiography.

## REASONS FOR CARDIAC CATHETERIZATION

### 1. Congenital heart disease

There are a number of congenital heart defects which require immediate surgery so that emergency catheterization and angiography are requested. Other defects are not detected until much later in life and will possibly require no surgical correction if the patient is asymptomatic.

The following are examples of the type of congenital abnormalities that may be encountered in a cardiac suite, together with the angiographic projections which are necessary to demonstrate the abnormalities.

*Fallot's tetralogy*

This consists of four major defects (see Figs. 8.17 & 8.18):

a. pulmonary stenosis
b. right ventricular hypertrophy
c. large ventricular septal defect
d. over-riding aorta.

These patients often return in later years for re-catheterization and it is most important that comparable radiographs are taken in the same projections, even if more complex views are obtainable. It would be more difficult to assess surgical repair if the radiographs were not comparable.

*Persistent ductus arteriosus* (PDA)

A communication between the pulmonary artery and the upper thoracic descending aorta (see Figs. 8.19 & 8.20).

*Transposition of great vessels*

In this condition the pulmonary artery arises from the left ventricle with some communication such as a ventricular septal defect or a patent ductus arteriosus to ensure some mixing of blood, to allow partially saturated blood to flow round the body. Correction by surgery is usually carried out in late childhood. Bi-plane ciné angiography is generally performed but the abnormalities are difficult to demonstrate on a single frame and are more easily seen when the ciné film is viewed at normal speed.

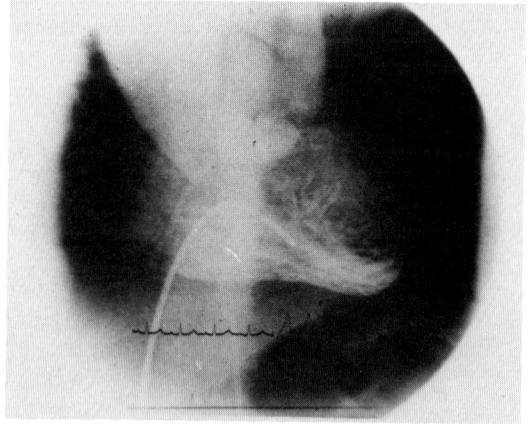

Fig. 8.17 Postero-anterior view.

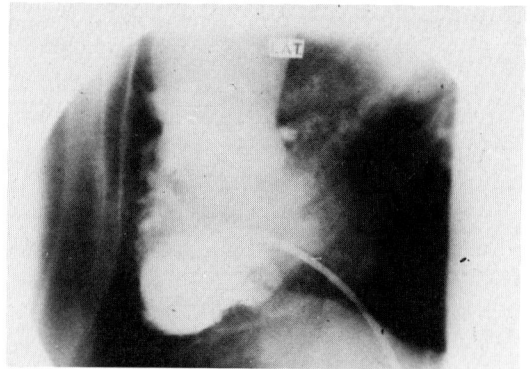

Fig. 8.18 Lateral view.

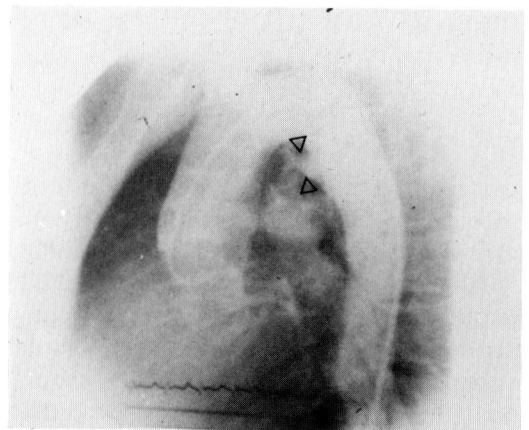

Fig. 8.19 Lateral view.

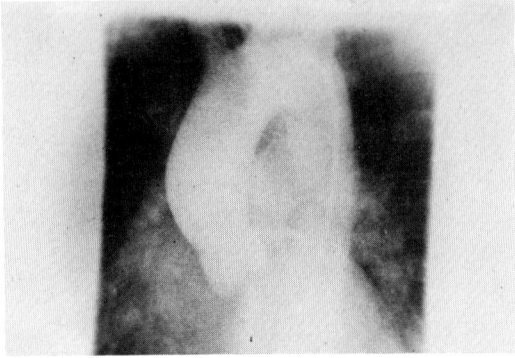

**Fig. 8.20** Left anterior oblique view.

## Coarctation of the aorta

The aorta is severely narrowed usually just beyond the origin of the left subclavian artery with possibly a bi-cuspid aortic valve instead of one with three cusps. These patients may not present until later life usually with heart failure or high blood pressure in the arms and rib notching is usually seen on the chest radiograph (Fig. 8.21).

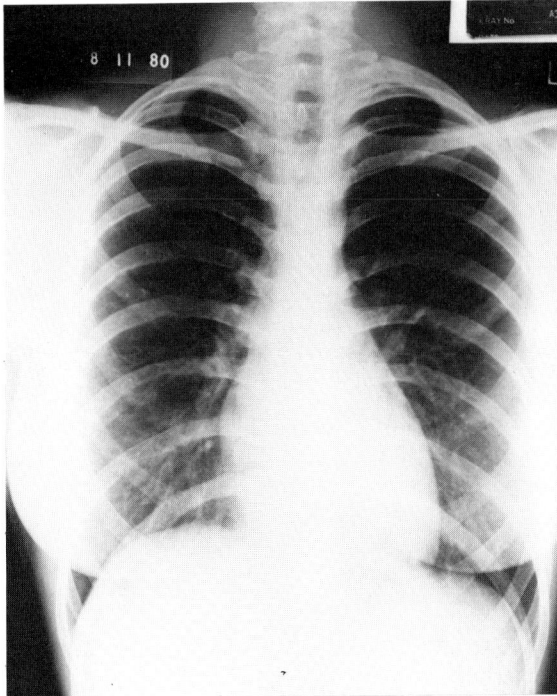

**Fig. 8.21** Postero-anterior chest radiograph.

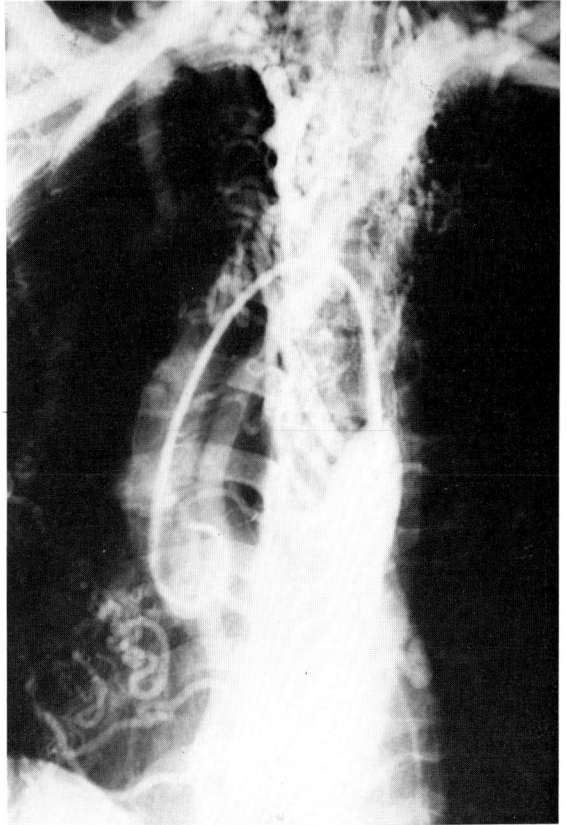

**Fig. 8.22** Right posterior oblique angiogram.

This is due to the expansion of the intercostal arteries being fed by a massive collateral supply to feed the aorta below the narrowing. This examination is generally carried out using large films rather than by ciné radiography.

## Atrial septal defect

This condition involves a communication between the right and left atria either an 'ostium primum' (an opening close to the valves) or an 'ostium secundum' (an opening away from the valves). Angiography is not always performed in this condition as diagnosis can be made on haemodynamic oxygen saturations only. When angiography is performed the detail is again difficult to demonstrate on a single frame but the contrast medium injected in the pulmonary artery can be detected passing from the left atrium to the right atrium when the ciné film is viewed at normal speed.

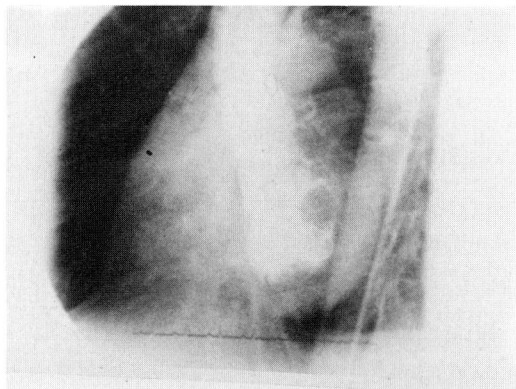

**Fig. 8.23** Left anterior oblique view.

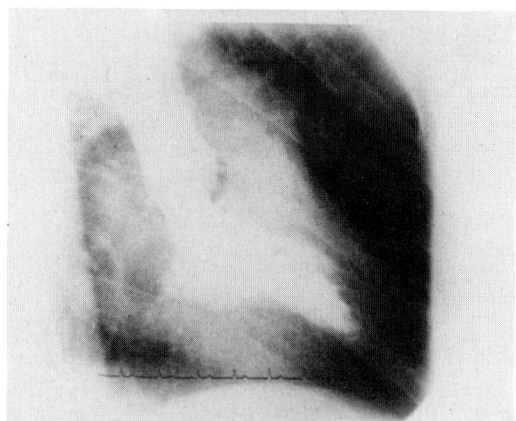

**Fig. 8.24** Right anterior oblique view.

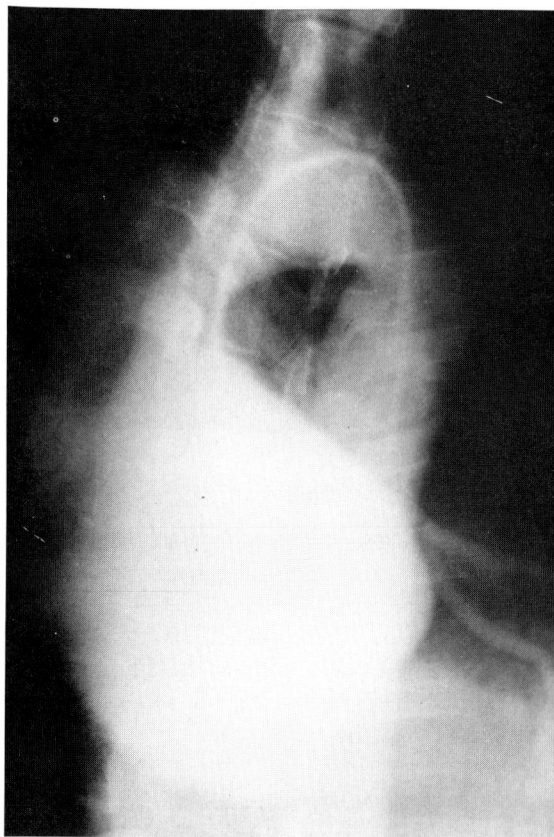

**Fig. 8.25** Right posterior oblique view. The sub-optimal appearance of the radiograph is due to the dilution of the contrast medium by the large aortic blood volume.

*Ventricular septal defect*

This defect concerns a communication between the right and left ventricles. The high pressure in the left ventricle causes a shunt of blood into the right ventricle and generally needs surgical correction whereas atrial septal defects may be relatively asymptomatic (see Figs 8.23 & 8.24).

*Patent foramen ovale*

This is necessary in fetal life to connect the left and right sides of the heart (as the fetal lungs are not in use) and it generally closes shortly after birth. However, in some patients it is possible to pass a catheter from the right atrium to the left atrium, through a foramen ovale which has not closed, or

is patent. Oxygen saturations may be normal signifying that no mixing of blood between the two chambers is taking place.

*Anomalous pulmonary venous drainage*

In this condition the pulmonary veins drain into the right atrium instead of the left atrium. If all four veins drain into the right atrium, then an atrial septal defect is essential for life.

*Pulmonary stenosis*

The stenosis if severe will cause right ventricular hypertrophy and post-stenotic dilatation of the pulmonary artery with a very high right ventricular pressure.

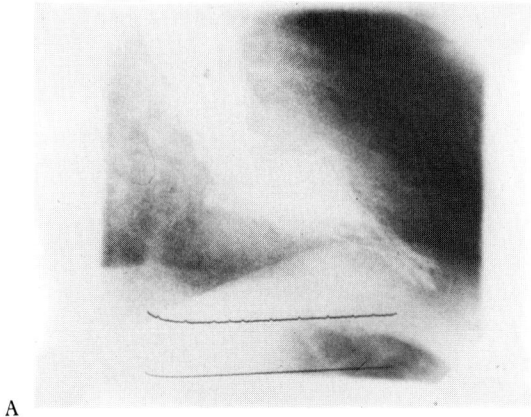

A

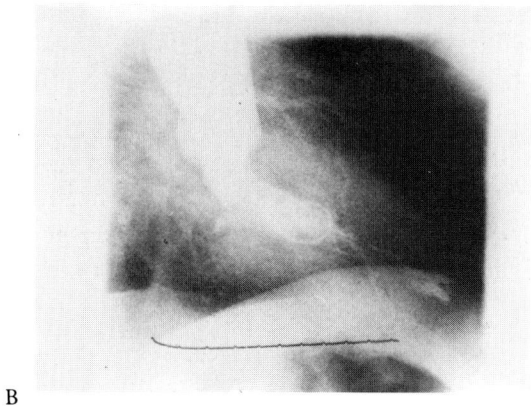

B

**Fig. 8.26** Right anterior oblique view. (A) Diastole. (B) Systole. The contrast medium remains in the apex of the left ventricle during systole.

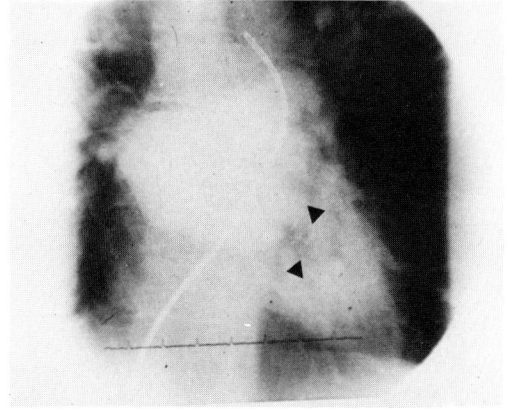

A

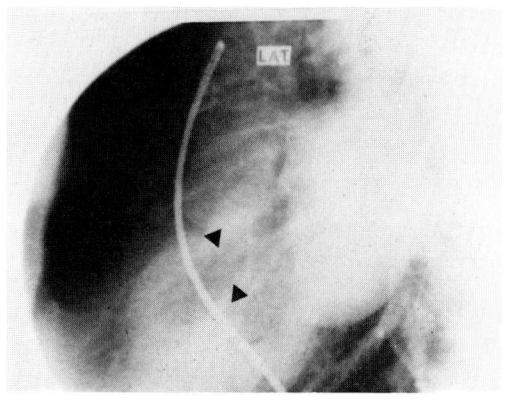

B

**Fig. 8.27** (A) Postero-anterior view. (B) Lateral view. Note the negative shadow of the myxoma in the contrast medium.

*Marfan's syndrome*

This is a congenital abnormality affecting the aorta. It normally runs in families (familial) and patients have abnormally long fingers and limbs. The spread of the limbs is proportionally greater than the patient's height. The patients also have dislocated lenses of the eyes, and abnormally high palate and poor muscular tone. Gross aortic dilatation occurs possibly leading to an aortic dissecting aneurysm. This may cause sudden death before investigation and a surgical repair can be performed. This examination is again normally carried out on large films (Fig. 8.25).

## 2. Acquired cardiac anomalies

*Hypertrophic obstructive cardiomyopathy* (HOCM)

This is constriction of the heart muscle usually at the level of the inter-ventricular septum causing a band of constriction and so creating a pressure difference across the left ventricle (see Figs 8.26A & B).

*Atrial myxoma*

A benign tumour usually in the left atrium. The patient usually presents with symptoms similar to mitral stenosis or infective endocarditis (see Figs 8.27A and B).

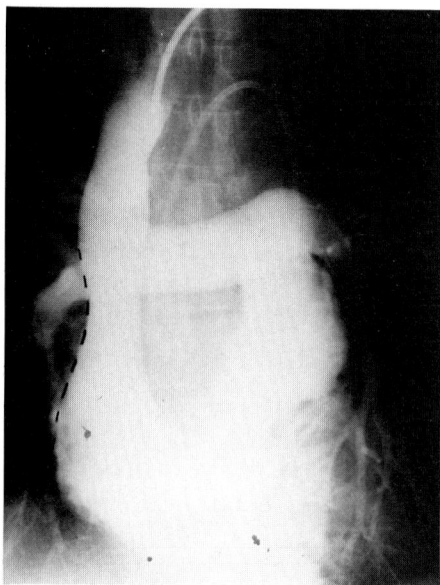

Fig. 8.28 Antero-posterior view. Contrast medium does not extend to edge of the right cardiac border.

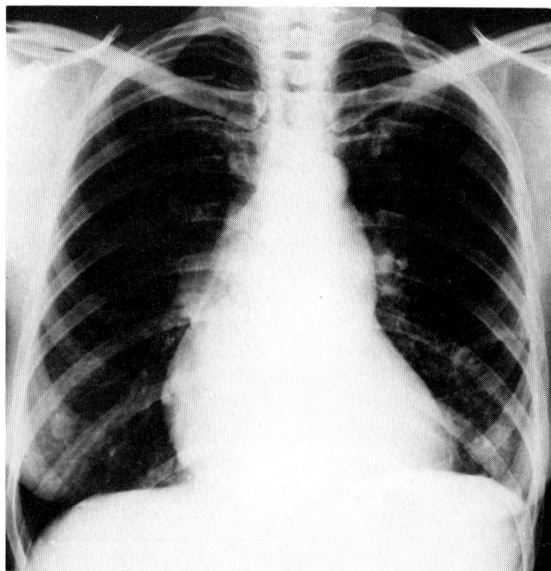

Fig. 8.29 Postero-anterior chest radiograph demonstrating classical enlarged cardiac outline with enlargement of the left atrium.

### Pericardial mass

In this condition the chest radiograph will show an enlarged right cardiac border and if angiography is performed a large mass in the right atrium will be demonstrated (see Fig. 8.28).

### 3. Valvular disease

The commonest valve to be affected by disease is the mitral valve in women and the aortic valve in men.

### Mitral valve disease

To examine the mitral valve, the left ventricle is catheterized via the aorta and a right heart catheter is positioned via the right atrium and right ventricle into the pulmonary artery where it is wedged in a distal vessel to record a pressure approximating to that of the left atrium. The difference, if any, between the wedge pressure and the end-diastolic left ventricular (LV-EDP) pressure gives an accurate measurement of the gradient across the mitral valve. If there is severe mitral stenosis then the pressure in the left atrium has to increase, to overcome marked mitral valve narrowing, in order to expel the volume of blood in the atrium. This increase in pressure is associated with dilatation (and thickening) of the left atrial cavity.

As mitral stenosis increases so increasing left atrial pressure, a backward reaction of increased pressures occurs to the right heart vessels, in order to keep a forward flow of blood (see Fig. 8.29). The stenotic valve can also be calcified, probably as a result of distorted pattern of blood flow (see Fig. 8.30). Left atrial enlargement can also occur in mitral incompetence (Fig. 8.31), due to blood regurgitating from the left ventricle during systole.

After obtaining all the haemodynamic (pressure) measurements, angiography can be performed to show the heart in the right anterior oblique projection. As demonstrated in Figure 8.1, this projection outlines the mitral valve optimally and will demonstrate regurgitation into the left atrium. An injection of 40–50 ml of contrast medium is made into the left ventricle.

### Aortic valve disease

Catheter studies are performed to try to measure

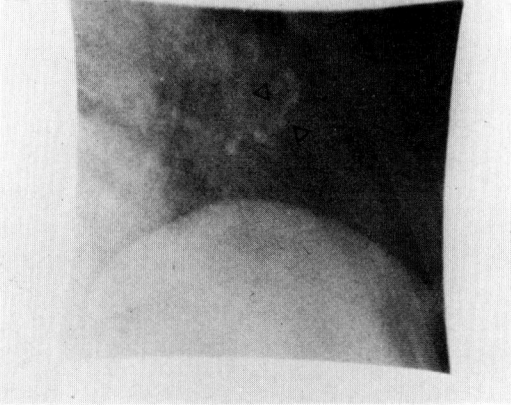

**Fig. 8.30** Calcified mitral valve. Right anterior oblique view.

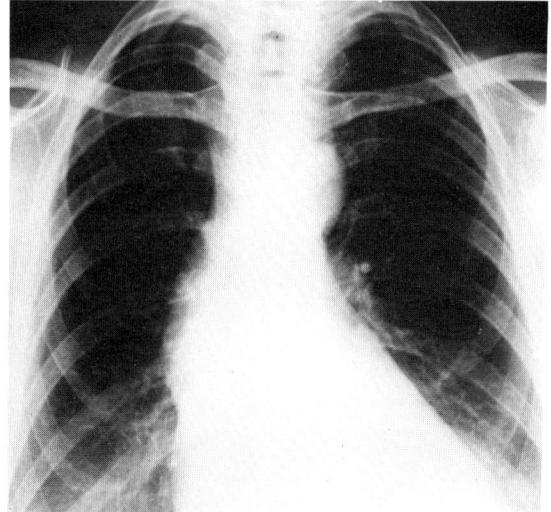

**Fig. 8.32** Chest radiograph showing aortic dilatation.

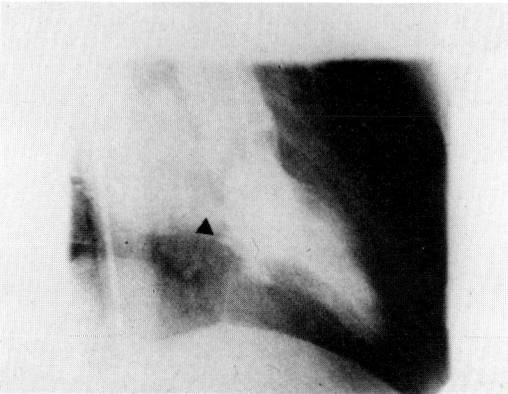

**Fig. 8.31** Left ventricular angiogram in right anterior oblique projection showing mitral regurgitation.

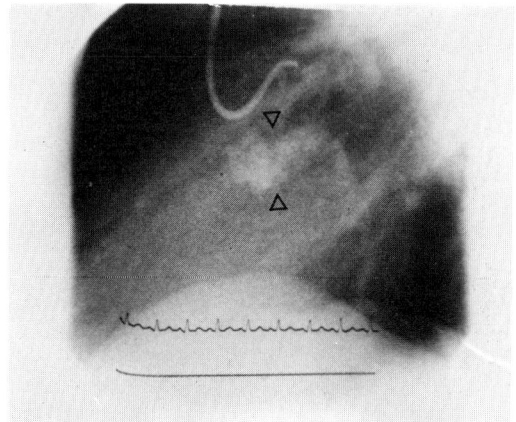

**Fig. 8.33** Calcified aortic valve, left lateral view.

pressures on either side of the aortic valve, i.e. in the aorta and the left ventricular cavity. In a case of severe aortic stenosis it may not be possible to cross the valve from the aorta. If this is so a left ventricular pressure can be obtained in two ways, one is by performing a trans-atrial septal puncture. The right femoral vein is catheterized and then a Brockenbrough needle is used passing the needle and its catheter across the septum from the right atrium into the left atrium. When the catheter is in the left atrium the needle is withdrawn and the catheter is passed through the mitral valve into the left ventricle. The second method of obtaining the pressure recording, involves direct left ventricular puncture through the chest wall, using a fine needle inserted into the left ventricle.

This method is used only for a pressure reading, not angiography.

The trans-atrial technique can also be used if the aortic valve is heavily calcified. In this case prolonged attempts to cross the valve may be hazardous because of the possibility of calcium deposits becoming dislodged and then finding their way into a coronary or cerebral artery, in either case producing most unfavourable effects. Left ventricular enlargement occurs when there is gross aortic regurgitation since blood leaks back across the valve during diastole, and also in stenosis, as the normal amount of blood cannot be expelled in each systolic contraction.

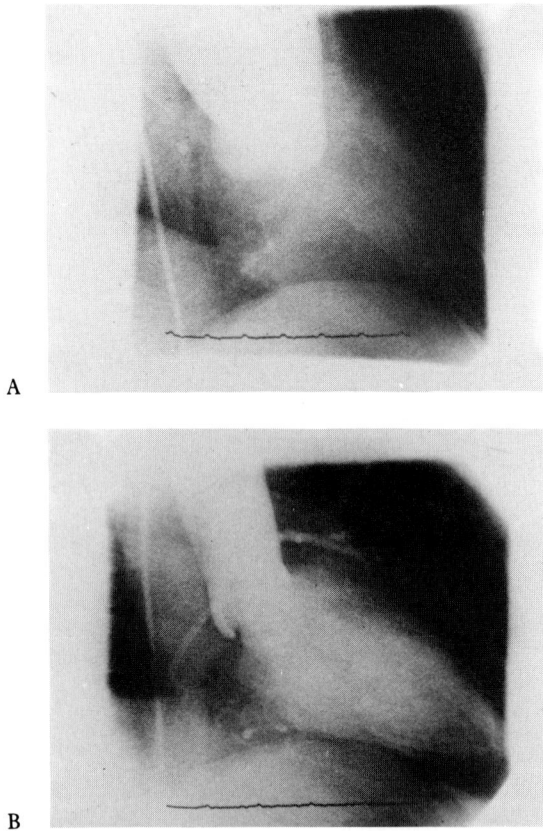

A

B

**Fig. 8.34** Right anterior oblique view, aortic root injecti· showing (A) post-stenotic dilatation of the aorta (B) left ventricle by aortic regurgitation.

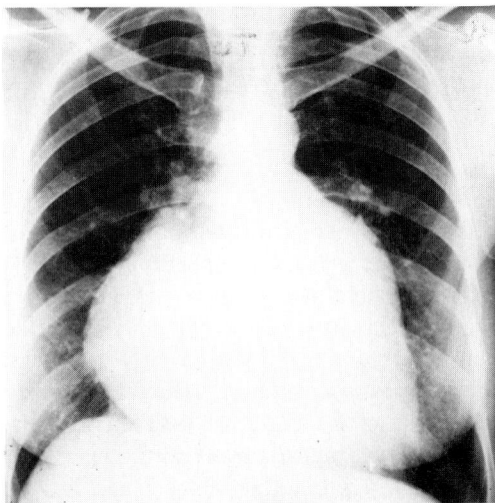

**Fig. 8.36** Antero-posterior view, right ventricular injection showing tricuspid regurgitation; the contrast medium in the right atrium is outlined.

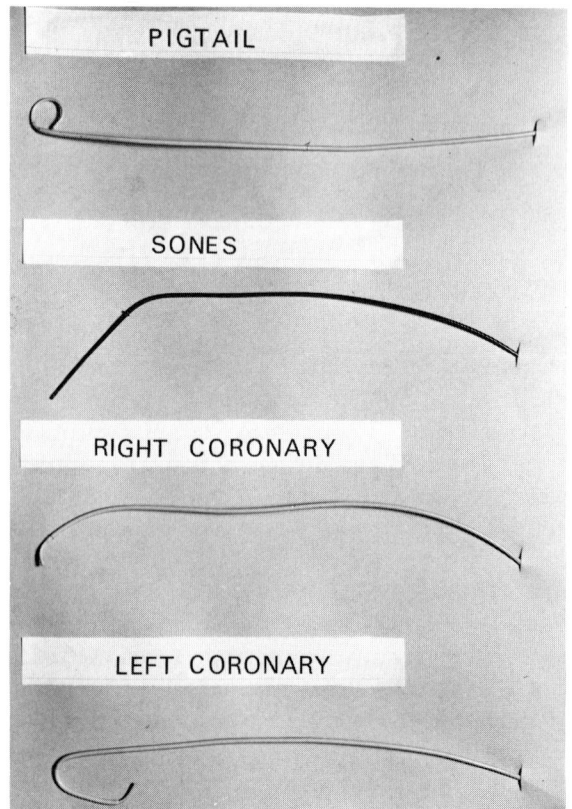

**Fig. 8.35** Chest radiograph demonstrating right atrial enlargement together with left atrial enlargement due to combined mitral and tricuspid valve disease.

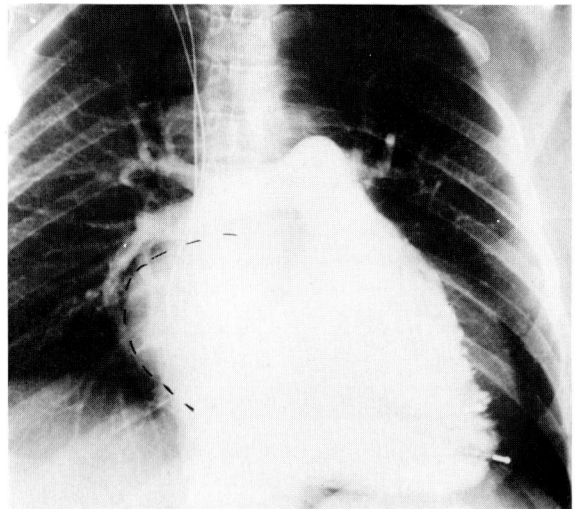

PIGTAIL

SONES

RIGHT CORONARY

LEFT CORONARY

**Fig. 8.37** Cardiac catheters.

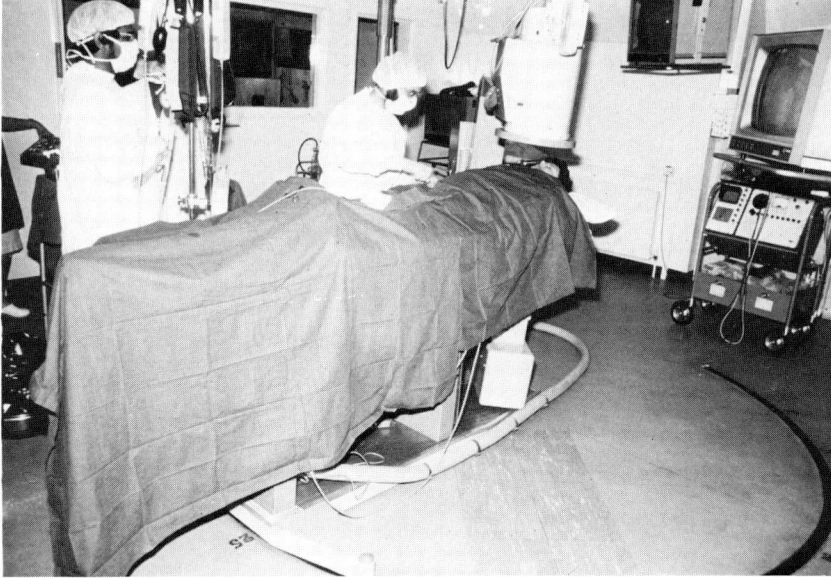

**Fig. 8.38** Judkins technique.

Gross aortic valve stenosis can also cause post-stenotic dilatation of the aorta (Fig. 8.34A), despite there being a reduction in the amount of blood being squeezed across the narrowed aortic valve. Angiography may be undertaken to outline the aortic valve which is equally well shown in the left lateral or the right anterior oblique projection but the latter is useful if the left ventricle is to be demonstrated by contrast medium regurgitating across the aortic valve (see Fig. 8.34B).

*Tricuspid valve disease*

The tricuspid valve is not commonly diseased but if so, it often occurs in conjunction with other valvular disorders which arise following rheumatic fever. Tricuspid regurgitation can be demonstrated angiographically by injecting contrast medium into the right ventricle and if tricuspid valve incompetence is present a considerable amount of contrast will regurgitate into the right atrium. Small amounts of regurgitation are open to conjecture as the catheter itself is across the valve thus causing a slight opening. As a result this is a slightly more difficult valve to demonstrate with confidence angiographically.

*Pulmonary valve disease*

This is generally connected with congenital heart defects rather than an acquired condition.

## 4. Coronary artery disease

The greatest proportion of work now, in most cardiac catheter suites, is the examination of the left ventricle and coronary arteries. The reason for performing coronary angiography may not be merely to demonstrate coronary artery disease but also to demonstrate the condition of the arteries if surgery is contemplated on one or more of the heart valves. There are *two* ways of performing left ventricular and coronary angiography:

*Judkins method.* Using a Seldinger technique into the femoral artery and using pre-shaped catheters for each part of the examination.

*Sones method.* By performing a brachial arteriotomy and using a Sones catheter for all parts of the examination. Great care must be taken when using this type of catheter for the left ventricular injection, as it is quite straight and open-ended. If it is not very free within the left ventricular cavity it may force its tip into the myocardium during a high force injection from a pressure

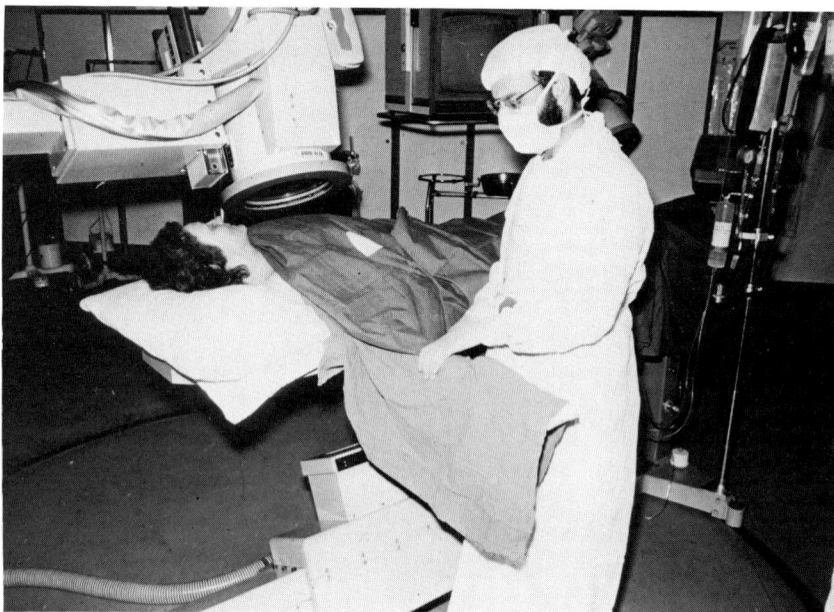

**Fig. 8.39** Sones technique. Note the close proximity of the operator to the X-ray tube.

pump. To avoid this, a slower flow rate of injection than that used with a pigtail catheter is advised and this also reduces the possibility of recoil of the catheter from the left ventricle to the aorta. The major hazard from using this technique is the high radiation dose to the operator, especially to the eyes and hands, as the operator must stand in close proximity to the X-ray tube and image intensifier throughout the procedure.

The choice of technique is entirely personal but there are limitations in the projections available in the Sones technique due to the abduction of the arm and so a lateral projection is not possible as the arms cannot be raised above the head.

The Sones technique is thought to be an advantage in the examination of patients with aortic stenosis, where manipulation of the catheter from above the valve may allow the catheter to cross the aortic valve more readily than manipulation of a catheter from the groin. In the Judkins method the femoral artery is haemostatically controlled by finger pressure in the groin, whereas the brachial artery in the Sones method must be sewn together. The operator must ensure there is a good radial pulse after suturing of the artery is complete.

Apart from the limitations due to the position of the arm in the Sones technique, the projections taken for either technique are very similar and as previously described depend very much on the type of equipment available. For this reason only one technique — the Judkins method — will be described.

*Judkins method of left ventricular and coronary angiography*

A pigtail catheter is used for left ventricular and aortic root injections. This can be used with greater safety because of its coiled shape at the end, and multiple side holes which produce a spray of contrast under high pressure rather than a direct jet. If the catheter is forced forwards under pressure during an injection the loop of catheter will touch the myocardium, rather than a point. It is also less likely to recoil through the aortic valve and so a faster flow rate of contrast medium can be injected without fear of trauma; a slower flow rate should however be used, if the ventricle is irritable, to reduce the likelihood of causing extrasystoles.

The left and right coronary catheters have a preformed shape and the curves come in various sizes in order to accommodate the varying shape and size of the patient's aorta and the aortic arch.

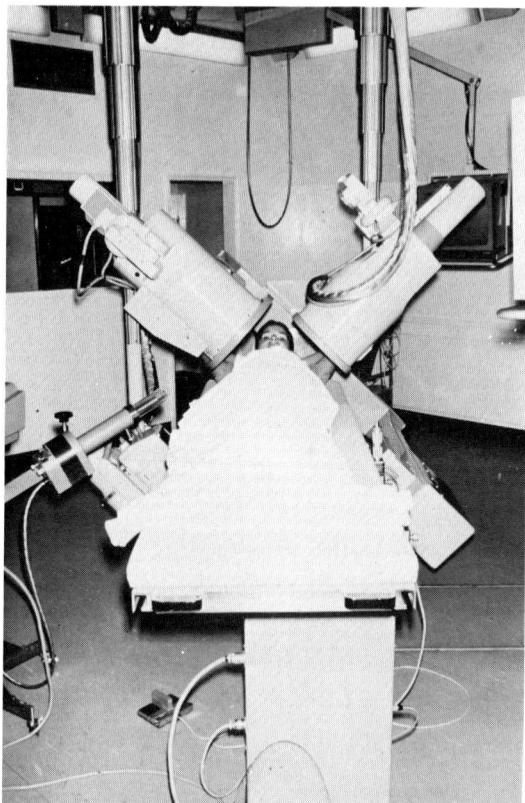

Fig. 8.40 Bi-plane left ventricular angiography.

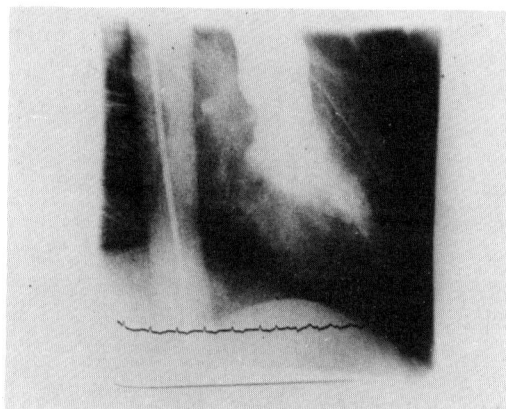

A

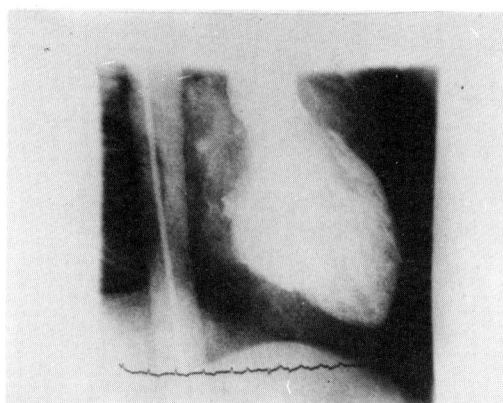

B

Fig. 8.41 Right anterior oblique views (A) Systole.
(B) Diastole.

Figs 8.41 & 8.42 A normally contracting heart.

Tapered tipped catheters are also available if pressure problems are encountered during selective coronary catheterization. Coronary bypass catheters, designed for cannulating coronary grafts, have also been useful for the right coronary artery when pressure readings drop considerably, due to a right coronary catheter that is fitting too tightly into the right coronary artery.

### Left ventriculogram

This technique is used to assess the size and function of the left ventricle and will also demonstrate mitral regurgitation or the presence of a left ventricular aneurysm. Bi-plane angiography can be used for all ventricular injections but it is of particular value in the assessment of an aneurysm since much of the ventricle can be overshadowed by the aneurysm in one view and an incorrect judgement made as to the function of the remaining well-contracting ventricle.

The best projection for the left ventricular angiogram is the right anterior oblique projection with the simultaneous bi-plane facility at 90° to this in the left anterior oblique projection. This is not always possible since some bi-plane facilities are fixed to give just a left lateral projection. If possible, however, it is advantageous to have the second channel mounted to give limited oblique projections as well as the lateral projection (see Fig. 8.40). The ultimate freedom of choice of projections can be achieved with the new Philips 'LARC' model, consisting of a Poly DIAGNOST C with an interlocking second channel mounted on a 'C' arm.

An injection of 30–50 ml of contrast medium is made into the left ventricle, using a high pressure pump. A hand injection of contrast medium is

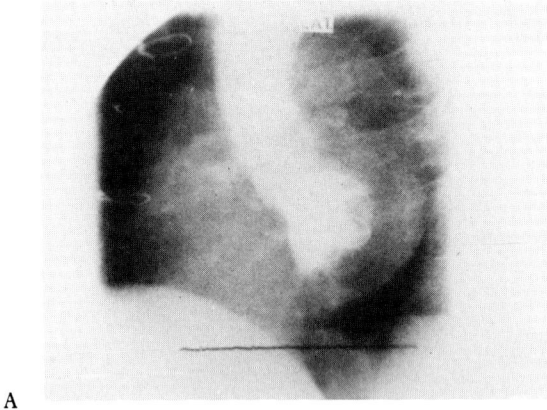

A

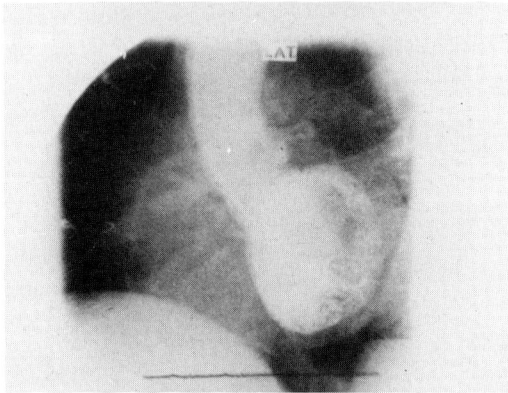

B

**Fig. 8.42** Left anterior oblique views (A) Systole. (B) Diastole.

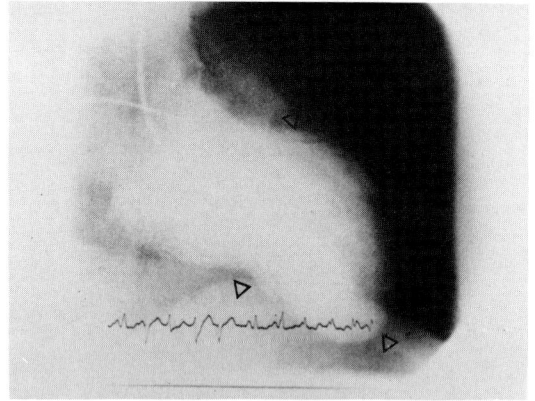

A

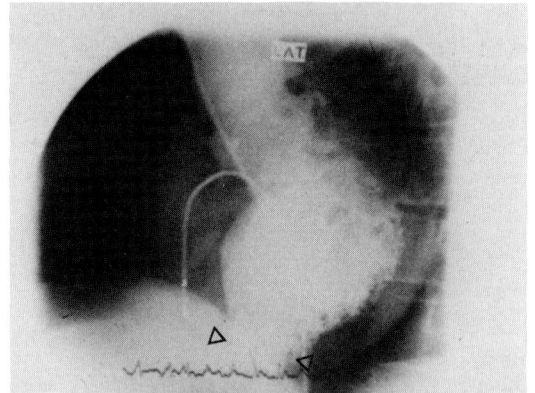

B

**Fig. 8.43** Left ventricular aneurysm (A) Right anterior oblique view (B) Left anterior oblique view.

made initially with the patient in inspiration to ensure that the catheter is free within the ventricle.

*Right coronary angiography*

The right coronary artery consists of one main vessel with a number of much smaller vessels arising from it, which feed the muscle fibres of the heart. Its origin is at the root of the aorta. The right coronary artery is catheterized by using the left anterior oblique projection or the lateral projection. The right coronary artery provides a blood supply to the right atrium, the inter-ventricular septum and both ventricles, and then anastomoses with the circumflex branch of the left coronary artery. In patients who have a normally large right coronary artery they have a relatively small circumflex artery on the left side, and vice versa.

In most patients, the right coronary artery supplies blood to the sino-atrial node so that when contrast medium fills the artery, instead of blood, during angiography it may give rise to severe bradycardia. To counteract this effect, it may be clinical policy to administer atropine (0.6 mg), via the catheter, before undertaking right coronary angiography. The anastomosis of the right coronary and the circumflex branch of the left coronary artery can be useful when the arteries are diseased. If the right coronary artery looks normal but retrograde filling is seen in the left coronary artery this is generally indicative of a serious lesion in the artery on the left side which is being filled retrogradely. If retrograde filling is seen, then it is important to take one projection to show this artery to its best advantage.

The average number of projections required to

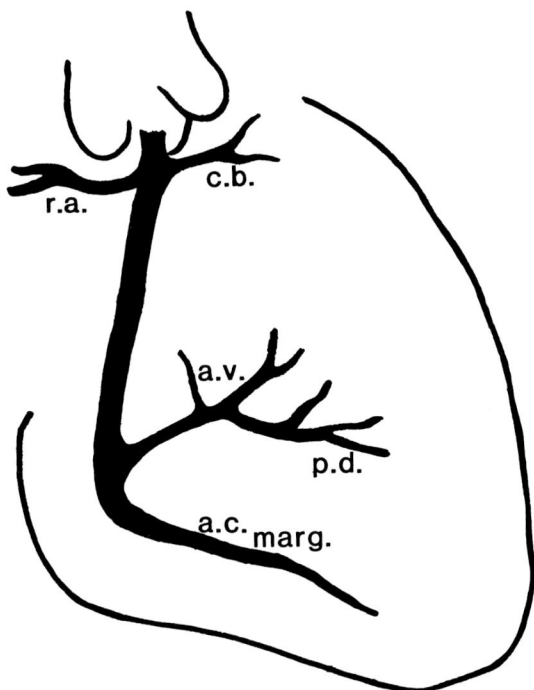

Fig. 8.44 Right coronary artery.

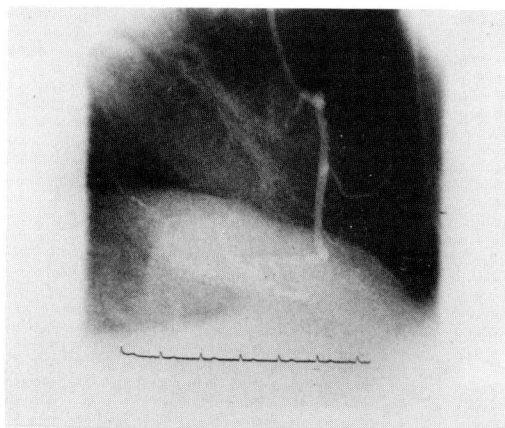

A

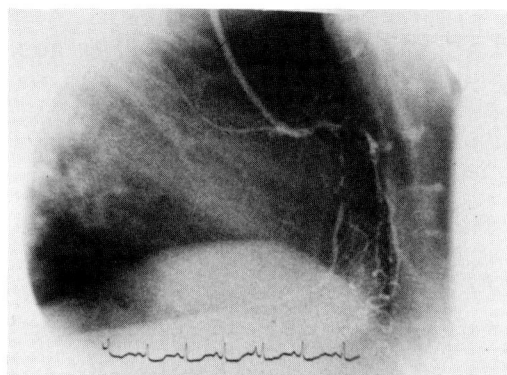

B

Fig. 8.45 Right lateral (unmagnified image) (A) Normal (B) Abnormal.

demonstrate adequately a normal right coronary artery is two or three; however, if the vessel is very tortuous or there is some doubt about the presence of a lesion, further views can be taken at different angulations to demonstrate the area in question properly. If a narrowing is seen and this is thought to be due to coronary artery spasm, the patient may be given either intra-arterial sorbitrate into the coronary artery itself or a glyceryl trinitrate (GTN) tablet to dissolve under the tongue. These act as vasodilators and will alleviate spasm that occurs in the artery. If a filling defect is still evident after the administration of a vasodilator, it is usually assumed to be a lesion. The only type of spasm in the artery that may not be alleviated by drugs is one that is induced by the tip of the catheter. One disadvantage of using GTN tablets is that they cause dilatation of all the vessels in the body and as a result the patient's blood pressure may drop, sometimes considerably, and a severe headache may develop.

The most important thing as far as projections for coronary angiography are concerned is versatility and where necessary alteration of angulations to suit the anatomy of the patient being examined.

It is, however, usual to start with pre-determined angles chosen to suit the equipment being used. The possible projections to demonstrate the right coronary artery and retrograde filling of the left coronary artery are as follows:

1. Right lateral using the unmagnified 9″ image intensifier field size, to give a complete cardiac outline (Figs. 8.45A & B).
2. 30° left anterior oblique (LAO) on a magnified 5″ image intensifier field size, this projects the origin and proximal portion of the artery away from the spine (see Figs. 8.46A & B).
3. Left anterior oblique with cranial tilt, this projection is achieved by rotating the table clockwise 45°, and rotating the 'U' arm 30° with the image intensifier towards the left

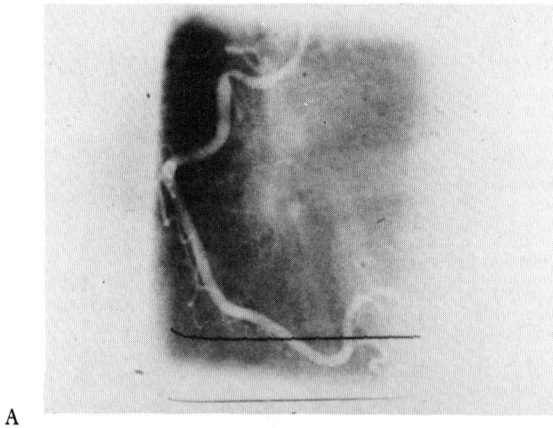

A

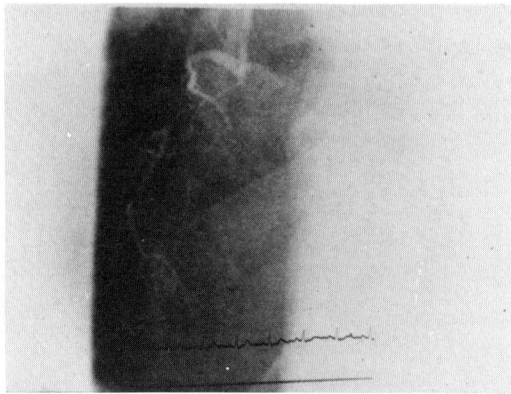

B

**Fig. 8.46** 30° LAO (magnified image) (A) Normal. (B) Abnormal.

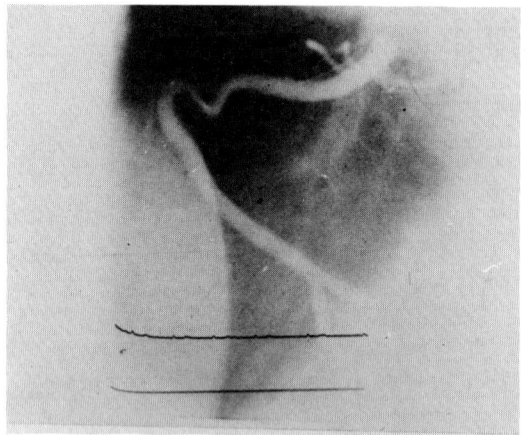

A

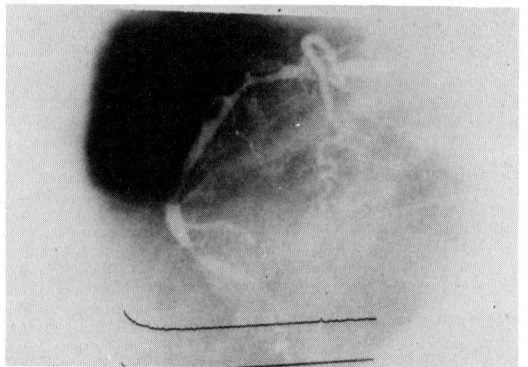

B

**Fig. 8.47** LAO cranial (magnified image) (A) Normal. (B) Abnormal.

shoulder; this view is also taken on a magnified 5″ field (see Figs 8.47A-D).

4. Left anterior oblique with caudal tilt, this projection is achieved by rotating the table anti-clockwise 45°, and rotating the 'U' arm 30° with the image intensifier towards the left hip; this view is also taken on a magnified 5″ field (see Fig. 8.48A–D).

5. Right anterior oblique with cranial tilt; rotate the table 45° anti-clockwise and rotate the 'U' arm 30° with the image intensifier towards the right shoulder; 5″ magnified field, this view demonstrates the distal part of the vessel (see Fig. 8.49A-D).

6. Right anterior oblique with caudal tilt; rotate the table 45° clockwise and rotate the 'U' arm 30° with the image intensifier towards the right hip; 5″ magnified field, this view demonstrates the proximal part of the vessel and retrograde filling of the left coronary artery (see Fig. 8.50A-D).

*Left coronary angiography*

The left coronary artery consists of one vessel (the left main stem) dividing after a short distance into two vessels: the anterior descending (LAD) branch which runs over the anterior surface of the heart with smaller branches feeding the ventricles and the inter-ventricular septum. The second branch is the circumflex, which traverses the left atrium and travels to anastomose with the right coronary artery. The circumflex branch also subdivides to give rise to another large vessel, known as the obtuse marginal branch.

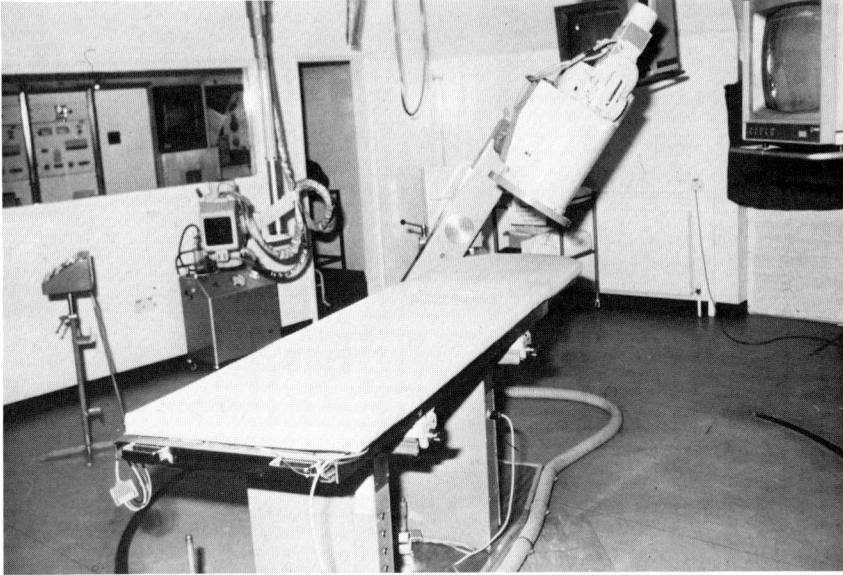

**Fig. 8.47** (C) U-arm and table rotation.

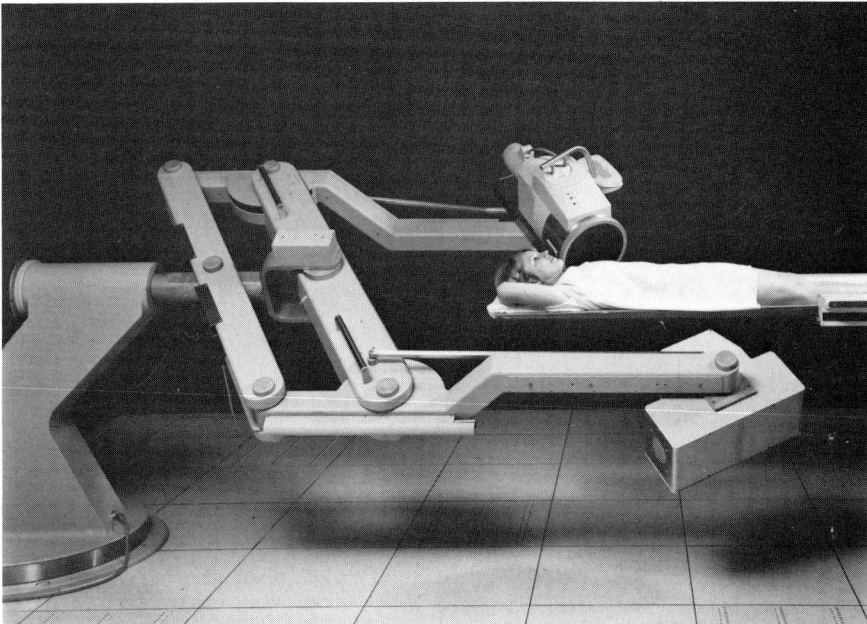

**Fig. 8.47** (D) Poly DIAGNOST C.

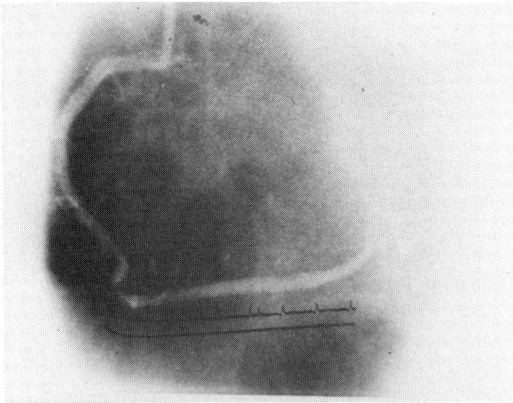

A

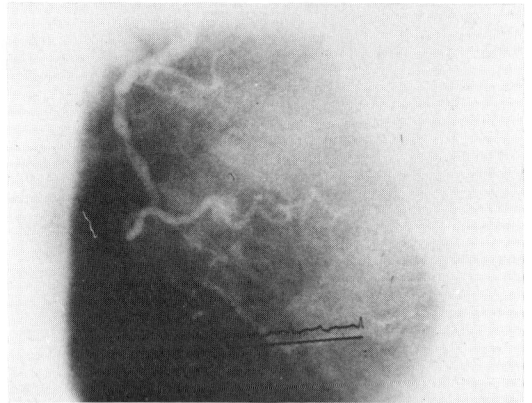

B

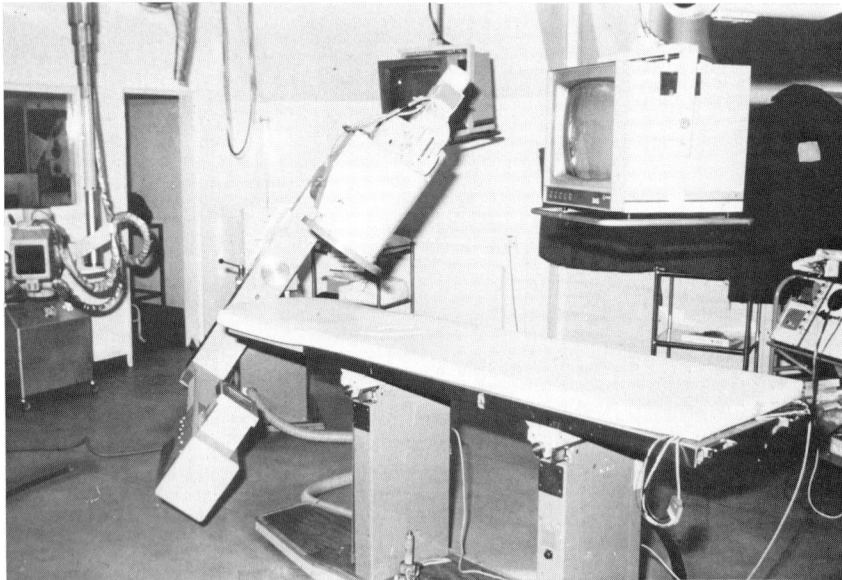

C

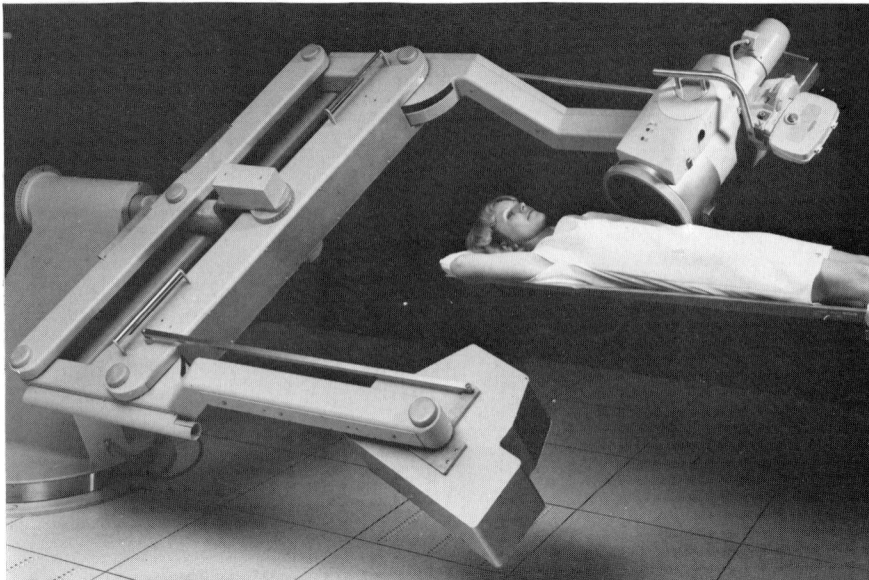

D

**Fig. 8.48** LAO caudal (magnified image) (A) Normal. (B) Abnormal. (C) U-arm and table rotation. (D) Poly DIAGNOST C.

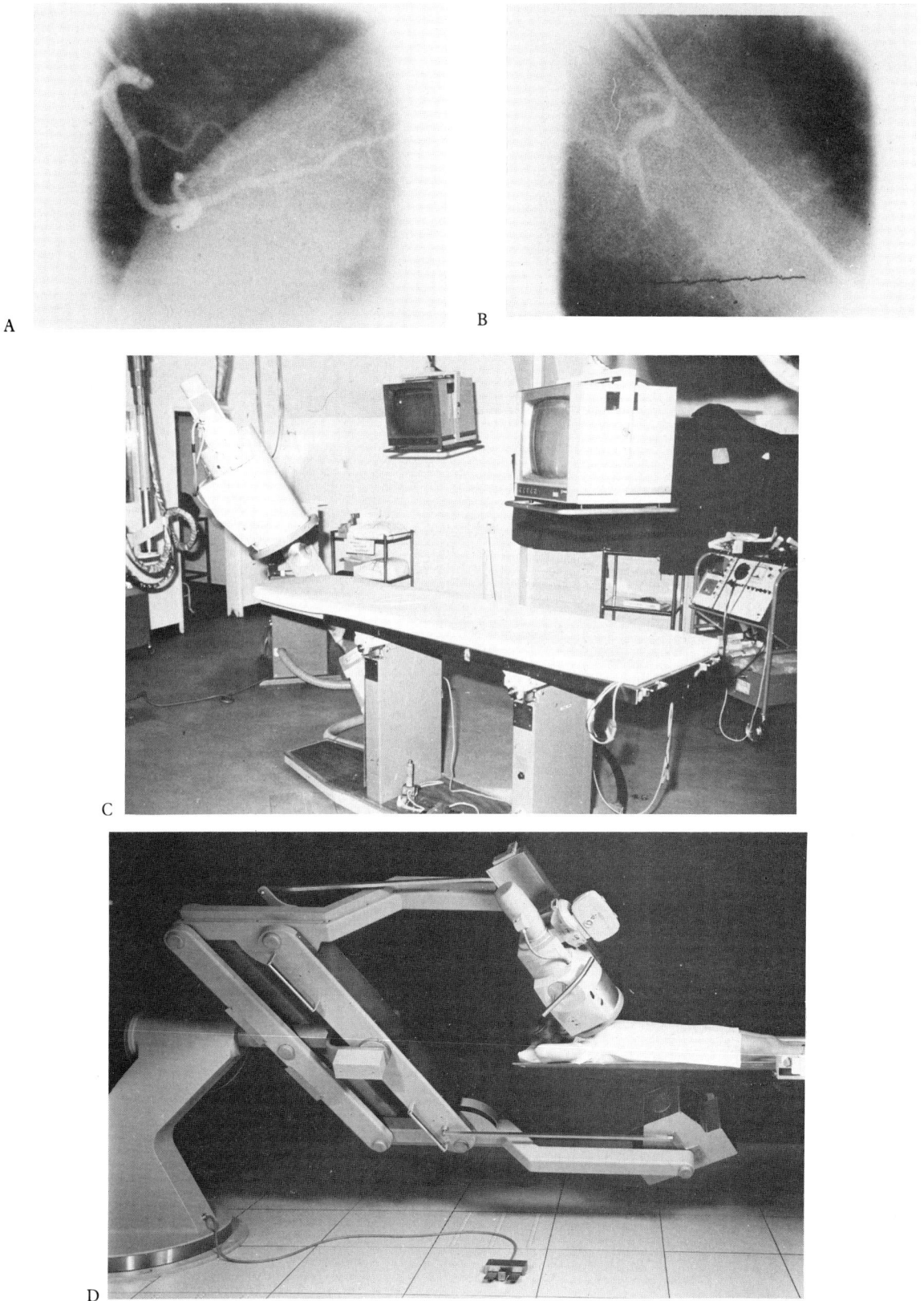

A

B

C

D

Fig. 8.49 Right anterior oblique (RAO) cranial (magnified image) (A) Normal. (B) Abnormal. (C) U-arm and table rotation. (D) Poly DIAGNOST C.

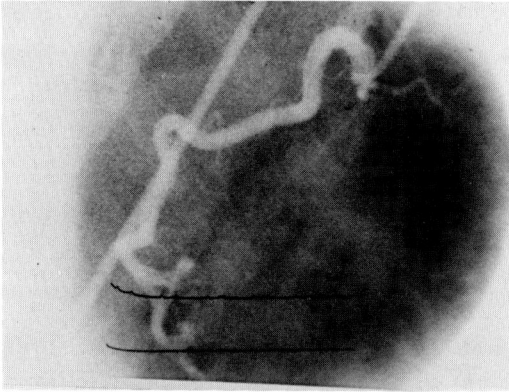

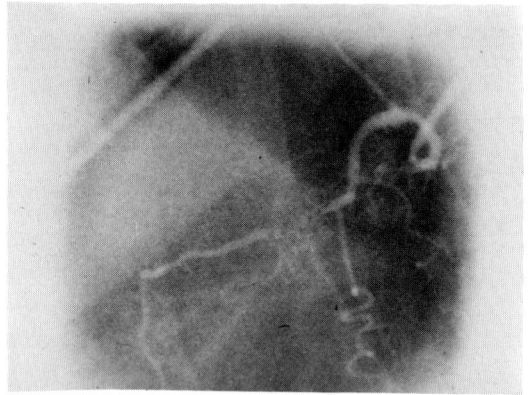

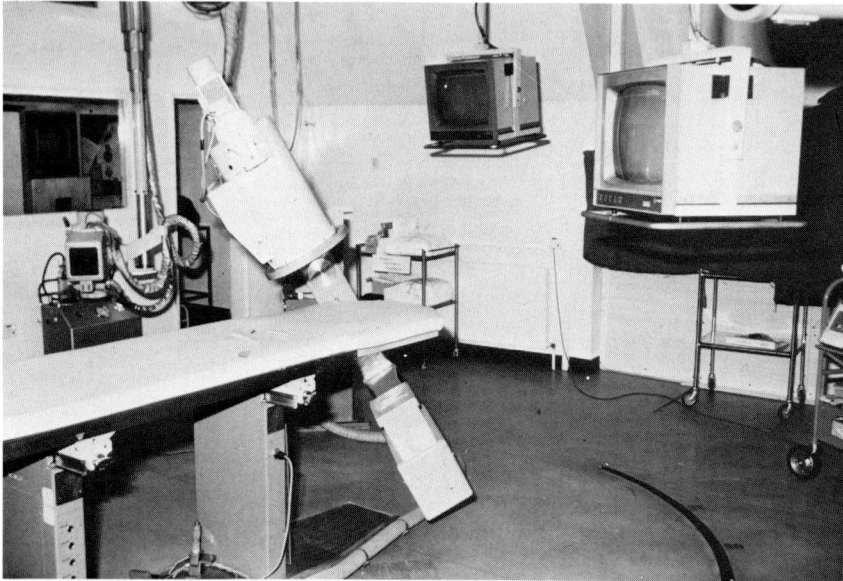

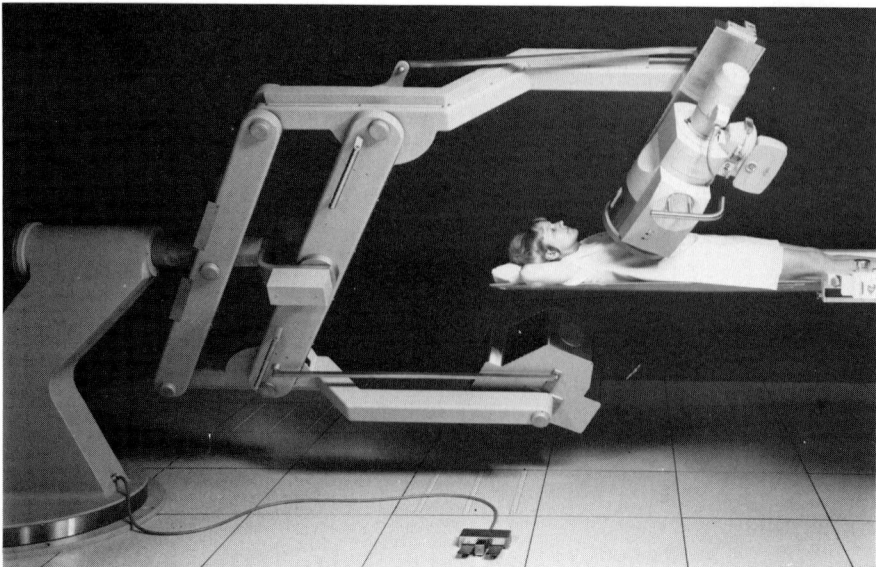

**Fig. 8.50** RAO (caudal) (magnified image) (A) Normal. (B) Abnormal. (C) U-arm and table rotation. (D) Poly DIAGNOST C.

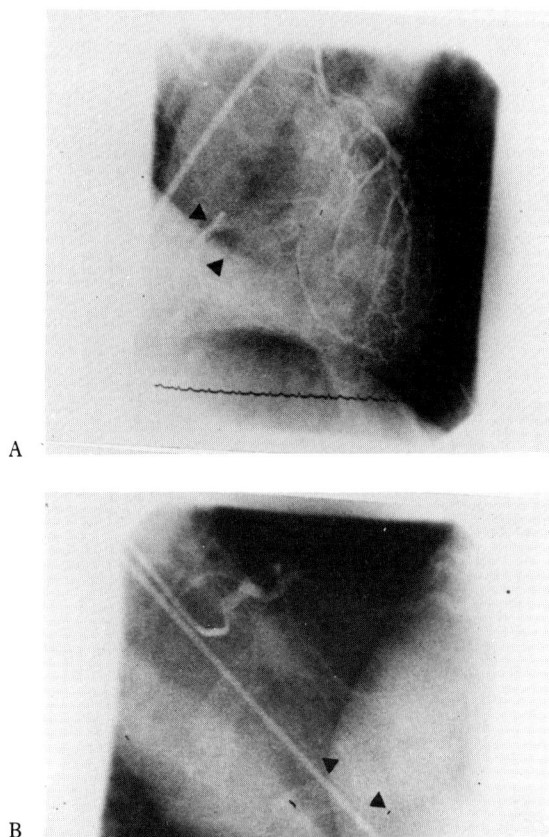

A

B

**Fig. 8.51** Retrograde filling of the right coronary artery from an injection in the left coronary artery. (A) RAO caudal projection unmagnified. (B) Magnified.

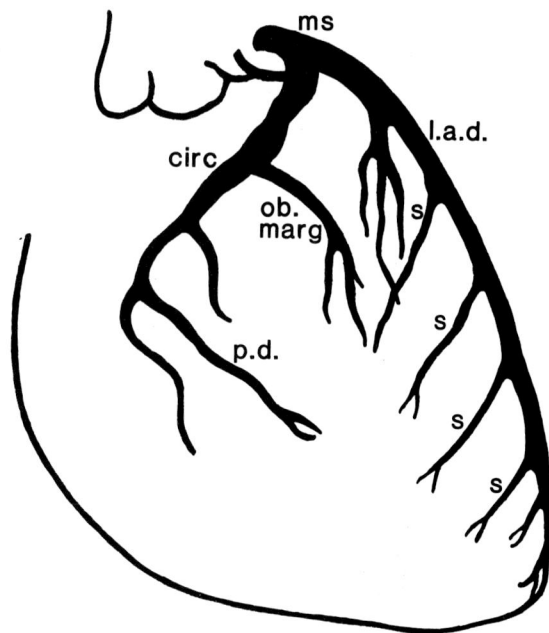

**Fig. 8.52** Left coronary artery.

Due to the much greater complexity of the anatomy of the left coronary artery, the number of projections for adequate demonstration of a lesion, in any major branch of the artery, must be greater than those for the right coronary. Six or seven projections are usually taken during the examination of the left coronary artery, possibly fewer may be taken if the patient had valvular disease and coronary arteriography is merely undertaken for exclusion of coronary artery disease. If the patient has a very short main stem, the catheter may preferentially fill only one of the branches, in which case the left anterior descending and circumflex branches will need to be examined separately.

The anastomosing of the left and right coronary arteries is invaluable to provide information on the right coronary artery if it is being filled retrogradely from the left coronary artery, due to a major lesion in the right coronary.

The routine projections found to be of greatest value for examination of the left coronary artery are as follows:

1. Right anterior oblique with caudal tilt; this projection is achieved by rotating the table clockwise 45° and rotating the 'U' arm 30° with the image intensifier towards the right hip, this view is taken on the magnified 5″ image intensifier field size (see Figs. 8.53A and B).
2. Left anterior oblique with cranial tilt, this projection is achieved by rotating the table clockwise 25° and rotating the 'U' arm 50° to 60° with the image intensifier towards the patient's left shoulder; this view is taken on a 5″ field size and is especially good to show the bifurcation of the circumflex and obtuse marginal branches (see Figs. 8.54A and B).
3. Left lateral projection with the table in the neutral position, this view is taken on an unmagnified 9″ image intensifier field size and gives a good overall picture and any retrograde filling to the right coronary artery that may be present (see Figs. 8.55A and B).
4. 25° right anterior oblique with the table in the neutral position; the magnified 5″ field is used here (see Figs. 8.56A and B).

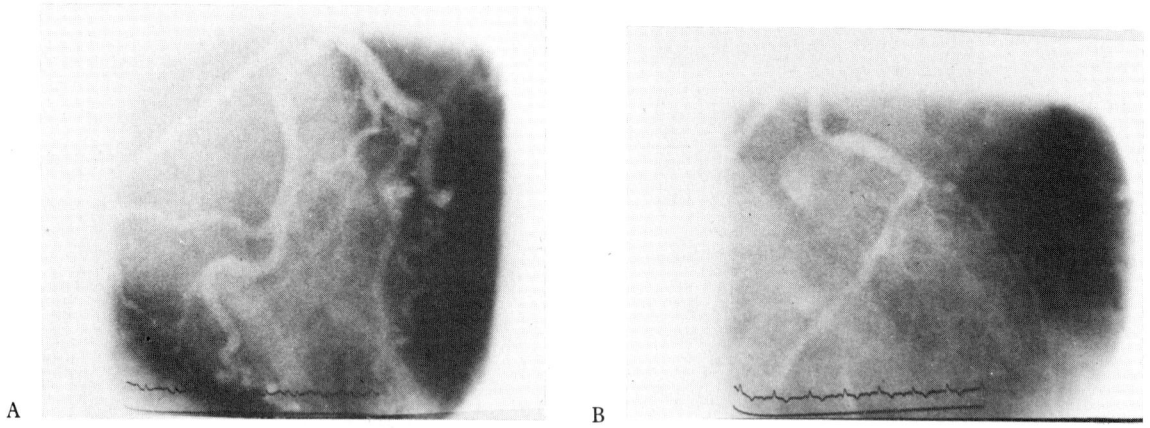

**Fig. 8.53** RAO caudal (magnified image) (A) Normal. (B) Abnormal.

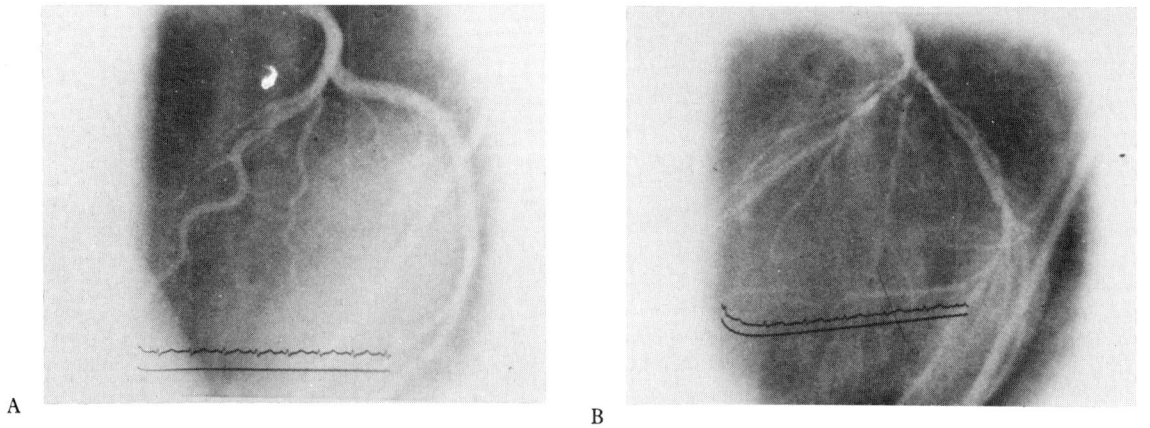

**Fig. 8.54** LAO cranial (magnified image) (A) Normal. (B) Abnormal.

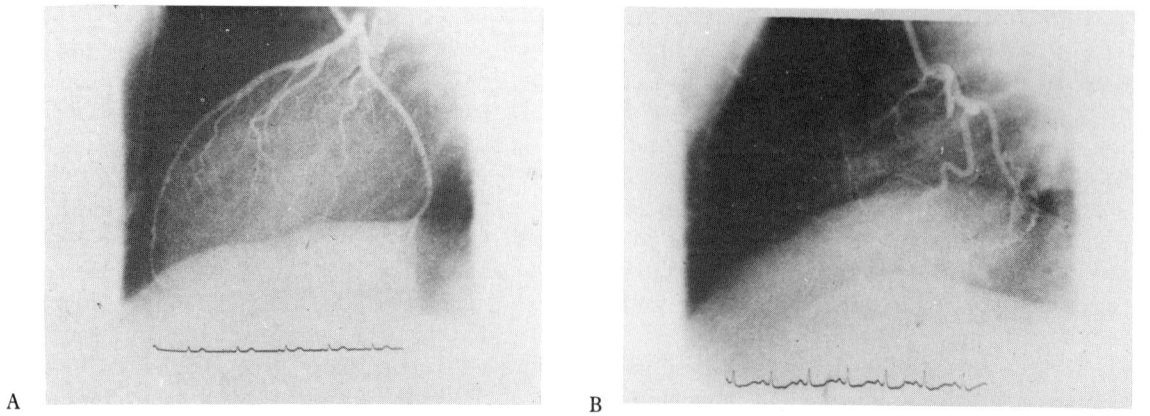

**Fig. 8.55** Left lateral (unmagnified image) (A) Normal. (B) Abnormal. Retrograde filling of the left coronary artery from a right coronary artery injection. (C) Left anterior oblique view. (D) Right lateral view.

C

D

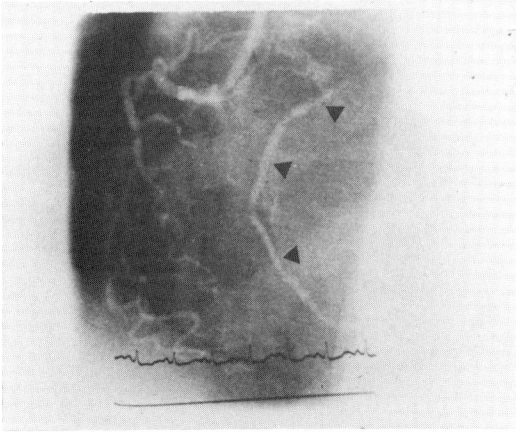

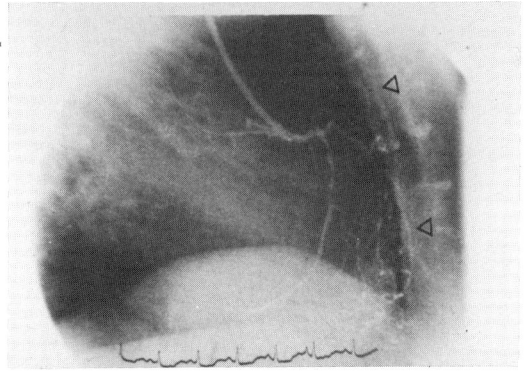

**Fig. 8.55** (contd)

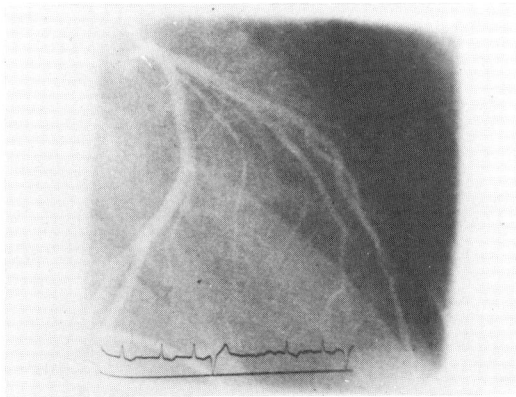

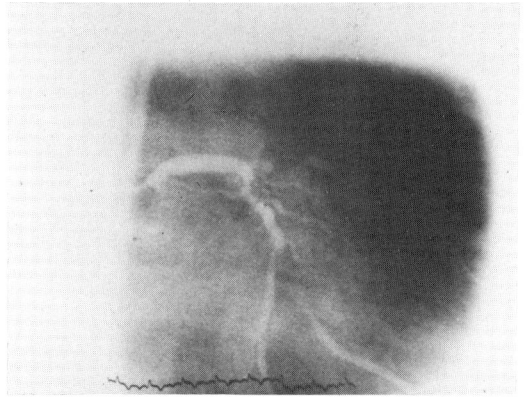

A

B

**Fig. 8.56** 25° RAO (magnified image). (A) Normal. (B) Abnormal.

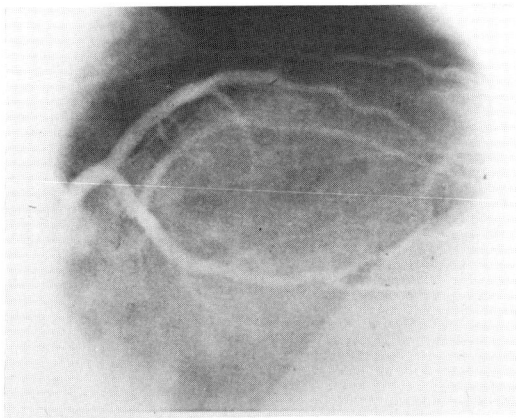

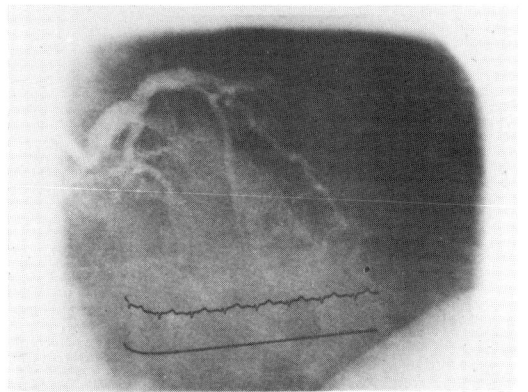

A

B

**Fig. 8.57** RAO cranial (magnified image). (A) Normal. (B) Abnormal.

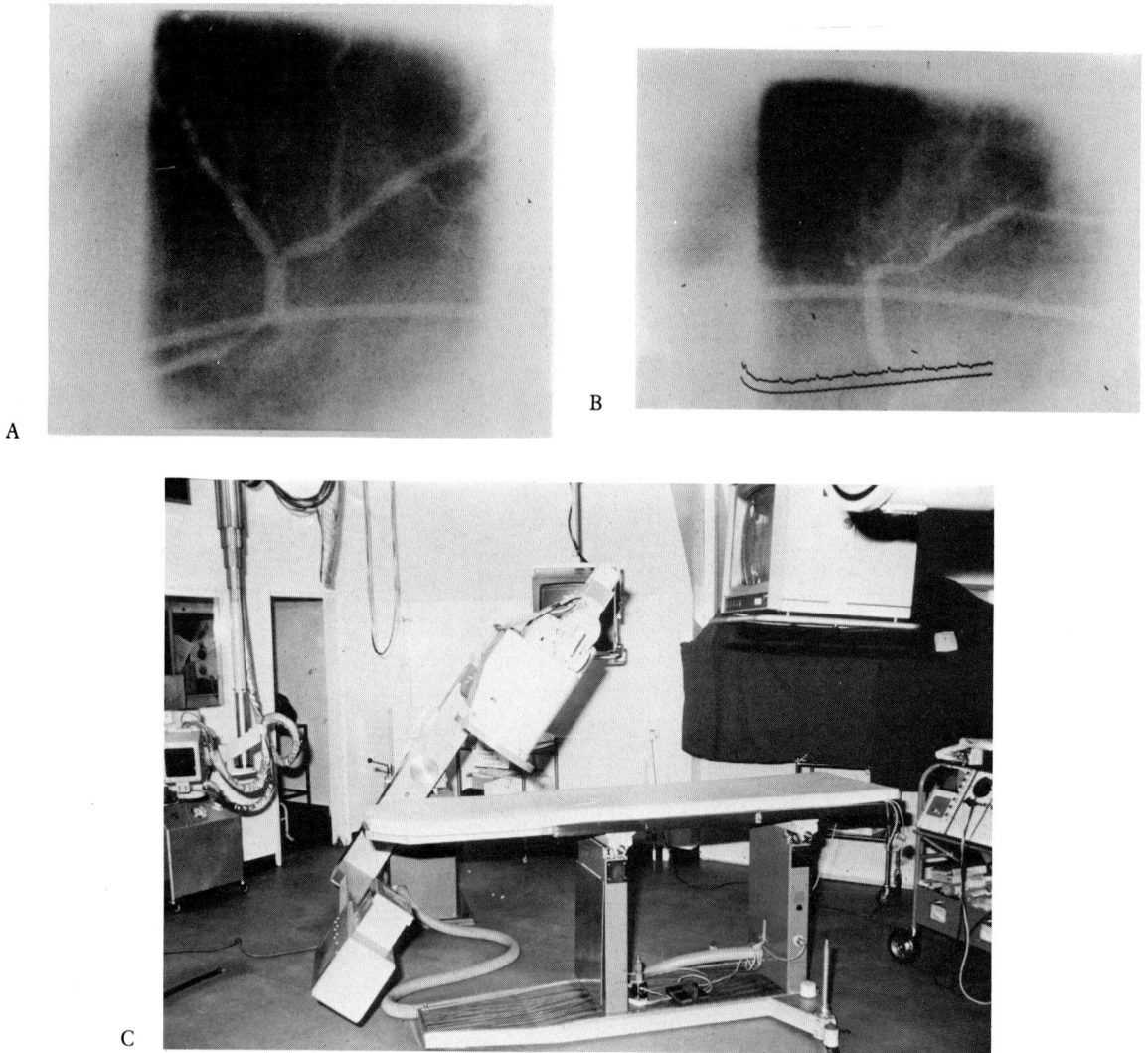

Fig. 8.58 Caudal (magnified image). (A) Normal. (B) Abnormal. (C) U-arm and table rotation.

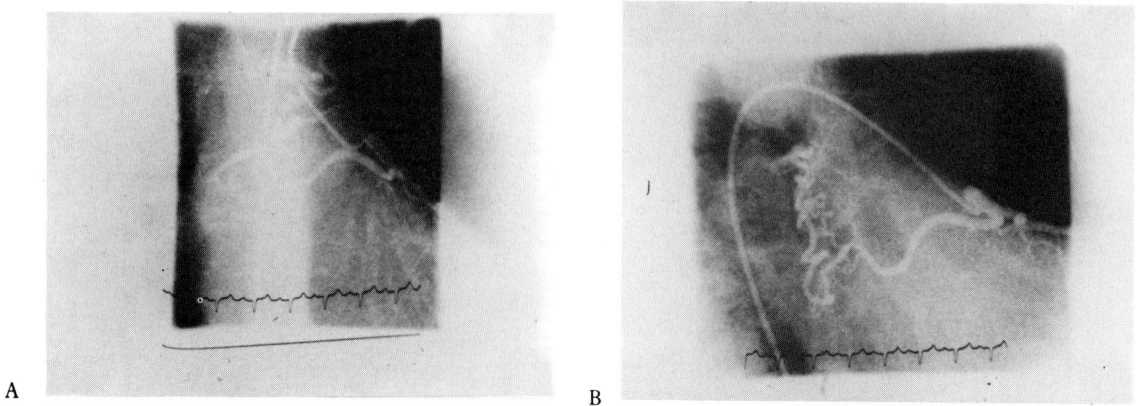

Fig. 8.59 (A) Postero-anterior view. (B) Right anterior oblique caudal view.

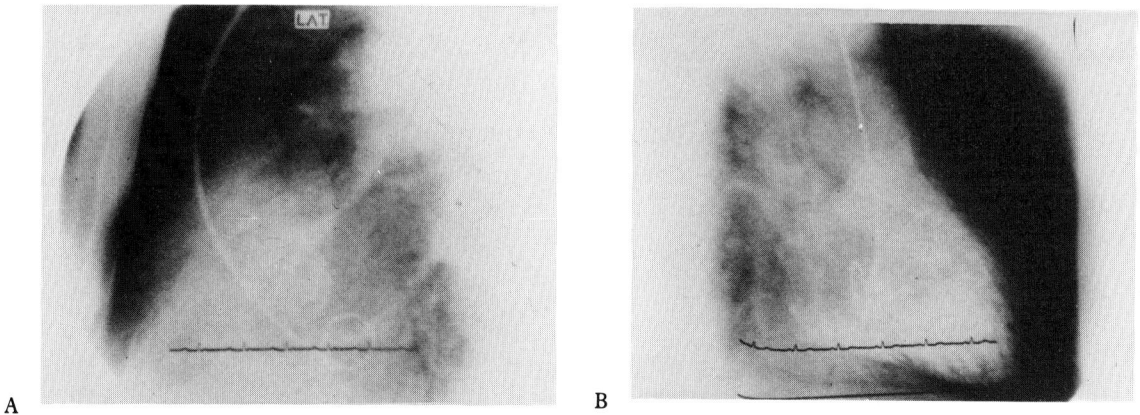

**Fig. 8.60** (A) Left lateral view. (B) Right anterior oblique view.

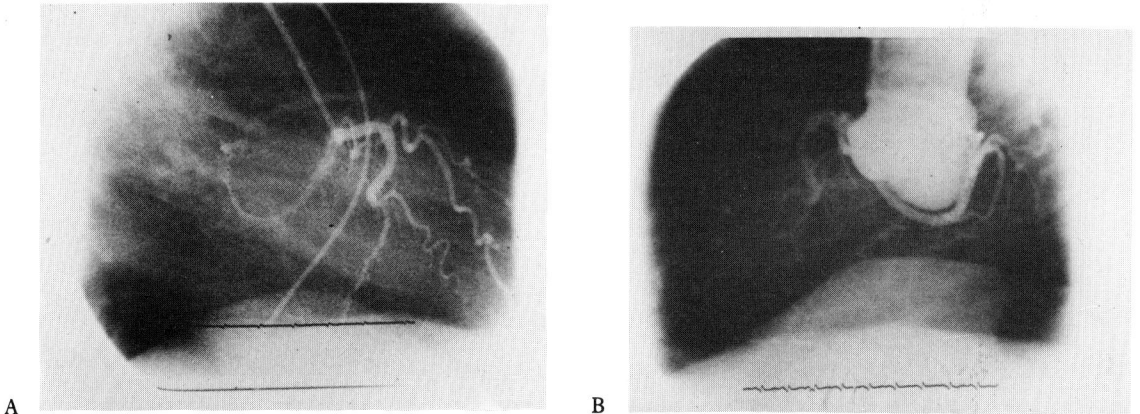

**Fig. 8.61** (A) Right anterior oblique caudal view. (B) Left lateral (aortic root injection).

5. Right anterior oblique with cranial tilt; to achieve this projection the table is rotated anti-clockwise 30° and the 'U' arm rotated 30° with the image intensifier towards the patient's right shoulder; a 5″ magnified field is used (see Figs. 8.57A and B).
6. True caudal projection, rotate the table 90° from the neutral position and rotate the 'U' arm 30° with the image intensifier towards the patient's abdomen, a magnified 5″ field is used and this view is particularly good to show the left main stem and proximal left anterior descending or circumflex lesions in the artery (see Figs 8.58A and B).

The angulations may need to be altered to suit the rotation of the heart or the varying anatomy of the vessel under examination. If coronary artery spasm is suspected then oral or intra-arterial nitrates may be given to alleviate this spasm. Repeat injections can then be taken to confirm spasm or an underlying lesion in the vessel.

From time to time unusual cardiac anomalies present themselves. Here are examples of some such anomalies:

This patient had a congenital heart defect and one branch arising from the left coronary artery was supplying the lungs, consequently each injection of contrast medium into the artery caused the patient to cough (see Figs. 8.59A and B).

This patient had a congenital defect which consisted of a single ventricle supplying the right and left sides of the heart (see Figs 8.60A and B).

This patient had a single origin for all three main vessels of the coronary artery circulation (see Figs 8.61A and B).

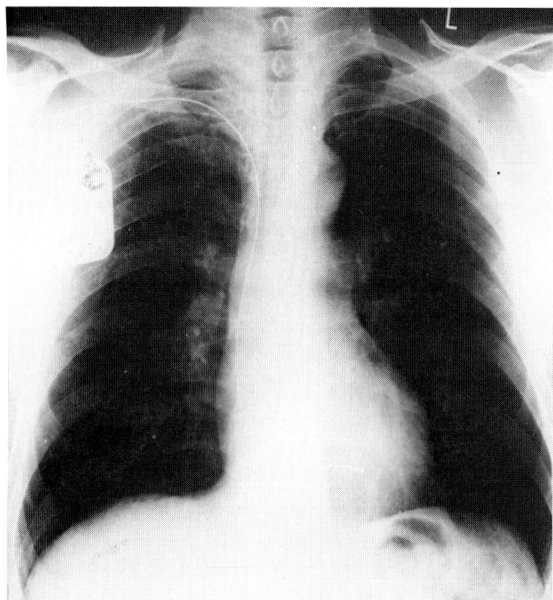
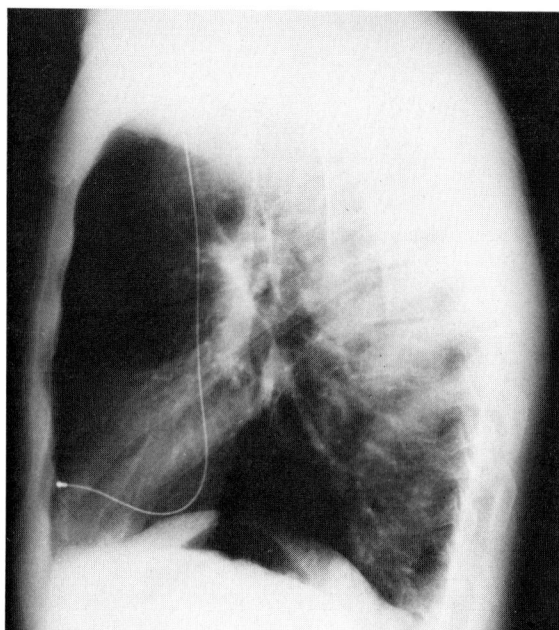

A                                                              B

**Fig. 8.62** Chest radiographs. (A) Postero-anterior. (B) Left lateral.

## NON-ANGIOGRAPHIC CARDIAC PROCEDURES

### Introduction of pacemakers

This technique constitutes a large part of cardiac work in certain centres. It tends to be a regional-ised service for the implantation of permanent pacemakers, whereas temporary pacemakers can be performed in a fluoroscopy room in any hospital which possesses suitable equipment and professional expertise.

It is preferable, however, when a permanent pacemaker implantation is carried out to have fluoroscopic facilities which include the availability of fluoroscopy in the lateral mode. This is to ensure that the catheter tip is positioned in the anterior section of the right ventricle and will therefore be less likely to dislodge by slipping back into the right atrium through a very mobile tricuspid valve. This procedure unlike cardiac catheterization with angiography involves a much more surgical approach, in order to implant the permanent pacemaker box within the muscles of the upper thorax. It is therefore much more important for a high standard of aseptic procedure

during this examination to reduce the risk of infec-tion. In some hospitals to achieve this a mobile fluoroscopic unit may be taken into an operating theatre to carry out the procedure there.

The patient may be kept in bed for 2 days following a permanent pacemaker implantation and a chest radiograph taken on the ward each day to ensure that the tip of the catheter is stable. On the third day the patient will probably be sent for departmental chest radiographs, both postero-anterior (PA) and left lateral projections, to provide permanent documentation of the tip of the catheter. A postero-anterior radiograph of the chest using a grid is preferable since this will best demonstrate the tip of the catheter.

Temporary pacemakers are generally performed as an 'acute' service, the speed at which it is arranged depending on the severity of the patient's heart block. The two approaches to insert a temporary pacemaker lead are:

(i) a cut-down procedure to the ante-cubital vein.
(ii) a sheathed needle approach to the subcla-vian vein.

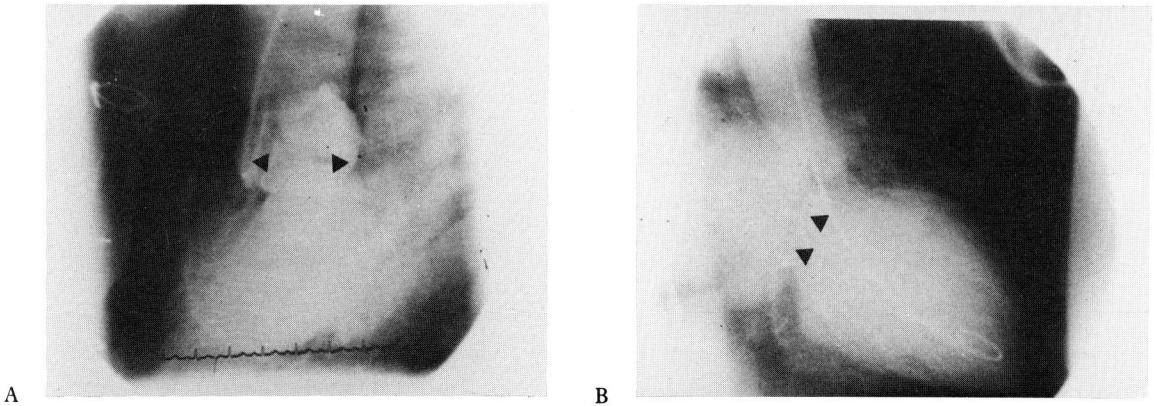

**Fig. 8.63** (A) Aortic valve replacement (left lateral). (B) Mitral valve replacement (right anterior oblique).

For the latter procedure, it is of great value to possess some means of tipping the patient's head down to increase the blood flow in the subclavian vein, for ease of cannulation.

*Pericardial tapping for pericardial effusions*

This procedure involves the introduction of a fine sheathed needle into the pericardium, under fluoroscopic control, to aspirate fluid which has accumulated, so causing the patient to have heart failure. These patients will present with an enlarged cardiac outline on a chest radiograph and will be sent for an echocardiogram to ascertain if this is fluid which can be aspirated. Pericardial effusions can occur following any kind of pericarditis. It can also occur in patients with renal failure, when as much as 1 litre of fluid may be aspirated from the pericardium.

**Post-surgical procedures**

*Angiography following valve replacements*

Angiography may be required as an 'acute' procedure or as a routine procedure, following valve replacement surgery. If an aortic valve replacement begins to leak badly, the patient will be admitted probably in heart failure and an emergency catheterization is performed with immediate corrective surgery. If, however, only a small leak is suspected the patient will be admitted as a cold case for cardiac catheterization and

multiple angiographic projections will be performed, using contrast medium to outline the exact position of the leak in the affected valve. The leak may be where the valve is stitched which is a perivalvular leak or through the replaced valve itself, i.e. a valvular leak.

This procedure can sometimes be extremely difficult and benefit from the new complex cardiac apparatus. Due to the need for bolus injections of contrast medium in multiple projections, it is necessary to achieve as many injections as possible in bi-plane angiography. This will reduce the overall number of injections and reduce the risk of contrast medium overload to the patient, which may have detrimental effects. These projections can only be ascertained by fluoroscopy, prior to angiography, to find the best profile and en face view of the valve in question.

It is always more difficult angiographically to demonstrate the tricuspid valve leaking as a catheter must be across the valve in order to make an injection into the right ventricle.

*Angiography following coronary bypass surgery*

When coronary bypass surgery has been performed the surgeon usually places radio-opaque markers at the origin of the bypass vessel. This is of great value if catheterization has to be undertaken at a later date if the patient's condition gives rise to concern as to the patency of the grafted vessels. Coronary angiography is performed by the routine method in the original right and left coronary

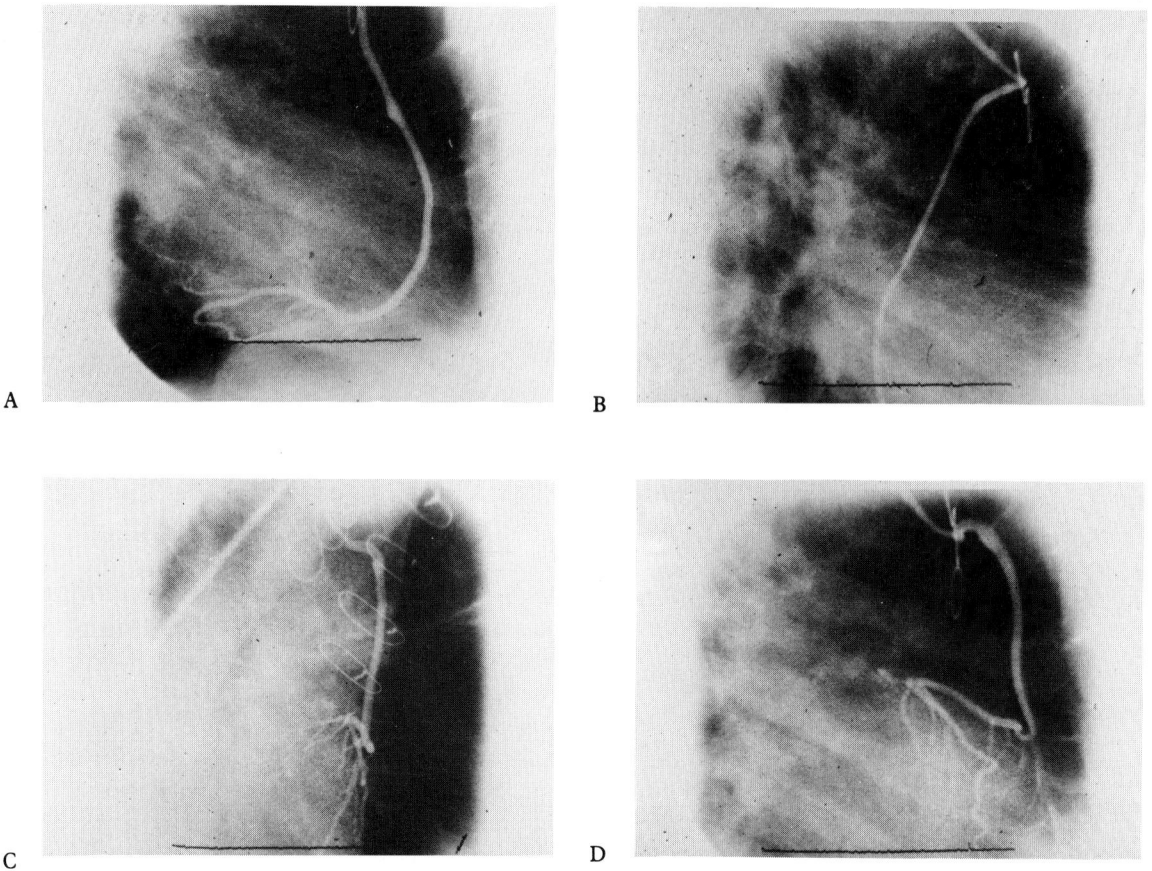

**Fig. 8.64** (A) Lateral view showing graft to right coronary. (B) RAO view showing graft to left obtuse marginal branch of the left coronary artery. (C) RAO caudal view showing graft to left anterior descending branch. (D) Lateral view showing graft to left anterior descending branch.

arteries. The procedure is then followed by catheterization of the grafts to determine any blocks which may have formed. The projections used for depicting the bypass graft vessel will depend upon the course it takes over the heart, i.e. to which vessel it is attached.

A technique that may be used in future to show graft patency which will not involve the costly and invasive procedure of selective angiography is digital vascular imaging using subtraction techniques (see Ch. 9). This will also be of great value in the dynamic study of the left ventricle.

## CONCLUSION

In conclusion it can be said that cardiac angiog-

raphy is an extremely interesting and rewarding part of radiography, and with the increase of heart disease in the western world it is becoming an increasingly busy and important aspect of radiographic work. A good working relationship with all parties concerned in cardiac catheterization (doctors, nurses, technicians and radiographers) is essential to the success and efficient service offered.

*Acknowledgement — for Chapters 8 and 9*

The author acknowledges with gratitude the assistance of the late Dr P.G. Keates, formerly Consultant Radiologist at The General Infirmary at Leeds; Dr N.P. Silverton, Senior Lecturer in Cardiology,

The University of Leeds; and Dr D.J. Lintott, Consultant Radiologist at The General Infirmary at Leeds. She also wishes to thank Mrs J. Archer for typing the manuscripts.

## REFERENCES

Cumberland D C 1981 Amipaque in coronary angiography and left ventriculography. British Journal of Radiology 54(639): 203–206

Petifier H, Crochet D, Fröst H 1980 Advantages of the biplane Angioskop system in angiocardiography and coronary angiography. Electromedica 4: 110–117

Pridie R B, Parnell B 1980 The importance of magnification in left ventriculography. British Journal of Radiology 53(631): 642–646

Sos T A, Baltaxe H A 1977 Cranial and caudal angulation for coronary angiography revisited. Circulation 56(1): 119–123

Walker J K 1981 Coronary collateral response and myocardial function. British Journal of Radiology 54(645): 731–735

Werménski K 1981 The use of the Angioskop-Koordinat 3D set-up for angiocardiography in children. Electromedica 1: 30–33

# 9

# New concepts in angiography — including therapeutic radiology

The Philips Medical UPI (U-arm, Puck, Intensifier) System was developed from the prototype in the Cardiovascular Suite at the General Infirmary at Leeds, and was described by Keates & Clarke (1977). The concept of this unit derives from a 35 cm × 35 cm square PUCK-U Film changer mounted on a 'U'-arm together with an image intensifier (Figs 9.1 & 9.2). The arrangement caters for in-line fluoroscopy and radiography around the patient through an arc of 180°. This will allow oblique and lateral projections to be taken during angiographic procedures by rotating the equipment instead of the patient, thus allowing the patient to lie comfortably and quietly on the back. It also removes the necessity for raising the patient onto foam pads to achieve oblique views.

The table must have a metal free extension in order to achieve these oblique radiographic projections. This extension should be in the order of 1 metre so that examinations from 'head to toe' can be achieved by positioning the patient's head towards the 'U'-arm for angiography above the symphysis pubis, and by positioning the feet towards the 'U'-arm for angiography of the lower limbs. The image intensifier and the Puck Changer are interchanged by a motor drive to achieve 'in-line' fluoroscopy and radiography. An electric motor is also used to drive the X-ray tube to the appropriate angle for achieving cranial and caudal projections. This X-ray tube angulation is especially useful for cerebral angiographic procedures.

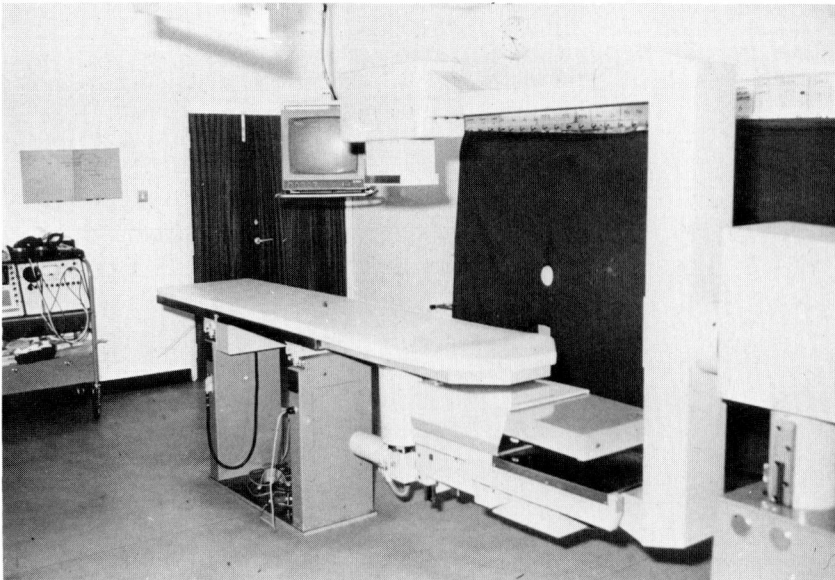

**Fig. 9.1A** The image intensifier is in line with the X-ray tube for fluoroscopy.

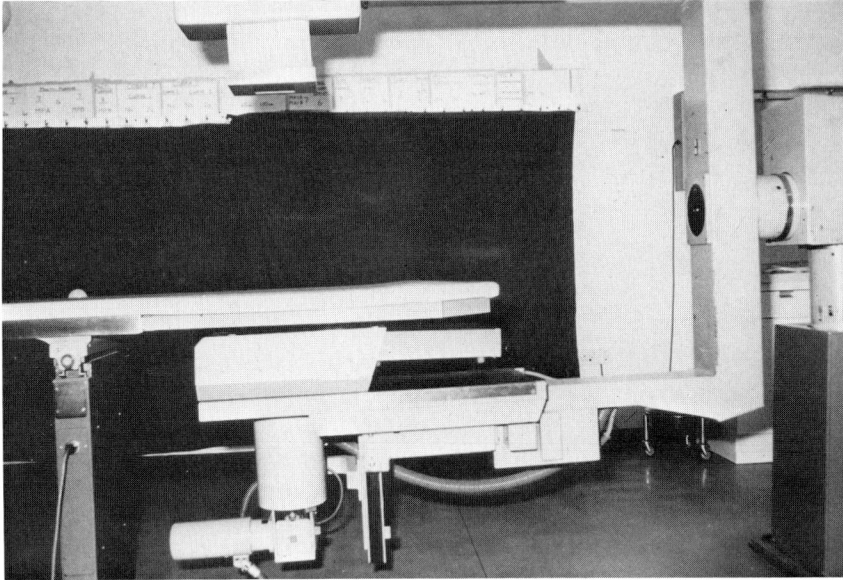

**Fig. 9.1B** The Puck changer is in line ready for angiography.

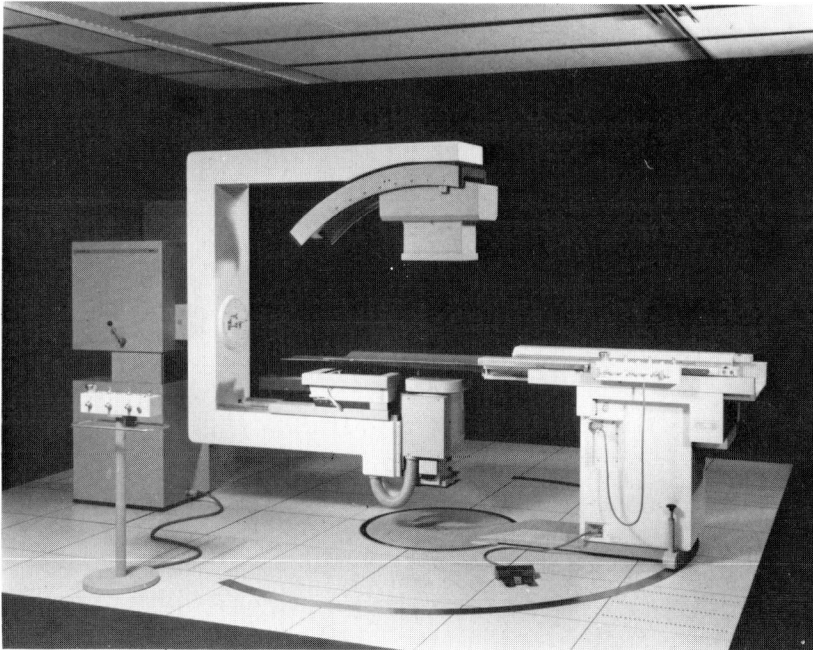

**Fig. 9.2** UPI (by courtesy of Philips Medical Systems).

The Angioskop (Siemens) is designed to produce similar results using a 'C'-arm and a Puck Changer (Fig. 9.3). This equipment does however have some limitations as the unit works in the postero-anterior position, and consequently a second stationary Puck Changer is needed for antero-posterior work which includes translumbar and femoral aortography.

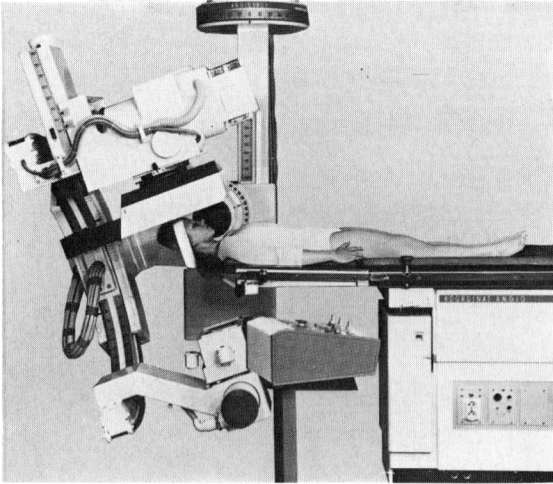

Fig. 9.3 Angioskop (by courtesy of Siemens).

The Angioskop is excellent for cerebral arteriography and for producing oblique and lateral projections of the head, thorax or abdomen. Again, the table must have a metal free portion to facilitate radiography in the oblique projections. This unit when fitted with a 35 mm cine camera is used for cardiac angiography. The Philips UPI can also be fitted with a cine camera to give limited cardiac angiography facilities.

These units could also have a spot film photofluorographic camera but of course the field size is then limited to the image intensifier field size which in many angiographic cases requires at least the 35 × 35 cm square facility offered by the Puck Changer. This problem may, of course, be alleviated in the future with the introduction of the larger field size image intensifiers (14″); which may have important consequences with the ongoing problem of large film format techniques and their high usage of silver.

This type of equipment has not dramatically changed angiography itself but has made the traumatic experience of an invasive study less unpleasant by removing the necessity to move the patient into uncomfortable positions which they may find distressing.

The 'U'-arm does, however, come into its own in the examination of the critically ill patient where oblique or lateral angiography may be of utmost importance and possibly could not be achieved using conventional film changers. The foremost of these procedures would be the examination of a patient suffering from a dissection or aneurysm of the thoracic aortic. These patients are quite often critically ill and require a right posterior (or left anterior) oblique projection to demonstrate the aortic arch to the best advantage.

Another feature of the prototype installation at the Leeds Infirmary which was found to be very helpful was a 'flash image' on the television

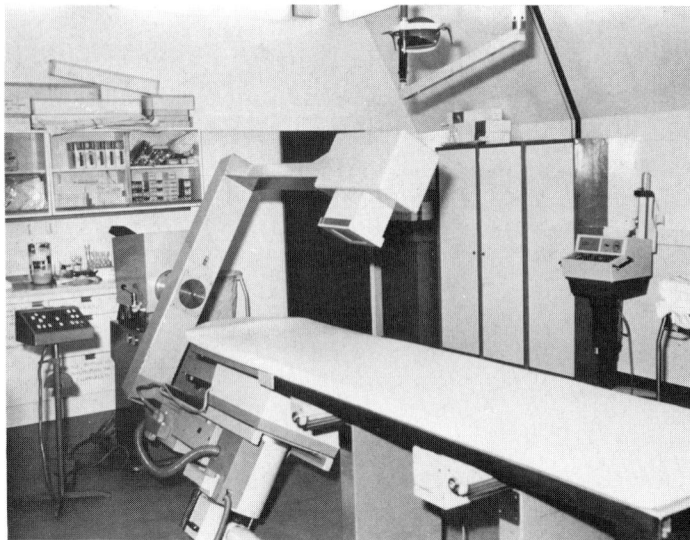

Fig. 9.4A The 'U' arm in a 30° right posterior oblique projection.

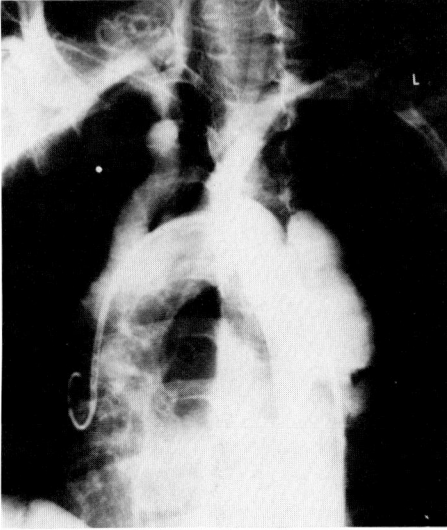

Fig. 9.4B Radiograph demonstrating an aneurysm of the descending aorta.

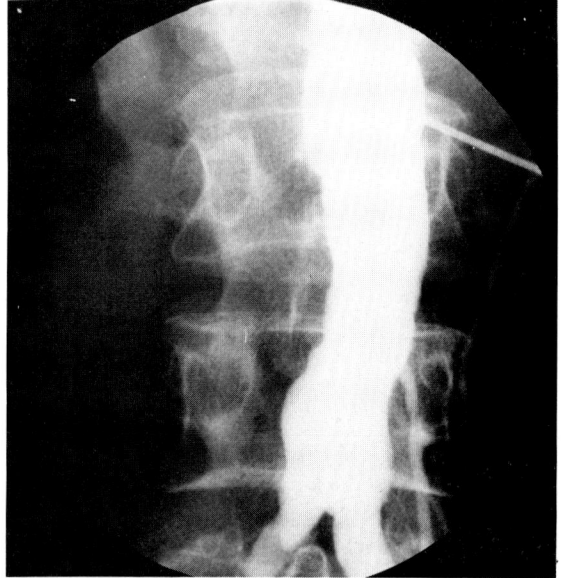

Fig. 9.5 This radiograph demonstrates the area shown on the television monitor during angiography; the magnification is due to the distance of the image intensifier from the patient.

monitor as each radiographic exposure occurs. This image shows only the centre 10 cm circle of the total image and it is made possible because of the image intensifier and the gain on the television tube, which can be suitably adjusted. Use of this feature has been made in translumbar aortography by withholding the transportation of the programme card for the Puck Changer and films taken until contrast medium has been seen to travel down the iliac arteries. It must be remembered that one is viewing only the centre 10 cm and there is a further 12.5 cm for the contrast medium to travel down the arteries shown on the 35 cm square film. When sufficient information has been acquired in the abdominal area, the programme card is then allowed to progress in pre-programmed form to demonstrate the full arterial system of the lower abdominal aorta to the midcalf arteries.

## TRANSLUMBAR AORTOGRAPHY

The table top movements are pre-set but the final movement has been overridden to give a greater overlap of films and to depict only the area to the mid-calf arteries. This is usually the extent of interest to the vascular surgeon. The feet are rotated outwards so that the division of the tibial arteries are projected over the centre of the tibia.

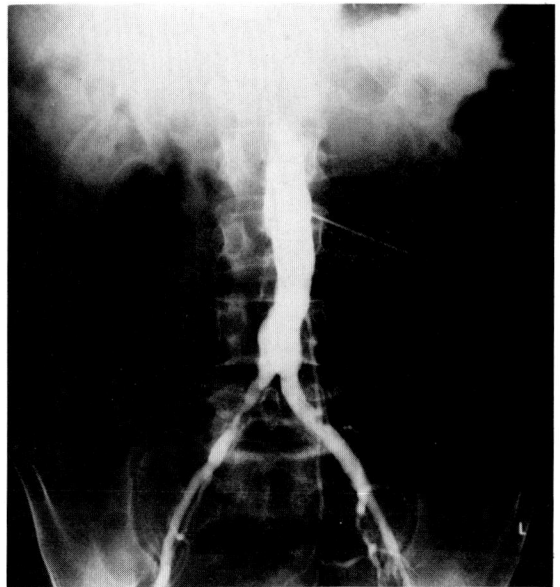

Fig. 9.6

Figs 9.6–9.9 Angiography demonstrated general irregularity of the abdominal aorta with narrowing of the right common iliac artery. The peripheral vessels are within normal limits.

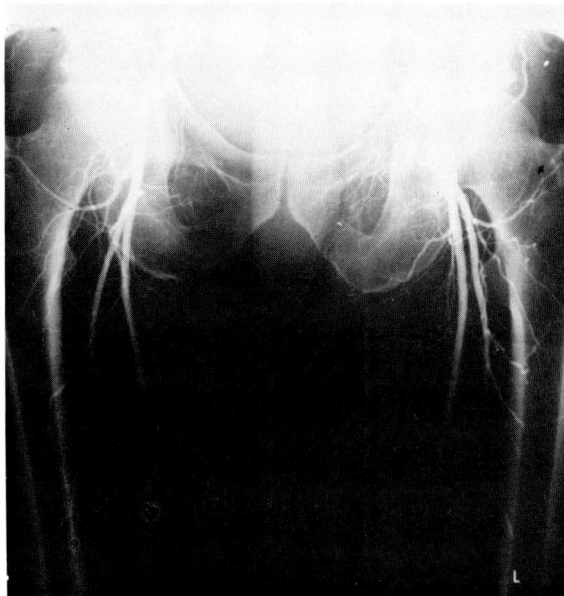

Fig. 9.7

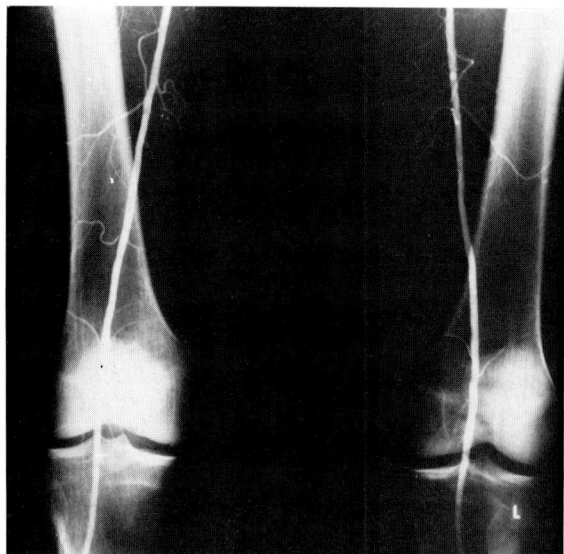

Fig. 9.9

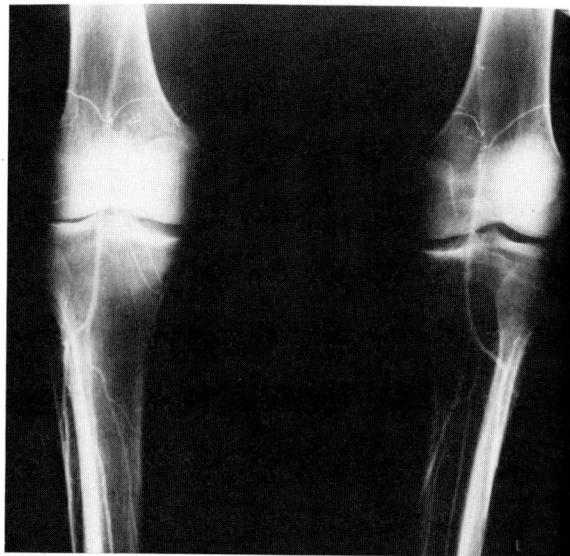

Fig. 9.8

Oblique projections of the profunda femoris artery or lateral projections of the abdominal aorta during a translumbar aortogram procedure are easily obtained by use of the UPI, whereas rotation of the patient with a translumbar needle in situ

would be inadvisable if using conventional angiographic equipment. It is of interest to note that the degree of obliquity needed using the UPI equipment for most angiographic procedures has proved to be less than was thought necessary when employing patient rotation techniques. For example, it has been found that 30° obliquity is quite adequate for demonstration to separate the origins of the common femoral and profunda femoris arteries. Similarly, as little as 20–30° is sufficient to produce oblique projections during renal arteriography. These degrees of obliquity can be accurately reproduced, whereas obliquity of a patient (when using conventional angiographic equipment) cannot be so accurately reproduced.

## ARCH AORTOGRAPHY

Thoracic aortography can be performed well on the UPI equipment. If the arterial system cannot be catheterised, then contrast medium may have to be injected into the right-sided system of the heart (e.g. main pulmonary artery) and follow-through films taken to demonstrate the thoracic aorta. Figures 9.12 and 9.13 illustrate 30° right posterior oblique projection.

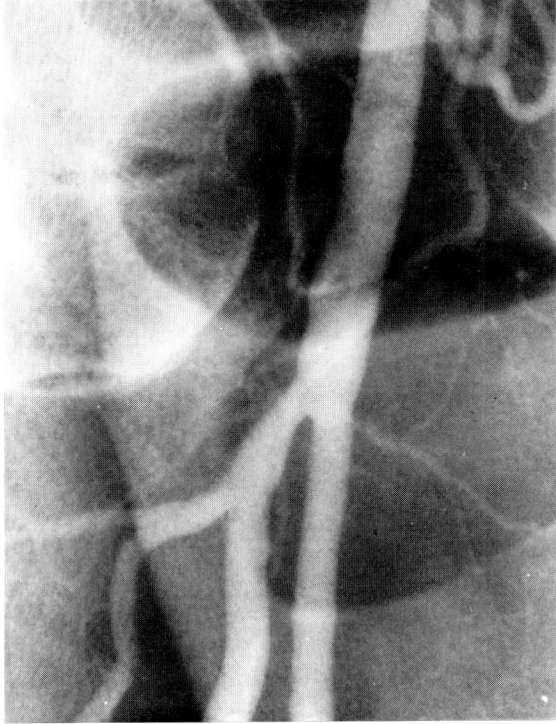

**Fig. 9.10** 30° left posterior oblique to 'open out' right common femoral and profunda arteries.

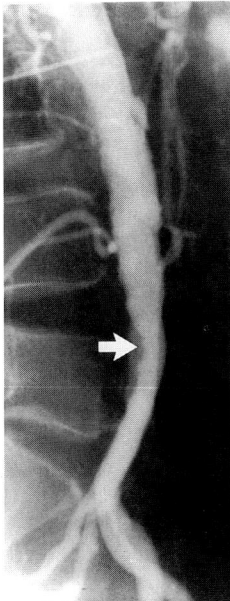

**Fig. 9.11** Lateral projection plaque and reduced flow.

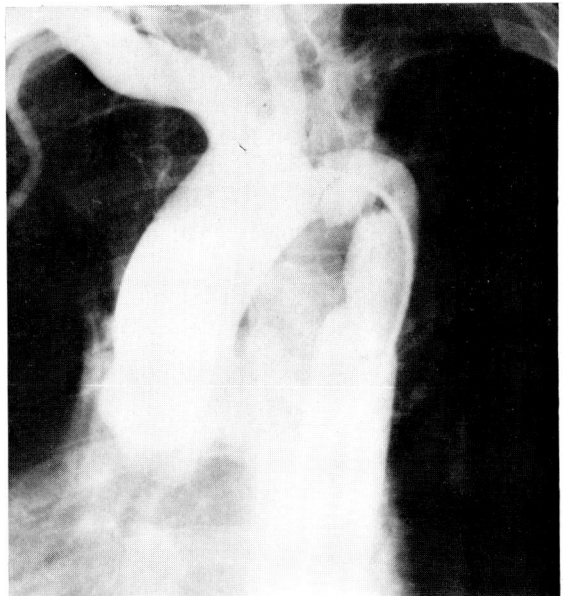

**Fig. 9.12** Left-sided injection; note aortic graft replacement.

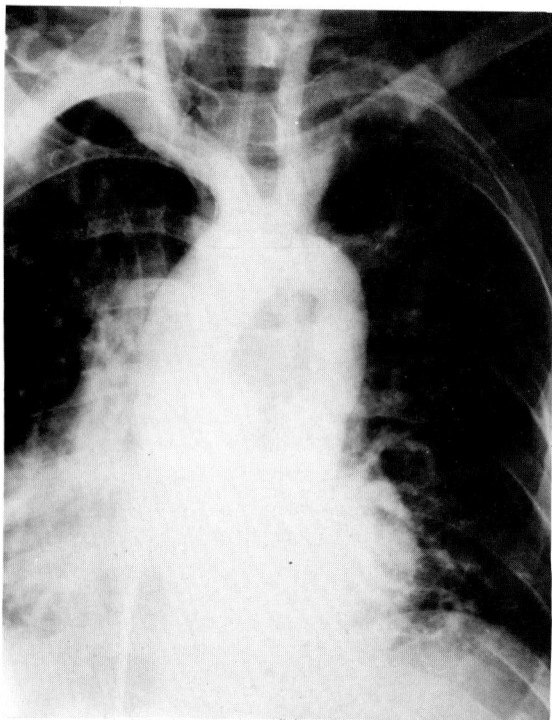

**Fig. 9.13** Right-sided injection — the graft is still clearly demonstrated.

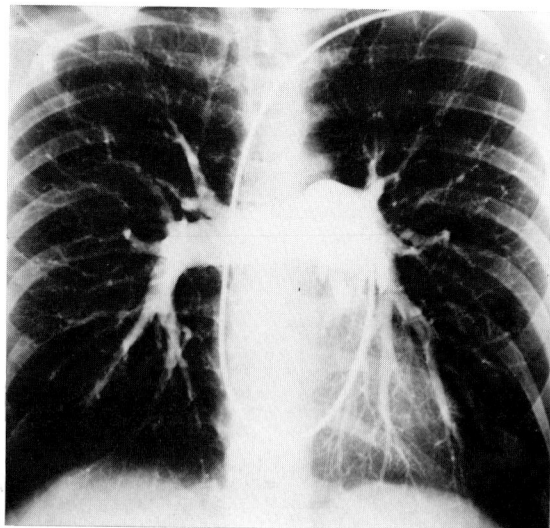

**Fig. 9.14** Normal

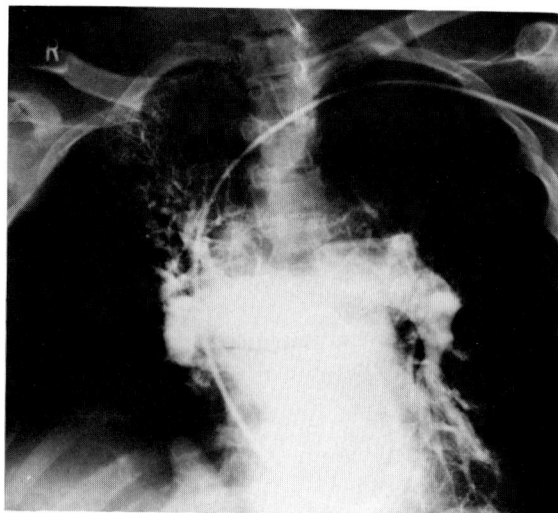

**Fig. 9.15** Abnormal, demonstrating gross pulmonary emboli.

## PULMONARY ANGIOGRAPHY

Pulmonary angiography can very often be an emergency examination where the quickest and least traumatic intervention to the patient is desirable. The facility for providing oblique projections has been invaluable in these situations. The right main pulmonary artery is demonstrated well in the antero-posterior projection because of its anatomical position, but the left main pulmonary artery travels posteriorly before dividing into its smaller branches and this presents difficulties in demonstrating its course. A significantly sized embolus may be obscured because of the course of this vessel. By rotation of the X-ray tube 10–15° to the patient's left, a small right posterior oblique projections is produced which will demonstrate the left pulmonary artery to better advantage without significantly creating a foreshortened appearance of the right pulmonary artery on the arteriogram (Figs 9.14 & 9.15).

## ANGIOGRAPHY OF THE GREAT VESSELS AND CEREBRAL ANGIOGRAPHY

Although the Philips UPI is not a unit dedicated to cerebral angiography a small Puck Changer can be fitted for cerebral work. If, however, the 35 × 35 cm Puck Changer is used there is enlargement in the lateral projection of the skull. This is due to the changer being positioned against the

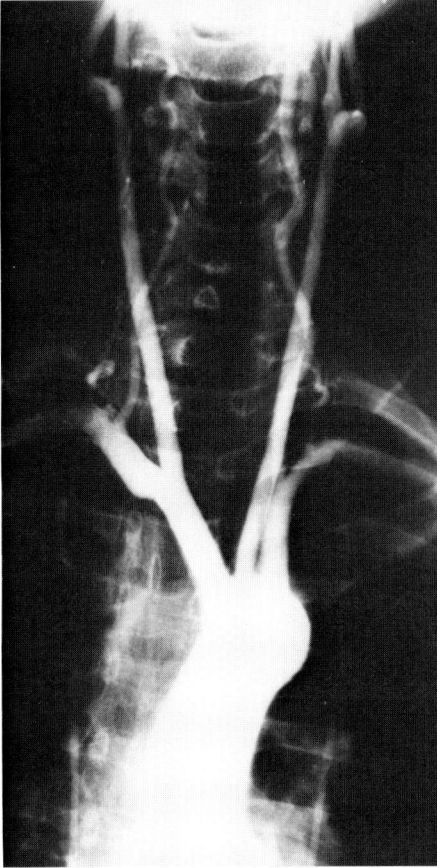

Fig. 9.16 Antero-posterior arch aortogram to demonstrate the great vessels

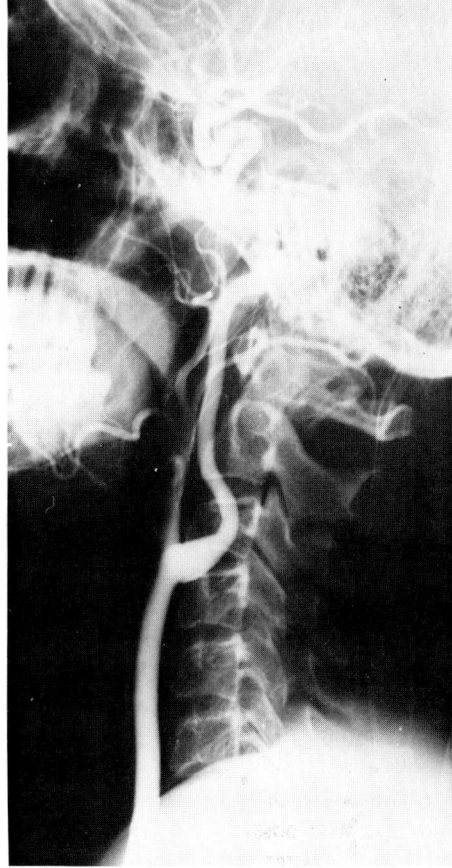

Fig. 9.17 Lateral projection — selective carotid arteriogram

shoulder in order that correct centring of the X-ray beam, anterior to the external auditory meatus can be achieved. The enlargement, may be advantageous providing that an appropriate focal spot size of the X-ray tube can be used to produce suitable quality of the radiograph. Radiographic detail of the blood vessels should then make diagnosis easier. The larger Puck Changer is especially good for depicting the arch of the aorta and the distribution of the great vessels.

## DIGITAL VASCULAR IMAGING

This is the newest concept in angiography using video images instead of film and allows the arterial system to be examined non-selectively and relatively non-invasively by the injection of contrast medium via a short catheter introduced into the antecubital vein. The resulting very low contrast video images are digitised, stored, computer-enhanced then converted back to an analog signal to be recorded onto video tape or disc and viewed on a television monitor.

There are essentially two modes of operation:

Radiographic or Serial Mode
Fluoroscopic or Continuous Mode

### Radiographic or serial mode

This is a similar technique to conventional angiographic subtraction except that all the arterial system in the area of view will be demonstrated rather than a selected vessel. The serial mode can be used to demonstrate vessels in any part of the

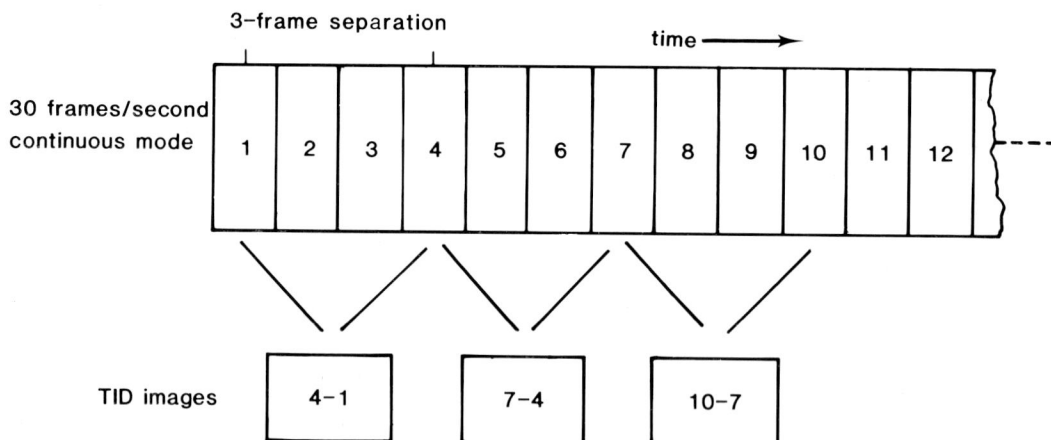

Fig. 9.18 TID mode

body with the exception of the heart and pulmonary vessels because of their movement. Extra filtration may be added to the X-ray tube, up to a total of 5 mm aluminium, to reduce soft tissue absorption thereby increasing the contrast of the iodine in the blood vessels. The exposure values are in the region of those used for conventional small format fluorography (e.g. 100 mm film) i.e. 100 mR/exposure.

The patient must be made comfortable on the table to reduce the possibility of movement blur. A short angio-catheter is first inserted into the ante-cubital vein and then connected to a pressure injector. The patient is positioned in order to visualise the area under examination. Following cessation of respiration, a bolus injection of 20–50 ml of contrast medium is injected at a rate of approximately 14 ml/second.

An initial exposure is taken at the start of the injection of contrast medium, before it reaches the arteries to be visualised, and this video picture is digitised and stored to be used as the 'mask' for the subsequent contrast images. A series of images at 1 second intervals are then taken during arterial opacification. The final images are subtracted 'real-time' images and are recorded onto video tape or disc. (Final images can be viewed a fraction of a second later than live action and are termed 'real-time'.)

If patient movement does occur, causing artefacts on the image, post processing can be undertaken using an alternative 'mask', i.e. any image from the series or another contrast free image made at the end of the series. Image enhancement can be achieved using a 'windowing' technique similar to that in computerised axial tomography (CAT).

This mode is particularly successful for visualising the cerebrovascular system.

### Fluoroscopic or continuous mode

Continuous mode is used for the visualisation of the heart and pulmonary vessels and can use either a low fluoroscopy dose or for better image enhancement the higher doses normally associated with cine angiography. The 'mask' is again taken after cessation of respiration before the contrast medium reaches the area under examination, but this time several frames are recorded and integrated to produce a blurred 'mask' during the motion of the cardiac cycle. The image of the moving organs in this 'mask', although having no detail, allows easier handling of the data by the computer. The contrast images are then subtracted from the 'mask', displayed and recorded on video tape. This technique is successful in demonstrating left ventricular wall movement. However, if patient movement occurs causing artefactual images or visualisation of coronary bypass grafts is desired, a third mode called Time Interval Difference (TID) can be applied.

### Time interval difference (TID)

This involves post processing of the pre-recorded continuous mode images preferably recorded at cine angiography exposure levels (i.e. 70–80 kV and 50–300 mA) in order to reduce noise level. Instead of the blurred 'mask' being used, the 'mask' is continuously updated, for instance, the first frame of the contrast image series is used as a 'mask' and is subtracted from the fourth frame. Then the fourth frame is used as the next 'mask' and is subtracted from the seventh frame and so on.

To obtain permanent hard copies of the vascular system a selection of the best subtraction images in the series may be transferred from the tape or disc to film via a multiformat camera, allowing the disc to be re-used.

These techniques are still in the relatively early stages of development in Europe, but a great deal of work has been undertaken in the United States of America. Hopefully some useful, less invasive angiography will be performed in the near future in this country. It is not intended to be a replacement for conventional angiography as it has limitations, but it could be extremely valuable for patients on whom conventional angiography is medically undesirable or physiologically prohibitive.

To conclude this section, abdominal arteriography has not been greatly advanced by the use of the 'U' arm except in achieving with greater ease the accuracy and reproducibility of oblique and lateral angiographic projections. The special advancements made in the field of abdominal examinations will now be discussed.

## THERAPEUTIC RADIOLOGY

The areas of therapeutic radiology that will be described are:

1. The abdominal bleeder
2. Percutaneous transhepatic drainage
3. Retrieval of stones from the bile ducts
4. Percutaneous transluminal angioplasty (balloon dilatation of arteries).

### 1. The abdominal bleeder

This patient will always be referred as an acute emergency since an abdominal bleeding site can only be shown angiographically at the time bleeding is occurring, and if it is greater than 1 ml per minute. The surgeon will generally have endoscoped and colonoscoped the patient and found no visual bleeding site, so he will then enlist the aid of the radiologist to attempt to show the bleeding point by angiography. As the site is unknown, it is usual to proceed with angiography of the inferior mesenteric artery. The distal territory of this artery will be obscured by the contrast medium in the bladder as it is excreted by the kidneys during the course of the examination. Progression is then made to the superior mesenteric artery and then the coeliac axis until a bleeding site is shown. It may be necessary to obtain subtraction films, if this is possible on an emergency basis, to aid diagnosis.

If a bleeding site is shown angiographically there are two courses of action open to the surgeon:

a. immediate surgery
b. therapeutic embolisation.

The latter is a radiological procedure which is being performed more widely and can be used either as a pre-surgical measure or in place of surgery. There are many methods of therapeutic embolisation but here are three that are more commonly used:

### Vasopressin infusion

This method is a drug infusion which causes temporary vasoconstriction. It is a safe method and can be used in any vessel but is not always successful. It is, however, often safe and effective in the inferior mesenteric artery, where an embolisation technique may be hazardous and cause extensive infarction of the large bowel, as the anastomotic network of arteries here is less extensive than in the small bowel.

The tip of the catheter is positioned as distally as possible in the vessel to be infused; this ensures maximal effect and less likelihood of catheter displacement. Vasopressin is administered at an accurate dose rate by using a vacupump, for 20 minutes, then a repeat angiogram is performed. If bleeding has not stopped the dose rate can be altered and administered over a period of 12–24 hours reducing the dose over that period of time. A further angiogram could be performed at the end of the infusion if this was thought to be desirable.

## Gelatin foam (Sterispon)

Sterispon is one of the easiest forms of embolisation device to use. The tip of the catheter is positioned as distally as possible in the vessel to be embolised to reduce the possibility of extensive infarction. The Sterispon foam is cut into strips and injected through the catheter; several small strips will be used to build up the occlusion of the vessel. When sufficient strips have been injected to cause satisfactory embolisation an angiogram is performed to check the occlusion angiographically.

## 'Gianturco' embolisation device

This device consists of a stainless steel coil with four wool strands, two inches long, at the end to encourage thrombus formation around the device. The coil can be made in varying sizes according to the clinical needs. It is generally used for occluding large vessels such as the renal artery.

The device is straightened out inside an introducer. A No. 7 teflon catheter is then introduced into the artery to be occluded. The introducer is then inserted into the catheter and the device advanced along the catheter using a special guide wire. It is pushed out of the end of the catheter, into the artery, where it resumes its spiral shape and wedges into position with the steel core and wool strands slightly intertwined. After approximately 10 minutes, clot will begin to form around the coil, to form a more complete occlusion of the vessel.

This technique is generally very successful and permanent.

coiled spring →

cotton strands

**Fig. 9.19** 'Gianturco' coil.

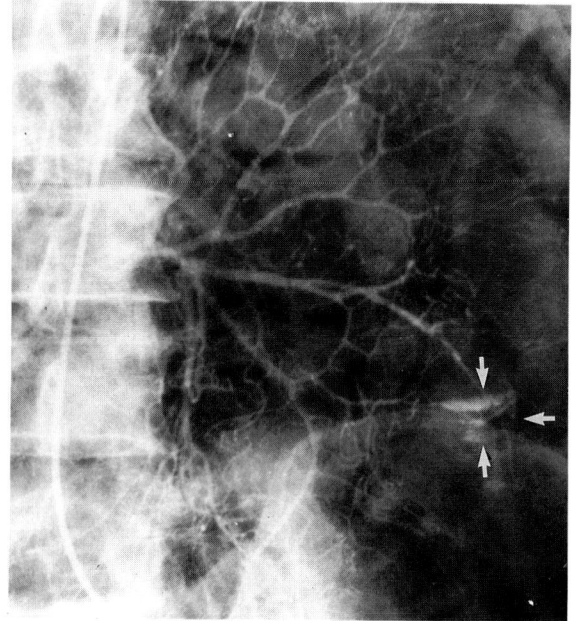

**Fig. 9.20** Superior mesenteric artery injection — bleeding site marked by arrows.

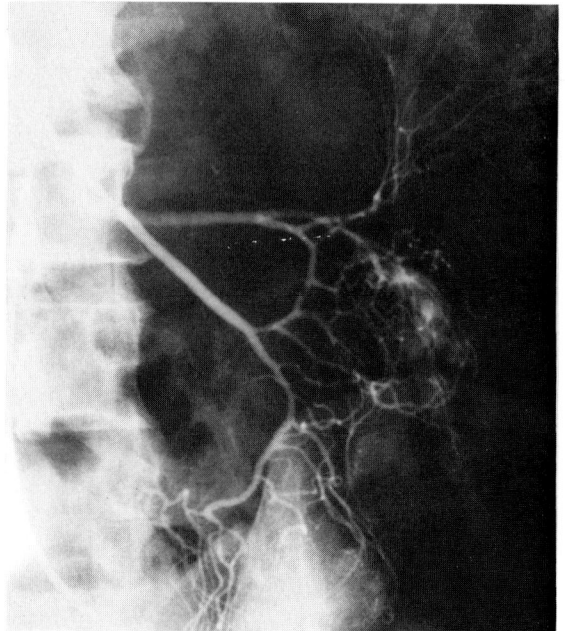

**Fig. 9.21** Super selective injection — the bleeding site is well demonstrated.

**Figs 9.20–9.23** demonstrate superior mesenteric artery bleeding followed by embolisation using Sterispon foam.

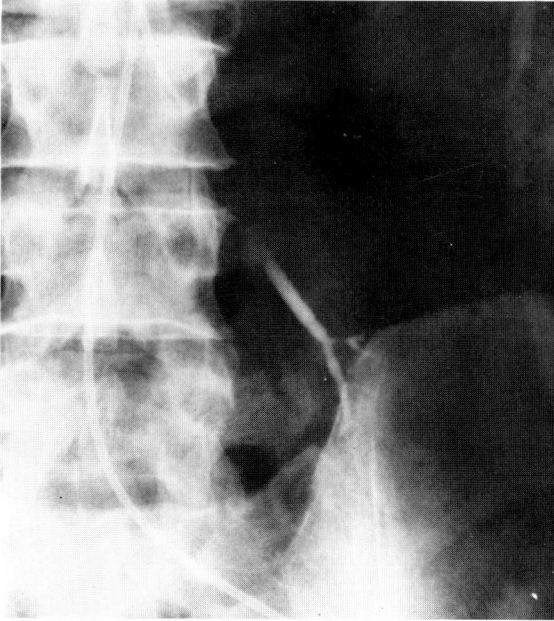

Fig. 9.22 Injection of Sterispon foam with contrast medium.

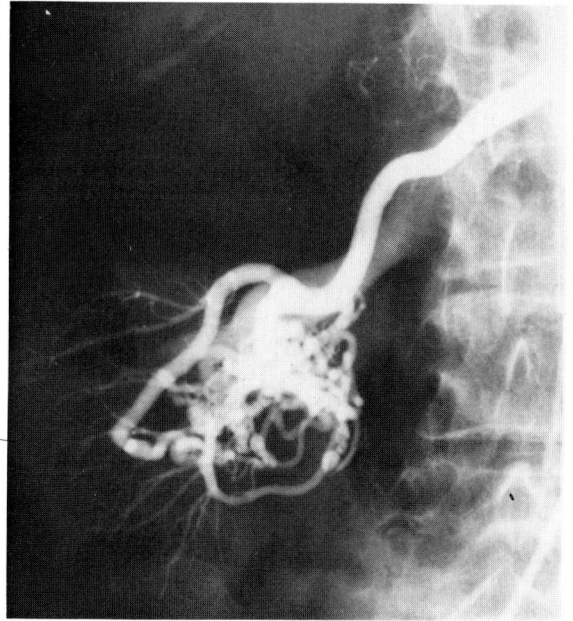

Fig. 9.24 Right selective renal arteriogram demonstrating an abnormal blood supply to a highly vascular tumour.

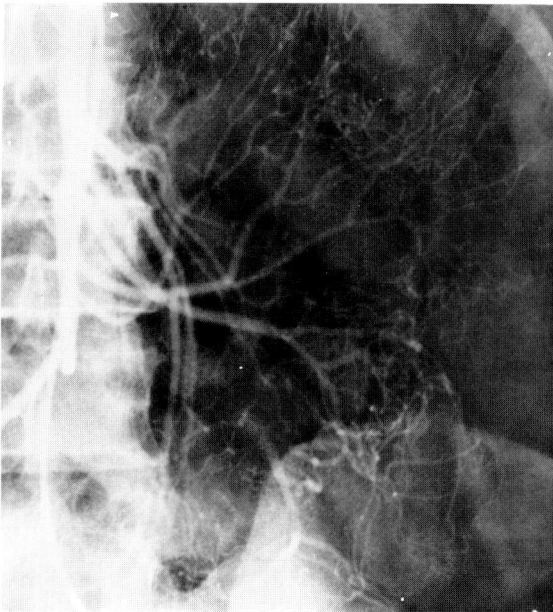

Fig. 9.23 Post-embolisation angiogram — there is no pooling of contrast medium, therefore the bleeding has stopped

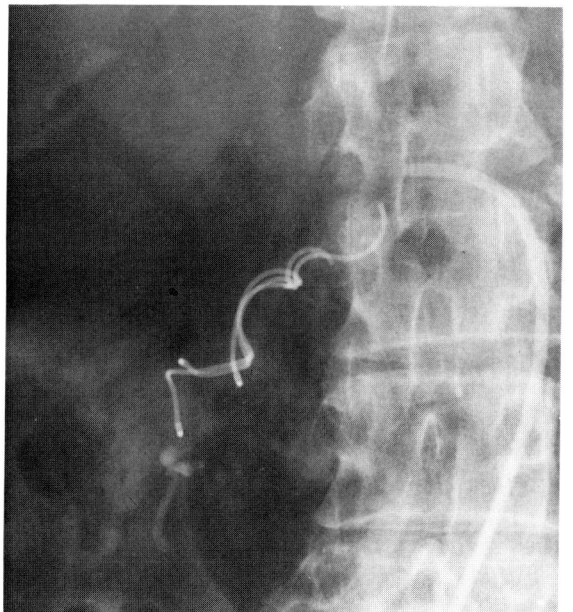

Fig. 9.25 Insertion of three 'Gianturco' devices.

Embolisation of right renal artery using 'Gianturco' devices is shown in Figures 9.24–9.26.

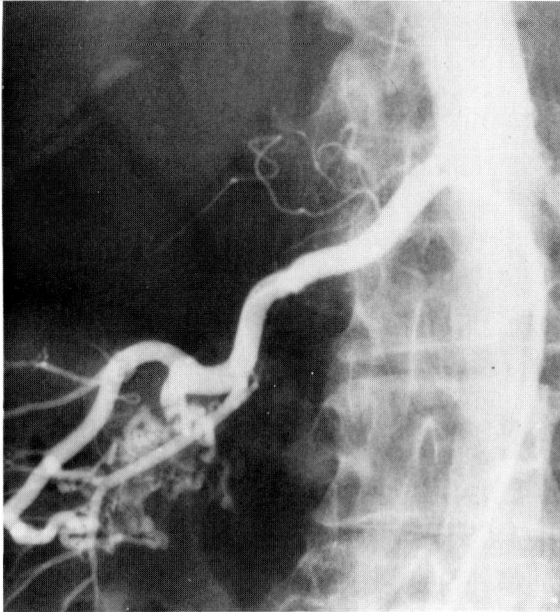

**Fig. 9.26** Post-embolisation angiogram — demonstrating reduced blood supply; eventually total embolisation will occur due to clot forming around the devices.

Fig. 9.27                                    stone

Fig. 9.28A                          total occlusion

Fig. 9.28B                          sub
                                    total occlusion

## 2. Percutaneous transhepatic drainage

The indications for drainage of the biliary tract following a percutaneous transhepatic cholangiogram (PTC) are as follows:

a. Pre-surgical temporary decompression, especially in patients with deep jaundice to improve their general condition prior to surgical intervention.

b. Permanent drainage for inoperable cases of bile duct malignancy.

A percutaneous transhepatic cholangiogram is performed to visualise the biliary tree and demonstrate the site of obstruction and its nature.

A duct showing a concave end will generally signify stone obstruction (Fig. 9.27).

A duct showing a cigar-shaped end possibly with narrowing will generally signify a malignancy encroaching on the duct (Fig. 9.28A & B).

If the system is then to be drained a sheathed needle is inserted into a suitable large duct. Facility for fluoroscopy in the lateral position may be useful at this stage to determine the antero-posterior position of the duct to be catheterised.

There are three types of drainage procedures:

a. External
b. External/Internal
c. Internal.

Wherever possible internal drainage is preferable, but is not always possible. It does however solve the patient's fluid balance problems which can be encountered with prolonged external drainage.

### a. External drainage

This technique is very useful as a temporary pre-surgical measure but may be necessary as long-term drainage where it is impossible to cross a total occlusion in the case of malignancy. A catheter with multiple side holes (e.g. pigtail) is left in a suitable duct and secured at the skin with tape, and the external portion of the catheter is placed in a drainage bag attached to the skin. If the catheter becomes blocked it can be syringed through with saline.

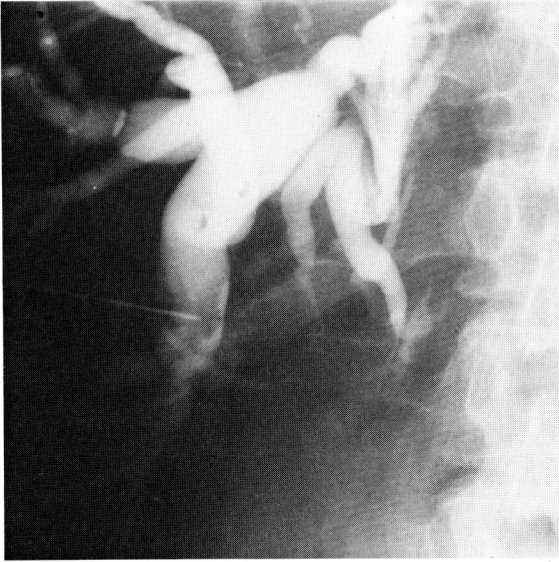

Fig. 9.29 Percutaneous transhepatic cholangiogram using a 'Chiba' needle demonstrating severe narrowing of the common bile duct.

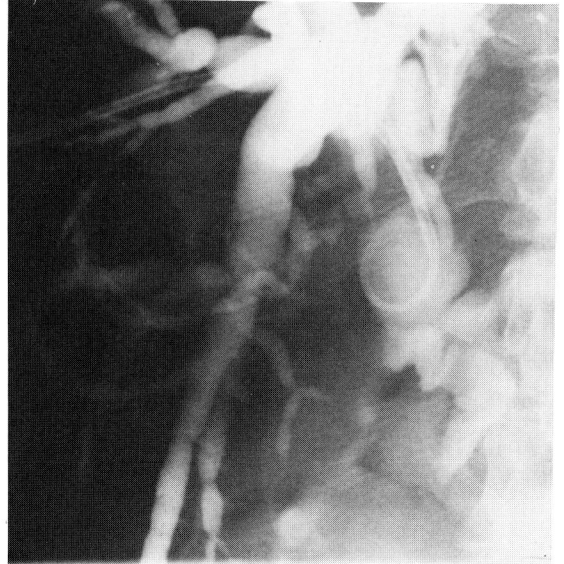

Fig. 9.31 External drainage. Pigtail catheter in situ.

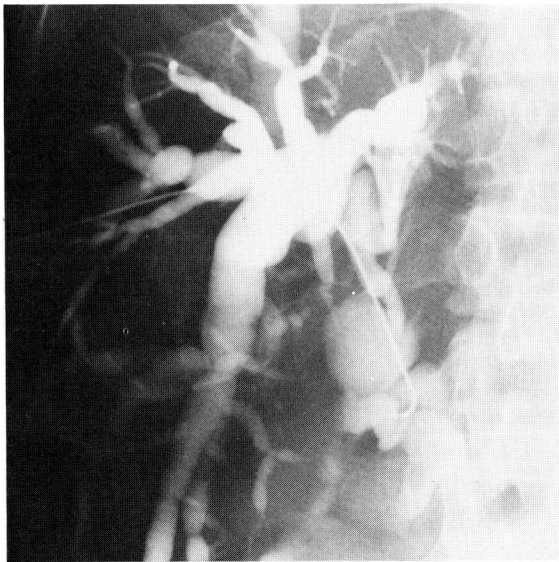

Fig. 9.30 A stiff guide wire is positioned in the bile duct to be drained.

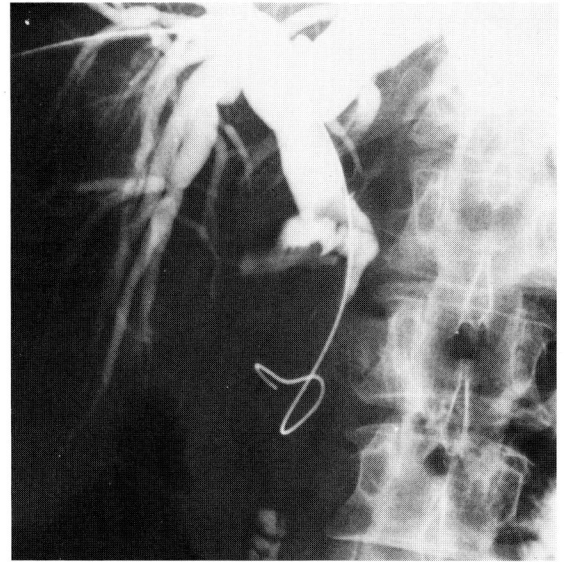

Fig. 9.32 Radiograph demonstrating a guide wire across the narrowing in the bile duct.

### b. External/internal drainage

This method is used where possible for long-term drainage. A sheathed needle is inserted into a duct and a guide wire is passed through the sheath which is then replaced by a longer catheter. The normal guide wire is then replaced by a special stiffer guide wire to aid advancement through the biliary system and across the obstruction into the duodenum. A multi-side hole catheter. is then

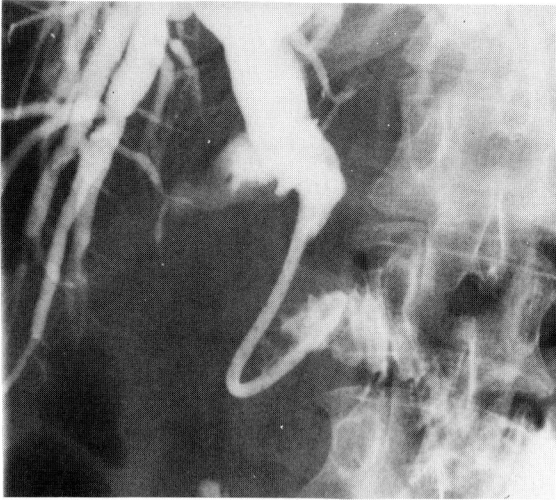

**Fig. 9.33** External/internal drainage. A special multi-hole pigtail catheter is used with the pigtail end in the duodenum and the drainage holes in the catheter on either side of the stricture.

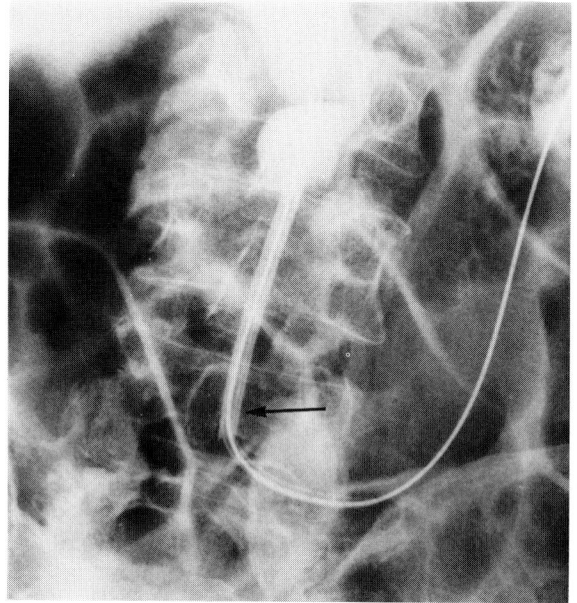

**Fig. 9.34** The guide wire is placed through to the duodenum, over which is passed a fine dilator and then a coarse dilator (arrowed).

inserted over the guide wire ensuring there are sufficient holes, in the catheter, on either side of the obstruction. A pigtail type catheter is useful for this procedure as it tends to anchor itself better in the duodenum and so prevent the catheter dislodging. The catheter should be taped at the skin surface and a tap attached to the end. The catheter should be flushed through with saline every 2 days. This type of system can be left permanently, with occasional catheter changes.

### c. Internal drainage

This is a complete internal drainage system which is the better permanent drainage for the patient as there is no catheter outside the body. The procedure is the same as for the external/internal drainage but instead of using a pigtail catheter, a special short catheter is placed across the obstruction to act as a 'bridge' across to the duodenum.

24-hour bed rest is recommended after any drainage procedure.

The radiographs in Figures 9.29–9.32 show external drainage, then a few days later an external/internal drainage was performed (Fig. 9.33)

Internal drainage performed on a second patient is shown in Figures 9.34–9.36.

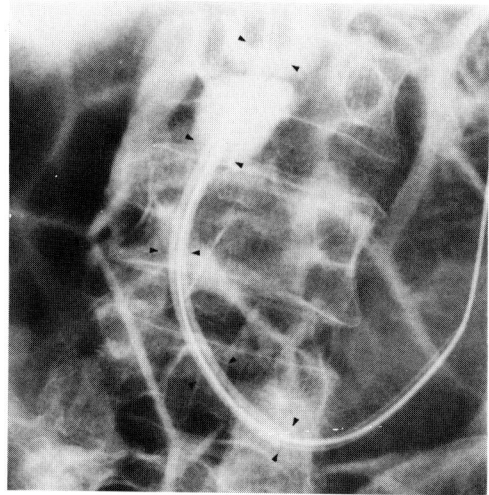

**Fig. 9.35** The coarse dilator is then removed and replaced by the short prosthesis as indicated. The guide wire and fine dilator are then removed.

### 3. Retrieval of stones from the bile-ducts

The retrieval of stones is generally requested as a postoperative procedure and can only be performed if a suitably wide T-tube has been inserted. The

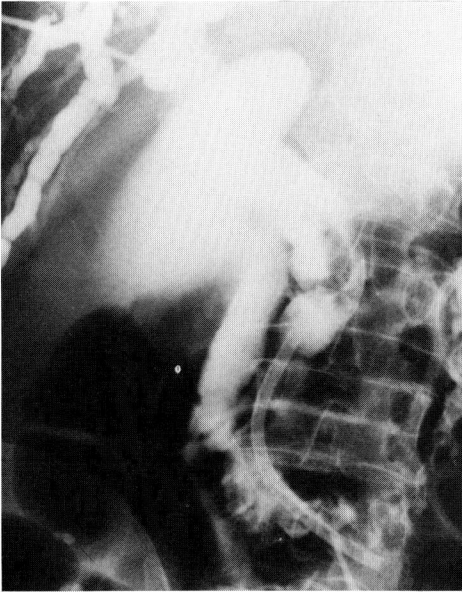

Fig. 9.36 Short internal drainage prosthesis in situ across the stricture; a final cholangiogram was performed via the external drainage catheter.

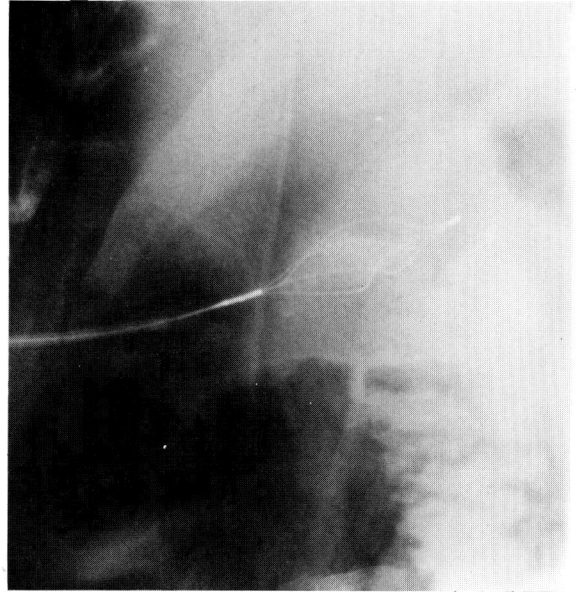

Fig. 9.37 A non-opaque gallstone inside the dormier basket is demonstrated and removed together with the T-tube.

T-tube must be left in situ for 4–6 weeks before retrieval can be performed, so that a fibrous track has had time to form. This enables the duct to be re-entered if the T-tube is removed during the procedure.

A catheter is advanced, over a 'J-tipped' guide wire to just beyond the stone. The guide wire is then replaced by the dormia basket which opens as it advances out of the catheter. The stone is caught in the basket and is removed either through the T-tube or if it is a large stone, the T-tube is removed as well.

If more than one stone is to be retrieved the procedure can be repeated easily, by re-catheterising the duct, possibly with the aid of a tip-deflecting guide wire, which can then be replaced by the dormier basket.

## 4. Percutaneous transluminal angioplasty

Angioplasty has been performed successfully for a number of years in peripheral arteries. It is now a widely accepted treatment for lesions of the femoral and iliac arteries and has been used with dramatic effect in renal artery stenosis associated with severe hypertension. Blood pressures in

excess of 200 mmHg systolic pressure can gradually drop to a normal of 120 mmHg after renal angioplasty.

The catheter used for this procedure is a balloon catheter, with varying sizes of balloon available, according to the clinical needs. The balloon catheter is introduced directly into the vessel as in the case of renal or peripheral angioplasty. However, for percutaneous transluminal coronary angioplasty (PTCA) an extremely fine balloon catheter is introduced through a larger guiding catheter placed at the origin of the appropriate coronary artery.

Coronary angioplasty is a new and exciting dimension in balloon dilatation of arteries. This is, of course, a very hazardous procedure and must only be undertaken by a skilled radiologist or cardiologist familiar with coronary angiography.

Percutaneous transluminal coronary angioplasty (PTCA) must only be undertaken in association with full cardiothoracic surgical standby, to allow immediate intervention if coronary occlusion and impending myocardial infarction should occur. The ideal lesion for PTCA is isolated, relatively proximal, concentric and less than 1 cm in length. Calcification in the lesion, or a major branching at

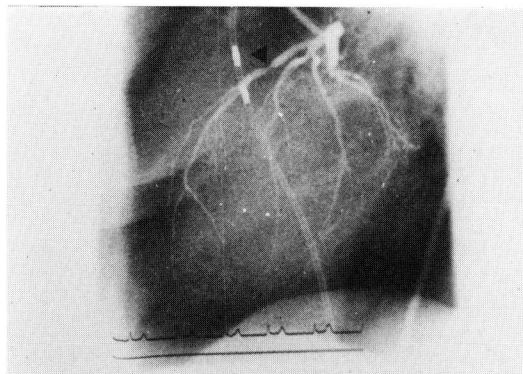

**Fig. 9.38** Before PTCA

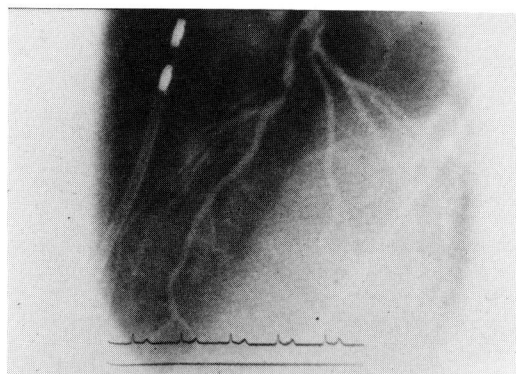

**Fig. 9.40** Before PTCA

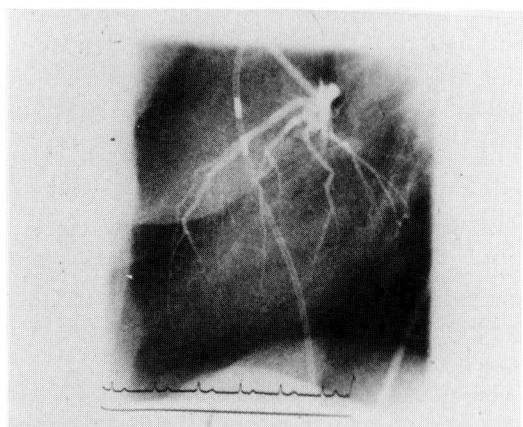

**Fig. 9.39** After PTCA

**Figs 9.38 & 9.39** Left lateral projection

**Figs 9.38–9.41** A severe lesion is demonstrated in the left anterior descending branch of the left coronary artery. (By courtesy of Dr N. P. Silverton.)

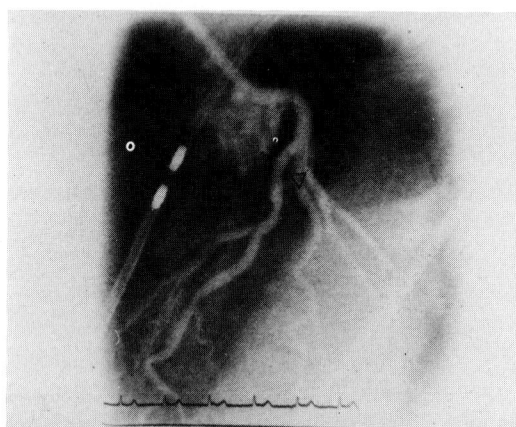

**Fig. 9.41** After PTCA

**Figs 9.40 & 9.41** Left anterior oblique (LAO) cranial projection

the point of obstruction are at present considered unfavourable factors. A few centres will undertake angioplasty when there is diffuse coronary disease, in order to dilate a particularly severe narrowing, which may be responsible for the patient's angina.

Prior to coronary angioplasty being undertaken, a special open-ended right heart catheter with bi-polar pacing electrodes a few centimetres from the tip is introduced into the pulmonary artery via the femoral vein. Through this catheter is infused Dextran 40 in dextrose to prevent red cell aggregation in the injured coronary artery, together with 1–5 mg of glyceryl trinitrate for the duration of the

procedure. An external pulse generator is connected to the pacemaker leads of this catheter and the pacing threshold established.

Coronary arteriography of the affected vessel is then performed to reassess the lesion in question. If the left ventriculography and coronary angiography have been very recently performed then left ventricular function and the unaffected coronary artery need not be re-examined on the day of the angioplasty. The normal coronary catheter is then exchanged for a much larger introducing catheter with a specific pre-formed shape according to the anatomy of the vessel to be dilated. Effective inflation of the balloon at the end of the dilatation catheter is checked and air removed by injecting a solution of 50/50 Urografin 370 and saline. The solution is then withdrawn from the balloon and the balloon catheter inserted through the guiding catheter. The fine guide wire at the very end of the balloon catheter guides it along the vessel and across the lesion.

Pressure recordings are taken from the ends of the guiding and balloon catheters, i.e. on either side of the lesion, and the difference in pressure (trans-stenotic gradient) is noted. An injection of contrast medium is made through the balloon catheter to ensure free flow and flushing away of contrast medium distal to the lesion. This is to ascertain that no atheroma has been dislodged and has caused occlusion of the vessel distal to the lesion which may result in myocardial infarction.

Once initial pressure recordings have been made, the first dilatation will be undertaken, inflating the balloon for up to 15 seconds. The contrast medium and saline solution used for inflating the balloon can be visualised fluoroscopically and recorded on ciné film to provide a permanent record of the dilatation. The solution is then immediately withdrawn, in order to deflate the balloon and the two pressures recorded. Intra-arterial glyceryl trinitrate is injected through the guiding catheter after the first dilatation and then as necessary to prevent coronary arterial spasm. An injection of contrast medium is made through the balloon catheter to check blood flow in the distal portion of the vessel.

Four or five dilatations are performed until the distal pressure approaches that of the proximal pressure.

Selective coronary angiography is then repeated in the same projections as prior to the dilatation, using the guiding catheter for the contrast injections. The success of this procedure may be confirmed either by changes in the trans-stenotic gradient, the angiography or both. A modest angiographic improvement may well be associated with enormous clinical and haemodynamic improvement for the patient.

PTCA is in its early stages of development in the United Kingdom, but hopefully in the future angioplasty may remove the necessity for some patients to be subjected to open heart surgery with its consequent trauma and expense.

## REFERENCES

Allison D J 1978 Therapeutic embolization. British Journal of Hospital Medicine December: 707–715

Allison D J 1980 Gastrointestinal bleeding, radiological diagnosis. British Journal of Hospital Medicine April: 358–365

Bean W J, Smith S L, Calonje M A 1974 Percutaneous removal of residual biliary tract stones. Radiology 113: 1–9

Christenson P C, Ovitt T W, Fisher III H D, Frost M M, Nudelman S, Roehrig H 1980 Intravenous angiography using digital video subtraction, intravenous cervicocerebrovascular angiography. American Journal of Roentgenology 135: 1145–1152

Crummy A B et al 1980 Computerised fluoroscopy, digital subtraction for intravenous angiocardiography and arteriography. American Journal of Roentgenology 135: 1131–1140

Dooley J S, Dick R, Olney J, Sherlock S 1979 Non-surgical treatment of biliary obstruction. The Lancet November: 1040–1043

Dooley J S, Dick R, Irving D, Olney J, Sherlock S 1981 Relief of bile duct obstruction by the percutaneous transhepatic insertion of an endoprosthesis. Clinical Radiology 32: 163–172

Fredzell G, Borggren A 1977 The Angioskop — a new multidirectional examination unit for advanced radiological examinations. Electromedica 3–4: 137–141

Keates P G, Clarke O F 1977 Free arc angiography. Medicamundi 22(1): 11–15

Ludwig W J, Engels P H C 1981 Digital vascular imaging (DVI). Medicamundi 26(2): 68–80

Meaney T F et al 1980 Digital subtraction angiography of the human cardiovascular system. American Journal of Roentgenology 135: 1153–1160

Mistretta C et al 1981 Digital vascular imaging. Medicamundi 26(1): 1–10

Nakayama T, Ikeda A, Okuda K 1978 Percutaneous transhepatic drainage of the biliary tract Gastroenterology 74: 554–559

Ovitt T W et al 1980 Intravenous angiography using digital video subtraction, X-ray imaging system. American Journal of Roentgenology 135: 1141–1144

Ring E J, Husted J W, Oleaga J A, Freeman D B 1979 A multihole catheter for maintaining long term percutaneous antegrade biliary drainage. Radiology 132: 752–754

Walker W J, Goldin A R, Shaff M I, Allibone G W 1980 Per catheter control of haemorrhage from the superior and inferior mesenteric arteries. Clinical Radiology 31: 71–80

Wallace S, Gianturco C, Anderson J H, Goldstein H M, Davis L J, Bree R L 1976 Therapeutic vascular occlusion utilizing steel coil technique clinical applications. American Journal of Roentgenology 127: 381–387

Wheeler P G et al 1979 Non-operative arterial embolization in primary liver tumours. British Medical Journal July: 242–244

# 10

# Research

The word research implies a scientific search or study for the cause of some problem or for a solution to a problem. Research is therefore the use of existing knowledge to establish new information. This is particularly so in the case of medicine and science. In the seventeenth century Francis Bacon had much to say about research and indeed set the scene for modern scientific research by recommending that observations of facts should be recorded and conclusions drawn from these facts.

A true researcher will be engaged in the task all his life for his thirst for knowledge and quest to investigate facts are never ended. There are some misconceptions about what work is actually involved in research, but the size of the job may range from a mini-research exercise based on the investigation of one or two small factors through to a full-scale research exercise involved in the production of a thesis for a philosophy degree (Ph.D). This may involve a person in a few hours of work per week for a mini-research project through to 3 years full-time study for a Ph.D degree.

## PREPARATION FOR RESEARCH

Generally speaking one observes phenomena throughout one's working life. One has an idea that this particular thing works because of that particular thing. Or one may think that is the case but wish to make further investigation. It might be that one wonders what would happen if such and such a procedure was undertaken. If one has a pet theory or subject then one reads all that is available on the subject. While reading textbooks,

**Table 10.1** Type of information to be recorded on index record card

Full name of author plus two initials
Full title of work — underlined
Place, publisher and date of publication
Edition (if available)
Particular page(s) of interest
Volume number

journals or manuscripts it is important to make a record of the source of the information (see Table 10.1). If this information is not catalogued at the time it is often lost. Over many years it is possible to build up a whole reference section about a particular speciality. One should not, however, confine the reading only to the special areas of interest but to the scientific field in general. In this way, general concepts of advances in the whole field are gained. Other ideas for future research often spring to mind through general reading. It is not possible to read every journal, nor indeed is it necessary. For a research project of any size however it would be necessary to read at least ten journals on a monthly basis.

It would also be important to go back through the immediate past 10 years of each of these journals to gain an appreciation of any recent trends which may have occurred or may be occurring in the particular field of interest.

One of the advantages of teaching is that the teacher is required constantly to keep up to date with modern trends and has to read widely to keep abreast of his subject. The research worker should not however commit to mind reams and reams of printed material. A quiet browse through books or journals is all that is necessary. Scientific facts, data, statistics etc should be stored on tape or on the computer or word processor. The mature

**Table 10.2** Biographical information to be recorded on index record card

| |
|---|
| Full name of author |
| Full title of article in inverted comas |
| Name of publication — underlined |
| Volume number |
| Date of issue |
| Number of pages — covered by article |

scientist knows where to locate information rather than being able to recite it off the cuff (Table 10.2).

## THE ACQUISITION OF KNOWLEDGE AND AIMS OF RESEARCH

### Personal experience

The researcher may draw upon his own observations and records of what has been experienced or performed, but the records may have been poorly made. Generalisations may be drawn on insufficient evidence or from too few examples. This may lead to incorrect conclusions being drawn because of personal prejudice and bias; evidence may be left out because it was not consistent with earlier experiences. Additionally there is always the danger of failing to recognise the salient features of the situation and those which are irrelevant. It can be dangerous to rely too heavily upon personal experience. The research worker should attend conferences, seminars and join discussion groups to broaden his outlook. On these occasions there is the opportunity to discuss matters with other workers in the same field and it is sometimes advantageous to discuss problems or findings with scientists in other fields of research. It is often useful to present a paper with data to a group of scientists to receive constructive criticism and comment. This may reveal deficiencies in the existing research programme or point to further channels for exploration.

### The authority of research findings

Man tends to look for authority in figures and literate man turns to the expert for authority, facts and advice. Knowledge may be handed down by teachers, other scientists, and often has to be taken on trust. It is virtually impossible to test the validity of every view held. Nevertheless, it should be remembered that because information has been handed down as generally accepted, it may not necessarily be true. In most cases the expert is better informed in his field than other people because of his intelligence level and his training experience. Differences of opinion may exist however, even amongst experts, and it is therefore important not to accept, without question, a particular viewpoint for all time.

### Deductive reasoning

This is a syllogism used for centuries as a method of acquiring knowledge. It is made up of three statements; the first two are:

*Premises* which provide a basis or evidence for conclusions of an argument given in a third statement.
*Categorical syllogism* which is a statement composed of clear definite facts, e.g. Mammals cannot live without oxygen. Rats are mammals, therefore rats cannot live without oxygen.

A syllogism may be either *hypothetical* or *alternative* when it is not made from categorical statements.
A hypothetical syllogism may be, for example:

Should the rain continue, the gardens will be in danger of being flooded. The rain continues, therefore the gardens are in danger of being flooded.

An alternative syllogism may be:

The dog must be allowed to have food or he will die. The dog is not allowed to take food, therefore the dog will die.

This type of deductive reasoning is in use in everyday life; doctors deduce what is wrong with a patient by examining the signs, symptoms and biochemistry; a detective on a criminal investigation deduces the facts as he examines the scene of the crime. Deduction is a useful tool for obtaining certain kinds of information but it must not be relied upon exclusively.

*Inductive reasoning*

It is rarely possible to count all the instances of a given class of events or objects because of the time and effort involved. As it is not feasible to weigh all the boys of 5 years of age in the United Kingdom, a representative sample has to be taken. If it was practical to weigh all 5-year-old boys, conclusions could be drawn and this would be *perfect* induction; real-life situations do not allow this, so the result is *imperfect* induction. The size of samples and the extent to which they are representative of a whole population limit the degree to which conclusions drawn are sound.

Both deductive and inductive reasoning have advantages and disadvantages.

The seventeenth century men such as Isaac Newton and Francis Bacon made use of a synthesis of reason and observation to give a method now frequently used in scientific research. Dewey (1933) described various stages in the way we think about a problem:

1. a problem is encountered which cannot be solved or explained immediately.
2. observations are made and facts are gathered which appear to be relevant.
3. a solution or hypothesis is formulated. Intelligent guesses are made to explain relationships between facts.
4. deductive reasoning is applied to work out if the consequences of the hypotheses are true.

The researcher looks for evidence of consequences that should follow if the hypothesis(es) is/are true. It is necessary to establish which of the hypothesis(es) is/are congruent with observed facts. This is often called the 'scientific method'.

Accidental discovery sometimes occurs and may be the result of finding something different from that which was being looked for but is of equal, if not greater value. It was, for instance, quite by chance that Alexander Fleming discovered penicillin. Sometimes knowledge comes to the research worker through intuition, a hunch or an inspiration. However, before intuition operates there has usually been much preliminary thinking about the topic.

## A HYPOTHESIS

A hypothesis is a supposition used as a basis of reasoning or assumption. The discovery of the value of oestrogen in the protection of bone evolved from the observation of a number of patients suffering Colles' fractures and hip fractures occurring in women after the menopause. There is a definite loss of bone in women after the age of 40 years compared with men and this is now known to be due to oestrogen loss. As a result of this finding some women are given oestrogen therapy to improve their calcium metabolism.

If a hypothesis is formed it helps to stimulate new ideas, suggests new experiments or observations. Some experiments carried out for research are specifically to test hypotheses. Not all hypotheses are correct, but they may nevertheless be useful because in testing them and obtaining a negative result, another basis for future work is substituted. In pure sciences, experiments may be repeated on several occasions and if the same results are established on each occasion, this gives credence to the hypothesis. Another important point for the researcher to remember is that he must be ready to drop a hypothesis, or modify it, if the experiments carried out do not support it. It is very useful to discuss ideas with others and to submit discussion papers at fairly regular intervals to benefit from an outsider's independent reaction.

## PREPARING FOR RESEARCH

In the Health Service, a great deal of the research carried out involves patients. When this is so, a carefully prepared case has to be submitted to an Ethical Committee. The case will require details of what the particular project involves. There should be details of the number and sex of patients to be included and exactly what the experiment involves. All these details are then forwarded to the Ethical Committee.

## USE OF IONISING RADIATION IN RESEARCH

When using X-ray procedures in research it is necessary to submit a protocol which indicates clearly the precise radiographic technique to be employed, the size of the X-ray film and hence area of body to be irradiated, and the radiation dose to that part of the body as well as to the gonad areas. The radiation dose is measured using thermoluminescent dosemeters (TLDs). When the radiographic examination is not directly for the patient's benefit, it is important that the patient is given a full explanation of the procedure and is told of any hazards likely to occur in the procedure. For example, a series of experiments was carried out on the knee joints and lower limbs of young atheletes. The radiographs were measured to give some indicators for designing prosthetic joints. Although in the trial procedure the dose to the gonads was found to be minimal, strict X-ray beam delineation was employed and the patient was given a lead scrotal shield. This was, in fact, a scrotal shield from a cricketer's box which was covered with a sheet of lead and housed in a 'Litesome' belt.

In another research project the upper limb joints (wrist, elbow and shoulder) and lateral views of the cervical spine were taken to examine the radiographs for arthritis or other damage. None of these sites was near to the gonads and most of the men were in the over 45 year age group. Nevertheless, a full protocol had to be designed and submitted to the Ethical Committee.

Exposure factors (ffd, kVp, mA, exposure time), type of generator, type of film and intensifying screens, type of X-ray tube, focal spot size, should all be recorded. This is not only valuable for the present series of radiographs, but should follow-up radiographs be required, comparative densities can be produced. Additionally, the information is required when documenting the research report and should other groups wish to establish the technique, important details are available.

*Records to be kept*

It is important that each patient radiographed should be correctly and accurately recorded. An X-ray request card should be completed for each patient. Details of exposure factors, processing, size and number of films should be recorded in each case. A note should be made on the back of the card or film of any special items or facts involved in each patient's survey. Whenever possible all patients

in a series should be radiographed under identical conditions, i.e. same X-ray room, table, X-ray tube, grid or bucky, same factors, generator and identical processing facilities.

**Application to carry out research**

Numerous bodies exist for the purpose of research, including the Medical Research Council and national groups such as Heart Foundation, Cancer Research, Regional Health Authorities and the College of Radiographers. Full lists can be obtained from national, society or local libraries, medical libraries and university libraries. It may be that you wish to conduct a small project within a department or carry out a regional or national study. Perhaps this will simply be a small contribution to an already vast amount of previous work and will result in the production of a paper in a relevant journal. On the other hand you may undertake to conduct research as part of a research degree programme. One of the differences between degrees is the length of time usually spent on research, e.g. a B.Phil 2 years, M.Phil 3 years and D.Phil 4 years. These are usually minimum periods of time, but there may also be a maximum imposed. The time will normally be doubled if the work is to be undertaken on a part-time basis.

To register for a research degree it is necessary to complete an application form similar to the one shown in Figure 10.1. Useful documents to consult include the higher/research degree regulations of universities: The Open University booklet entitled 'Postgraduate Prospectus and Student Handbook 1984' and the CNAA booklet 'Regulations for the award of the Council's degrees of Master of Philosophy and Doctor of Philosphopy 1984'.

It is useful to consider certain details before you commence and these can help formulate the protocol (Warren, 1982). Consideration should be given to:

1. The purpose of the study. Is it to investigate current radiographic procedures and their effectiveness? Is it to discover a new technique? Is it part of a larger survey, e.g. muscle measurements of arms and thighs as part of a complete research programme on the effects of anabolic steroids in weight lifters?

COUNCIL FOR NATIONAL ACADEMIC AWARDS

FORM CNAA 9(R) (1983)

*For Official Use*
Date Rec.
Item No.

**Application to register for CNAA's Research Degree**

Submitted by............................................. *(Sponsoring Establishment)*
for the degree of *(delete as appropriate):* (i) Master of Philosophy (MPhil)
(ii) Master of Philosophy with transfer possibility to Doctor of Philosophy (MPhil/PhD)
(iii) Doctor of Philosophy (PhD)

---

**1 The Applicant**

Name: Male/Female* Date of Birth:

Private address:

Present post and place of work:

Particulars of any scholarship or other award held in connection with the proposed research programme:

Qualifications gained (Regulation 2 refers) *(include places(s) of higher education, courses completed, main subjects, classification of award, date and name of awarding body):*

Training and experience *(include details of activities (with dates) relevant to this application, and of any research or other relevant papers, books, etc, which have been published):*

**2 Academic Referees** (see Note (a)):

**3 Name of Collaborating Establishment** (see Note (b)):

**4 The Programme of Research**

**4.1** Title of the proposed investigation:

**delete as appropriate*

**Fig. 10.1** Application form to register for CNAA's research degree (by kind permission of the Council for National Academic Awards)

**4.2** Aim of the investigation:

**4.3** Proposed plan of work, including its relationship to previous work, with references (See Note (c)):

**4.4** Details of facilities available for the investigation (including funding and location):

**4.5** Relationship between work to be undertaken in the collaborating establishment and that to be undertaken at the sponsoring establishment or elsewhere (Regulation 3.6 refers):

**Fig. 10.1** (contd)

**5    The Programme of Related Studies** (Complete *either* 5.1 *or* 5.2)

**5.1** Details of programme of related studies to be undertaken (Regulations 3.8 and 3.9 refer):

**5.2** Where an integrated programme of study is proposed, details of the course of postgraduate study on which candidate's performance is to be formally assessed (Regulation 3.10 refers) are required:

**6    Supervision of Programme of Work** (Regulation 6 refers)

**6.1** Director of Studies (see Note (d)) *(include name, qualifications, post held and place of work):*

Experience of supervision of *registered* research degree candidates:
*Currently supervising* . . . . . . . . . . . . . . . . . . *CNAA and UK university candidates*
*Previously supervised* . . . . . . . . . . . . . . . . . . *CNAA and UK university candidates*
*(successfully completed supervision)*

**6.2** Second Supervisor(s) (see Note (d)) *(include name, qualifications, post held and place of work):*

Experience of supervision of *registered*  research degree candidates:
*Currently supervising* . . . . . . . . . . . . . . . . . . *CNAA and UK university candidates*
*Previously supervised* . . . . . . . . . . . . . . . . . . *CNAA and UK university candidates*
*(successfully completed supervision)*

**6.3** Details of any other person(s) who will act in an advisory capacity *(name, qualifications, post held and place of employment):*

**Fig. 10.1** (contd)

**7  Period of Time for Completion of Programme of Work** (Regulation 4 refers)

**7.1** Expected starting date for registration purposes (Appendix 1 of the Regulations refers):

**7.2** Mode of study (full-time or part-time):

**7.3** Amount of time (hours per week average) allowed for programme:

Expected duration of programme (in years) on the above basis to MPhil:

and additionally to PhD:

**8  Statement by the Applicant**

I wish to apply for registration for . . . . . . . . . . . . . . . . . . . . . . . . . . . . . . . on the basis of the proposals given in this application.

I confirm that the particulars given in Section 1 are correct.

I understand that, except with the specific permission of CNAA, I may not, during the period of my registration, be a candidate for another award of CNAA or of a university.

I understand that, except with the specific permission of CNAA, I must prepare and defend my thesis in English.

Signed . . . . . . . . . . . . . . . . . . . . . . . . . . . . . . . Date . . . . . . . . . . . . . . . . . . . . . . . . . . . . . .

**9  Recommendation by the Supervisors**

We support this application and believe that. . . . . . . . . . . . . . . . . . . . . . . . . . . . . . . . . . . . . . . . . has the potential to complete successfully the programme of work proposed.

We recommend that this applicant be registered as a candidate for CNAA's research degree.

Signed . . . . . . . . . . . . . . . . . . . . . . . . . . . . . . . Date . . . . . . . . . . . . . . . . . . . . . . . . . . . . . .

Signed . . . . . . . . . . . . . . . . . . . . . . . . . . . . . . . Date . . . . . . . . . . . . . . . . . . . . . . . . . . . . . .

**10  Recommendation by the Sponsoring Establishment** *(unless Section 11 below is completed)*

I support this application for registration of . . . . . . . . . . . . . . . . . . . . . . . . . . . . . . . . . . . . as a candidate for a research degree of CNAA

Signed . . . . . . . . . . . . . . . . . . . . . . . . . . . . . . . Date . . . . . . . . . . . . . . . . . . . . . . . . . . . . . .
*(Head/Principal Officer of the Establishment or authorised deputy or Chairman/Secretary of approved college Research Degrees Committee)*

**11  Notification of Registration on behalf of CNAA**

*Note: This section may be completed only by a college which has a Research Degrees Committee approved by CNAA.*

I confirm that the candidate was registered by this college for the degree of MPhil or MPhil/PhD.*

on . . . . . . . . . . . . . . . . . . . . . . . . . . . . . . . with effect from . . . . . . . . . . . . . . . . . . . . . . . . . . . . . . . . . . . . .

Signed . . . . . . . . . . . . . . . . . . . . . . . . . . . . . . . Date . . . . . . . . . . . . . . . . . . . . . . . . . . . . . .
*(Chairman/Secretary of approved college Research Degrees Committee) *delete as appropriate*

The sponsoring establishment should send the completed form and any attachments together with the appropriate application fee to CNAA's Assistant Registrar for Research Degrees at: 344–354 Gray's Inn Road, London WC1X 8BP.

**Fig. 10.1** (contd)

2. What is the present state of the literature? Has anything been written up already? There will be a need to consult documents in the university or medical school library, such as Index Medicus, Index Medica Radiographica, cumulative indices for various medical and radiological journals.

3. Patient details if they are involved. Details concerning section of population included in survey. It is also important to indicate how the sample of patients or subjects will be chosen, e.g. will it be a random sample?

4. Data collection. It is valuable to consider the factors or variables which will be involved in the survey. When any scientific research is to be undertaken it is of paramount importance that as many variables as possible are held steady. Ideally only one variable should be investigated at any one time. For example, exposure factors for the radiographic technique should be identical. The only differences in each case will then be the part under investigation.

5. Data processing. Consideration should be given, well in advance, to the amount of data which will need to be handled. A choice of computer or data processor will need to be made. It may be that certain statistical functions will need to be undertaken and a programme may be necessary for the computer.

6. Ethics of the programme. When any research programme is being contemplated it is necessary to consider the ethics of the procedure, particularly when ionising radiation is involved. Before commencing the project, permission or approval of the Ethical Committee in the hospital will have to be sought. It will be necessary to outline exactly the procedure for informing the subjects or patients about their right to refuse the X-ray examination. This is particularly important if the examination is of no direct benefit to the patient.

7. Finally, before submitting the programme for research, it will be important to have an estimate of the cost involved. This is particularly so if one is applying to a research group or foundation for funds to support the project.

Costs will include staff (radiographers, nurses, technicians, office personnel), materials (X-ray film, intensifying screens, grids) and special facilities, e.g. tools for research. There may be telephone or postal charges, cost of paper or printing etc.

There could be other points to consider and sometimes specific regulations are laid down by the body awarding the grant. In the case of theses for universities or learned bodies there will be the cost of examiners, various course fees, printing of the thesis and binding (three copies may have to be submitted at the end of the project).

## Research funds

If funds are required to carry out research, and usually they are, it is important to proceed in the specified way. In addition, the application must be clear and concise and should define precisely the course to be followed (Research Funds Guide, 1973).

It is important to apply well in advance of the estimated starting date since the application may need to be considered with several others, and by several committees. When completing an application form ensure that all details are given as requested and that statements made are accurate. Another important point regarding applications for funds is to make sure that it is sent to the appropriate body. It may be expedient to make enquiries locally at first. The Regional Health Authorities and Universities have personnel specially appointed to deal with such matters. Make as many useful contacts as possible in order to sound out the feasibility of the proposed work and to gain support. Not every application is accepted particularly in times of economic restraint, but the researcher should not be disheartened. It is usually possible to become engaged in some useful interim work whilst a second or third application is being considered. Possibly a limited amount of funding will be available and it may be necessary to trim one's proposals according to the funds available.

## WRITING UP THE PAPER OR THESIS

Similarities exist between writing a scientific paper and writing up a project or thesis. Much will depend upon the research involved. Having carried out the research and gathered the data, the most important and yet often the most laborious and difficult part is the writing down. The information and the facts must be put together in a succinct, precise and logical manner. Generally speaking the account will be dealt with under several headings, which may include:

*Title* — this should be short but specific and should describe the work precisely.

*Introduction* — a short description of the project and an outline of the proposed method of conducting the research.

*Synopsis/Historical background* — a synoptic record of the literature on the subject dealing with the events which have been dealt with so far and giving leaders to support the present proposed programme.

*Method* — an account of the method of approach and the experiments conducted. This will probably form the bulk of the written description.

*Results* — the results should be presented in a clear and concise fashion. Graphs, tables and experimental findings should be included here.

*Discussions and Conclusions* — the work should be discussed pointing out what has been achieved and what problems were encountered. The facts should be founded on previous work and conclusions should be well substantiated by statistical as well as scientific evidence.

*Future projects* — it is always useful at the end of a thesis to indicate where next the research may lead.

At the end of the thesis there should appear appendices. Here extracts of important documents may appear rather than in the body of the text. Illustrations will also be presented in this section.

Finally, a bibliography indicating clearly any textbooks or reference books which have been used in reading up the material and a list of references indicating the source(s) of material used to support statements in the text. References may appear in number order in the text, or in alphabetical order by title or principal author's name.

Acknowledgement should be made to those who have helped in the work in any way: firms or Research Foundations who have provided finance to support the programme and supervisors who have given advice and encouragement throughout.

## Presentation of the thesis

If this is a full-blown thesis or research project for a Masters or Doctorate degree, the university will require the manuscript to be bound (Hawkins, 1982). Details of the precise regulations should be available from the Higher Degrees Office of the particular university. This will indicate size of pages, margins, typescript, numbering of pages and colour and type of binding. In addition, details will be given for lettering of titles etc., on the spine and outside front cover.

### Use of statistics

What are statistics? According to Browning 1964: 'They are numerical facts systematically collected and set out.' If the research worker has never used statistics before it is useful to read around the subject. A very easy-going and lighthearted approach is adopted by Huff (1974). Terms which need to be explained include median, mean and mode, normal distribution, standard deviation and significance. Another useful book by Robson (1975) deals a little more seriously and scientifically with statistics and progresses to Chi squares and degrees of freedom, giving worked examples. Finally a textbook of statistics such as Ractliffe (1969) is of some value to the novice. It is also important to learn how to present tables of facts and to draw graphs which will help to outline relevant points to which reference has been made in the text. It may be that you need to seek advice from a statistician in the hospital. Whatever happens do not present tables or graphs or use statistical evidence which you cannot interpret if asked in your thesis viva voce.

## EXAMPLES OF THE USE OF RADIOGRAPHY IN SOME RESEARCH PROJECTS

### Bone mineral measurement by morphometry

#### Metacarpals

Experiments were conducted to show that small rotations of metacarpals about their long axes produce small systematic changes in cortical dimensions as measured by radiographic morphometry. The hand is frequently radiographed for morphometric purposes to show bone loss which occurs with advancing age in both men and women. In tubular bones like the metacarpal this results in an increase in medullary width and a reduction in cortical width (Barnett & Nordin, 1960; Garn, 1970). Cortical bone loss can be determined by measurements of total and medullary width on radiographs using Vernier calipers (Barnett & Nordin, 1960).

Measurements of the second metacarpal at the mid-shaft have been, and still are, used in such studies. A more accurate estimate of bone loss can be obtained if measurements are made of the total width, and medullary width at the midshaft of metacarpals 2, 3 and 4 of both hands (Horsman & Simpson, 1975). These measurements can be taken using a semi-automatic computer-controlled morphometer.

Errors are introduced into these measurements however if slight rotations of the hands occurs. The effects of such rotations were, therefore, investigated. 10 right 2nd, 3rd and 4th metacarpals were purchased from an anatomical dealer. Each bone was mounted in turn in a jig in which it pivoted about two fixed points on the proximal and distal metaphyses. The bones were radiographed in the jig placed directly on top of the cassette.

Each bone was radiographed in the neutral position — as in the neutral position in the hand (Meschan, 1959, Clark, 1979). The bone was rotated through 5 degree intervals from 10 degrees medial to 10 degrees lateral, and radiographs were taken in each position. Measurements were made of each bone from the radiographs and the results showed that medial rotation of the second metacarpal causes very small systematic increases in total bone width of the order of + 0.025 mm/degree accompanied by a similar change in medullary width of about + 0.021 mm/degree. Lateral rotation of the second metacarpal causes decreases of total width in the order of 0.013 and medullary width of the order of 0.017 mm/degree (Armes et al, 1979). These small changes are inconsequential in cross-sectional studies. However, in sequential studies involving measurements on six metacarpals, small rotational positional errors of the hand

are of importance. Thus this radiographic experiment was of value in establishing the importance of hand positioning of patients for radiographic morphometry.

*Morphometric measurements of the femur*

Changes in the cortex of the femur related to age were noted by Atkinson et al (1962) and Barnett & Nordin (1961). Resorption of cortical bone of the femur is greater with increasing age both in men and women. Women lose approximately 0.4% per annum and men 0.2% per annum of their cortical bone by intracortical resorption (Horsman, 1976). The rate of decrease in thickness of the lateral mid-femoral cortex is approximately 0.05 mm per annum in females and approximately 0.01 per annum in males (Arnold et al, 1966). Minute changes in bone geometry can only be detected with greater accuracy in positioning of the patient's femur for radiography.

To investigate the effects of small changes in radiographic technique the following experiment was carried out (Bentley, 1978): a jig similar to the one used in the metacarpal experiments was used to fix the head of the femur at one end and the distal condyles at the other (Fig. 10.2). In the jig it was possible to rotate the femur through 360 degrees by steps of 5 degrees. Radiographs were made with the femur in the vertical position and then in various stages of internal and external rotation.

From the radiographs measurements of the total width, cortical width and medullary width were made. It was found that greatest differences attributable to rotation of the femur occur in medial rotation for the total width and cortical width. Lateral rotation causes a decrease in total width associated with an increase in medullary width which together produce a decrease in cortical width.

This is another instance of radiographic research establishing important parameters for measurement techniques.

## Other projects

In other research projects radiography has been used in the studies of the knee joint to enable

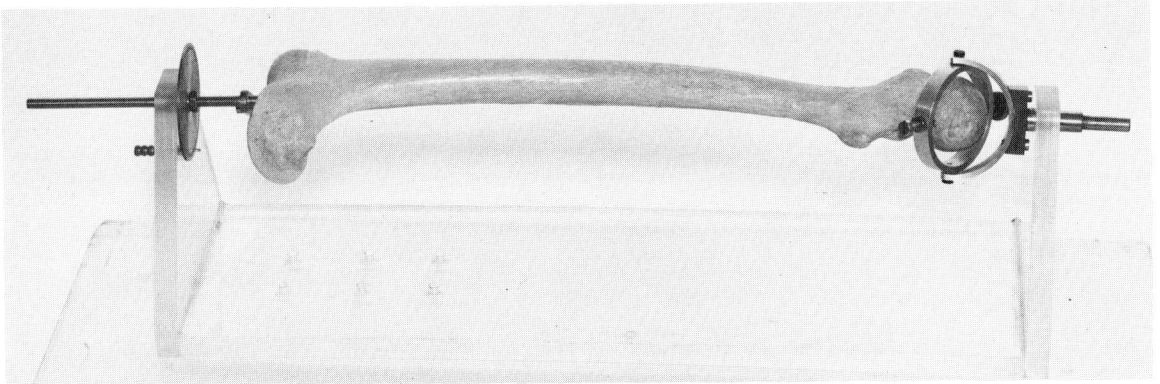

Fig. 10.2 Femur in jig for experiments on effects of rotation of leg (femur)

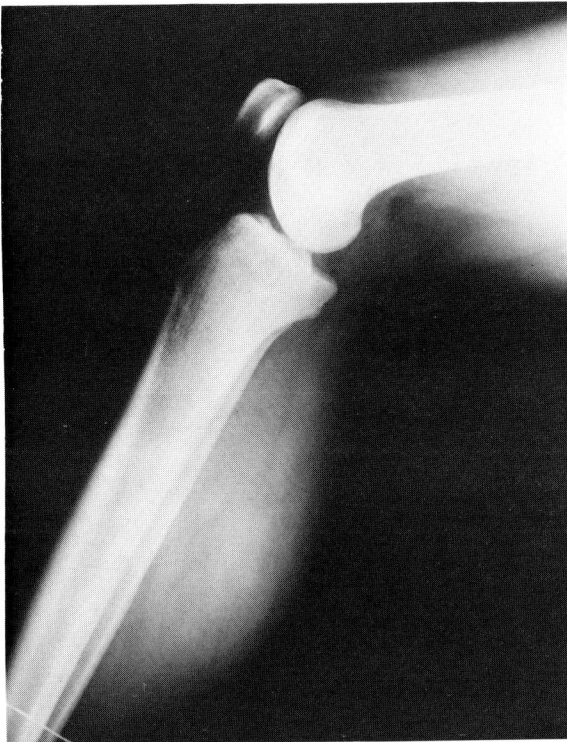

Fig. 10.3 Radiograph of knee and leg for measurement purposes

procedures and have formed part of other theses or dissertations.

Research then is not something taken on without consideration and careful planning. The research worker needs to be constantly reading and searching literature and attending seminars and conferences. Information must be carefully collected and correlated and research work must be carefully and accurately recorded. The researcher must have patience and be prepared to accept that sometimes a particular avenue explored may not yield great reward. However, once one has adapted to the system of research and has become disciplined in the routine, the rewards of research lie in the sense of a job well done and the constant excitement of new avenues awaiting exploration.

*Acknowledgement*

The author wishes to acknowledge the assistance of Dr G.J.S. Parkin, Consultant Radiologist at The General Infirmary at Leeds and Wharfedale General Hospital, Otley, for kindly reading the script and offering constructive criticism.

calculations of the forces acting in the knee joint (Fig. 10.3). Further studies of the cadaveric knee joint have helped to evaluate the tensions along the quadriceps tendon and the patellar ligament. This project was used in establishing joint prostheses for knee joint replacement surgery. Numerous other small projects have used radiography in their

REFERENCES

Armes F M, Horsman A, Bentley H B 1979 Effects of rotation on radiographic dimensions of metacarpals. Radiography XLV (536): 172–175
Arnold J S, Bartley M H, Tont S A L, Jenkins D P 1966 Skeletal changes in aging and disease. Clinical Orthopaedics xlix: 17–38

Atkinson P J, Weatherall J A, Weidmann S M 1962 Changes in density of the human femoral cortex with age. Journal of Bone and Joint Surgery 44B:496

Barnett E, Nordin B E C 1960 Radiological assessment of osteoporosis. Clinical Radiology IX: 166–174

Barnett E, Nordin B E C 1961 Radiological assessment of bone density. The clinical and radiological problem of thin bone. British Journal of Radiology 34: 683–692

Bentley H B 1978 Radiographic morphometry of the femoral shaft. Radiography XLIV (526): 233–237

Brunning D C 1964 Everyman's English dictionary. Dent, London

Clark K C 1974 Positioning in radiography. Heinemann, London

Garn S M 1970 The earlier gain and later loss of cortical bone. In: Nutritional perspective. Thomas, Springfield, Illinois

Hawkins C 1982 In: How to do it. British Medical Journal Publication, p 52

Horsman A, Simpson M 1975 The measurement of sequential changes in cortical bone geometry. British Journal of Radiology 48: 471–476

Horsman A 1976 Radiographic morphometry of the second metacarpal 572 In: Nordin B E C (ed) Calcium, phosphate and magnesium metabolism. Churchill Livingstone, Edinburgh

Huff D 1974 How to lie with statistics. Pelican, Harmondsworth

Meschan I 1959 An atlas of normal radiographic anatomy. Saunders, Philadelphia

Ractliffe J F 1969 Elements of mathematical statistics, 2nd edn. OUP, Oxford

Research Funds Guide 1973 BMA, London

Robson C 1975 Experiment, design and statistics in psychology. Penguin, Harmondsworth

Warren M D 1982 In: How to do it. British Medical Journal Publication, p 91–94

# Index

Abdomen radiography
  in A and E, 82
    common pathologies, 82
  bleeders, therapy, 189–91, 192
  paediatric *see* Paediatric radiography
Accident and emergency radiography,
    74–86
  equipment, 76–9
    accessories, 77–8
    alternatives, 78
    generator, high tension, 77
    patient support/transport system,
      76–7
    X-ray tube supports, 77
  foreign bodies, 86
  multiple injuries, 86
  of particular body regions, 79–86
  popularity with radiographers, 74
  rooms
    location, 75–6
    planning, 74–5
    radiodiagnostic, 77
  standards, maintenance, 74
Acromegaly, 130–1
  radiographic features, 130–1
    hand and wrist, 130
    heel, 131
    skull, 131
Aluminium filter, 64
Angiography, new concepts, 180–209
  (*see also* Coronary
    angiography; Heart
    radiography and angiography;
    Ventricle)
  equipment, 180–3
    digital vascular imaging *see*
      Digital vascular imaging
    filming, 182
    flash image TV monitoring, 182, 183
Angioplasty, percutaneous
    transluminal, 195–7
Angioskop, 181–2
  C-arm, 181
  Puck changer, 181
Ankylosing spondylosis *see* Spondylosis
Ano-rectal malformation, 62
  atresia, 63
Anus, plugging in urodynamic studies,
    110–11

Aorta
  abdominal, irregularities, 183, 184
  coarctation, 155
  descending, aneurysm, 183
  dilatation, 159
  great vessels, distribution, 186–7
  valve disease, 158–61
    calcification, 159
    regurgitation, 160, 161
    replacement, 117
    stenosis, 160, 161
Apgar score, 56
Apnoea, neonatal, 54
Aortography (arteriography)
  arch (thoracic), 184, 185, 186
  translumbar, 183–4
    oblique views, 184, 185
Arm radiography, 79–81
Arteriography *see* Aortography
Atrium
  myxoma, 157, 158
  septal defect, 155
Auditory canals, 30, 31–2
  external, 30
  internal, 32

Barium examination, paediatric, 69–70
Battered baby syndrome *see* Paediatric
    radiography
Bile ducts, stone retrieval, 194–5
Bladder
  dysfunction, 112–17
    neurogenic, 101
    in paraplegics, 111, 115
    sequence after injury, 114
    trabeculation, 110, 116
    types, 112–13
    urodynamic/radiographic profiles,
      115–16
  neck
    female, 104
    male, 103–4
  nerve supply, 102
  pressure
    intrinsic, 109
    measurement, 117–18
  radiographic image, 108
    contrast medium filled, 111

total tension, 117
  wall, 101–2
Bone
  age, 68–9
  density, reduction *see* Osteoporosis
  erosion, 121
  formation, 120
    disturbances *see*
      Osteochondrodystrophies
  mineral measurement, 207–8
  scan *see* radionuclide imaging
    in osteomalacia, 129
    in Paget's disease, 133–4
  tumours, primary, 121–3
Breast cancer, 121–2
Bucky holder, 51, 61, 76, 77

Cardiac procedures, non angiographic,
    176–7
Cardiac radiography *see* Heart
    radiography and angiography
Cardiomyopathy, hypertrophic
    obstructive, 157
Carotid artery angiography, 187
Cartilage production, 134–6
Catheters
  balloon, in angioplasty, 195–7
    dilatation, 197
    pressure recording tests, 196–7
  cardiac, 153, 160, 196
    in Judkins method, 160, 162–3
    in reasons for, 154
    Sones, 160, 161–2
  hepatic drainage, 192–5
  urethral, 109–10
    choice, 110
Cerebro spinal fluid leaking, 10, 11
Cervical spine (general features), 1–27
  (*see also* Cervical vertebrae;
    Spinal cord)
  anatomy, 1–5
    divisions, 1–2
  cadavers, use in experimental
    models, 2–5
    preparational requirements, 4
  injury *see* Cervical spine injury
  normal movements, 5–6
    neck rotation, 5

Cervical spine injury (*see also* Spinal cord injury)
  facet dislocation *see* Facets
  flexion, 12–13, 19
  hypertension, 13, 14, 19
  initial treatment and examination, 4, 9–13
    diagnostic errors, 4–5
    fluid control, 10
    skull traction, 10–11
  radiography *see* Cervical spine radiography
  in A and E, 85–6
  stable fractures
    definitions, 11–12
    types, 12
  sudden failure, 4
  unstable fractures, 11
    late presentation, 11
    types, 12–13
  unsuspected features, 9
    pre- and post-manipulation, 8
    use of radioisotopes, 26–7
  urodynamic/radiographic combined studies, 115–17
  vertebral injury *see* Cervical vertebrae
Cervical spine radiography, 13–27, 85–6
  diagnostic errors, 4–5
  discography, diagnostic value, 24
  exposure times, 15
  in flexion and extension, 21–22
    procedure, 21
  45° oblique view, 16, 18
  45° supine view, 16, 17
  lateral ('first site') view, 10–11, 14–15
  lateral oblique view, 17–18, 22
    caudad and cephalad tilting, 17, 18
    comparison to 45° oblique view, 18
    facets shown, 18
  myelography, diagnostic value, 24, 25
  off-lateral view, 16
  research techniques, 4–5
  stationary grids, use, 15
  'swimmers' view, 15–16
  tomography, computed, 25–7
    future use, 26
    improvement, 26
    problems, 25
  tomography, transverse axial, 24–5
  vertebrae *see* Cervical vertebrae
Cervical vertebrae (C1–7)
  bony arrangement, 2–5
  injuries, radiographic views
    in A and E ward, 85–6
  burst, 9, 12
  C1, C2, differences from other vertebrae, 18–20
  facet dislocation *see* Facets
  Hangman's fracture, 15, 19
  tear-drop fracture, 13, 14, 16

wedge fracture, 11
ligamentous arrangement, 2–5
transverse section, 3
Chest radiography
  paediatric *see* Paediatric radiography
  post pacemaker implementation, 176
Choanal, atresia, contrast examination, 59–60
Cholangiography, 192–5
Chondrodystrophies, 134–6
Cochlea, 31
Coeliac disease, 63
Computed tomography *see* Cervical spine radiography; Spine, radiography; Tomography
Contrast media (*see also* Metrizamide)
  in angiography, 153
    doses, 70
    minimization of amount, 147
    test dose, 153
  neuroradiography, 95–6
    accidental discovery, 87
  in urography, 71
    dosage, 70
    introduction, 71, 109–10
Cord, sacral, lesion, 115
Cord, spinal *see* Spinal cord
Coronary angiography, 145–53, 161–78
  bi-planar facilities, 147, 163
  equipment, 152–3
    ancillary, 152
    C arm, 145–6 (*see also* Angioskop)
    image intensifier, 152
    'LARC' model, 147, 151, 163
    poly- DIAGNOST C, 147, 148, 167, 168, 169, 170
    TV/video recorder, 152
    U arm 145–7, 148, 167, 168, 169, 170, 174
  Judkins method, 161, 162, 164–75
  left artery, required views, 167–75
    left anterior oblique, 171, 172
    left lateral, 171, 172, 175, 196
    postero-anterior, 174
    right anterior oblique, 171, 172, 173, 175
    true caudal, 174, 175
  oblique views, typical equipment positions, 145–7
    45° right anterior, 148
    30° caudal, 149
    30° left anterior cranial, 150, 196
    30° right anterior cranial, 149
    right and left anterior, 150
    right anterior, 151
  right artery, required views, 164–6
    left anterior oblique, 165, 166, 168
    right anterior oblique, 166, 169, 170
  Sones method, 161–2
Coronary arteries (*see also specific vessels*)
  bypass surgery, 177–8
    post-operative views, 178

disease, 161–75
  left, 166–71
    complex anatomy, 171
    retrograde filling, 172
  plastic surgery, 195–6
  position, 143
  right, 164, 165
    retrograde filling, 171
  transposition, 154
Council for National Academic Awards, 201, 202–5
Croup, lateral pharyngiogram, 59
Cushing's syndrome, 126
Cystogram, micturating, 72, 102

Defibrillator, 152
Deliberate injury *see* Paediatric radiography
Detrusor muscle, 102
  activity, effect of injury, 112–13, 115–16
  relationship to sphincter, 113
Diaphragm movement, demonstration, 81–2
Digital vascular imaging, 187–9
  modes of operation
    fluoroscopic or continuous, 188
    radiographic or serial, 187–8
    time interval difference, 188–9
Disc disease, 90
Discography, spinal injury, 24

Ear, *see also* Petrous bone
  anatomy, 30–2
  development, 29–30
  inner (internal), 31
  middle *see* Tympanic cavity
  tomography *see* Petrous bone tomography
Electrocardiography, 152
Electromyography, 110–11
Embolisation, therapeutic, of bleeders, 189–91, 192
Emergency radiography *see* Accident and emergency radiography
Epiphyses, multiple dysplasia, 136
Eye protection in tomography, 34

Face bone radiography, 83–4
  projections
    additional, 84, 85
    basic, 84
Facets, vertebral, dislocation, 12
  bilateral, 13
    treatment, 17
      radiography, 17–18
  unilateral, 13
Fallot's tetralogy, 154
Femoral artery, 185
Femoral epiphysis, slipped capital, 66
Femur, measurement, 209–10
Fibrillation, ventricular, 152

Fibrous dysplasia, 137
Film
 cassetteless, 78–9
 changers
  ADT, 106–7, 143
  Puck, 180–1
 for paediatric radiography, 51–2
 problems, 33
 processing, 78
Filter, aluminium, 64
Fluorography, bladder, 113
Fluoroscopy, heart, 144
 pre-pacemaker implantation, 176
Foot radiography, 79
 paediatric radiography, 68
Foramen ovale, patent, 156
Foreign bodies, 86

Gelatin foam, embolisation see
  Sterispon
Generator, high tension, 77
Genito-urinary radiography, paediatric
  see Paediatric radiography
Gonad protection, child, 64–5
Gianturco embolisation, 190, 192
 insertion, 191
Great vessels, angiography, 186–7
Grey matter, sensitivity to trauma, 6–7

Haematomas, pre-vertebral, 9
Haemodialysis, skeletal disorders
  arising from, 129–30
Hand radiography, 79
Hangman's fracture, 15, 19
Head injuries see Skull
Heart
 cavities, 141–2
 defects, unusual anomalies, 175
 non-angiography procedures, 176–7
 radiography see Heart radiography
Heart radiography and angiography,
  141–78 (see also Angiography,
  new concepts)
 catheterization, reasons for, 154–75
  acquired heart anomalies, 157–8
  congenital heart disease, 154–7
  coronary artery disease see
   Coronary arteries
  valve disease, 158–61
 ciné angiography, 145
  X-ray tube and generators, 143
 coronary arteries see Coronary
  angiography
 large film techniques, 143–4
  cut film changer, 143
  roll film changer, 143, 144
  X-ray tubes and generators, 143
 oblique views, 145
 patient, preparation, 153
  contrast medium, 153
  non-invasive tests, 153
 post-operative, 177–9
 ventricles see Ventricles

Hepatic drainage, percutaneous trans-,
  192–4
 types, 192
  external, 192–3
  external/internal, 193–4
  internal, 194, 195
Hip radiography, 81
 congenital dislocation, 85
Hirschsprung's disease, 62
Hurler's syndrome, 136
Hydromyelia, 94
Hyperparathyroidism
 primary, 126–7, 128
 secondary, 126, 127
Hyperthyroidism, 126
Hypogonadism, 126

Iliac artery, narrowing, 183
Illness, child, complications and
  requirements, 58–60
Image intensifiers, 152, 181, 182
Incus, 30
Injuries in A and E wards, 79–86 (see
  also Cervical spine injury;
  Spinal cord injury)
Intussusception, 62

Jejunum, biopsy, 63

Knee joint radiography, 81
 measurement of forces, 208–9

Leg radiography, 80–1
 length measurement, 68–9
Ligaments, cervical spine, 3, 18–19
 transverse, 5
Limb radiography, 79–81
Liver see Hepatic drainage

Magnification radiography, 61
 chest, 61
 coronary, 164–5, 166, 168, 169, 170
 petrous bone, 34
 spine, 88, 89
Malignant disease, 120–4
 secondary deposits, 121
Malleus, 30
Mandible radiography, 84–5
Mannitol therapy, 7
Marfan's syndrome, 137, 157
Mesenteric artery, embolisation, 189,
  190, 191
Metabolism, disorders, 125–37
Metacarpals, bone mineral
  measurement, 207–8
Metrizamide, use, 95
 advantages and disadvantages, 95–6
  avoidance of major reactions, 96
  psychological side effects, 95–6
Metro-cysto-ureterogram, 106

Micturating cystogram, 72, 102
Micturition, 101, 102
 controls allowing/inhibiting, 104
Mitral valve
 disease, 158
  calcification, 159
  regurgitation, 159
 replacement, 177
Morphometry, bone mineral
  measurement, 207–8
Morquio-Brailsford syndrome, 134–6
Myelography, spinal injury, 24, 25
Myeloma, multiple, 123–4
 punched out lesions, 123, 124

Nasal (choanal) atresia, contrast
  examination, 59–60
Neck
 radiography
  soft tissue, importance, 20–1
  xero-, 21, 27
 rotation, 5
Neurofibromatosis, 137
Neuroradiography, new techniques,
  87–99
 historical development, 87–8
 tomography, computed see Spine
  radiography
Non-accidental injury see Paediatric
  radiography
Nuclear magnetic resonance
  (Zeumatography), 86, 96–9
 advantages and disadvantages, 97,
  98, 99
 application, 97
  soft tissue definition, 97
 artefacts, 98
 high cost, 98–9
 historical developments, 97
 patient care, 98
 principles, 96

Odontoid peg fracture, 19, 20
Oesophagus, blockages, 58–9
Optic foramina, view of, 67
Ossicles, auditory, 30, 34
 tomographic demonstration, 39
Osteitis deformans see Paget's disease
Osteochondritis, femoral head, 65
Osteochondrodystrophies, 134–7
 main types
  chondroid production, 134–6
  osteoid production, 136–7
  others, 137
Osteodystrophy, renal
 cause, 129
 radiological features, 129–30
Osteogenesis imperfecta, 138
Osteogenic sarcoma, 122–3
Osteoid formation see Bone
Osteoma, osteoid, 123

Osteomalacia and rickets, 127–9
  cause, 127, 128–9
    haemodialysis associated, 130
  characteristic features, 127–8
    Looser zones, 127, 128
    saucer shaped deformity, 128
Osteomyelitis, CT diagnosed, 91
Osteopetrosis, 135, 136
Osteopoikilosis, 136
Osteoporosis, 129
  primary, 124–6
    biochemical cause, 125
    radiographic views, 124
  secondary, 126–31
Oxygen, neonatal monitoring and
    delivery, 55


Pacemakers, introduction, 176–7
  temporary, 176–7
Paediatric radiography (general
    features), 51–73
  abdomen, 60–3
    alternative positioning, 63
    baby, 61
    erect view, 61
    indications for, 61–3
    infant/child, 61
    lateral view, inverted, 61, 62–3
    preparation, 52
    supine views, 60–1
  barium examination, 69–70
  bone age, determination, 68–9
  chest (thorax), 57–60
    baby, 57
    child, 58
    illness, complications and
      requirements, 58–60
    magnification, 60
    toddler, 57
  equipment, 51–2
    dose reduction, methods enabling,
      51–2
    immobilization aids, 52
    support cradles, 51
    X-ray tube, 51, 52
  feet, 68
  non-accidental injury, 69, 137–8
    conditions simulating, 138–9
    signs, 69
    types, 69
  patient care see Paediatric
    radiography, patient care
  pelvis, 64–6
    Andrén-von Rosen view, 65
    frog lateral view, 65
  radiation protection, 53
  sinuses, 67–8
  skull
    antero-posterior view, 66, 67
    bone size ratios, differences from
      adult, 66
    lateral view, 67
    occipital (Towne's) view, 66–7
    postero-anterior 15° view, 67

submento-vertical view, 67
  spine, 63–4
    lateral view, 64
    postero-anterior view (erect), 64
  ultrasound, advantages, 72–3
  urography, intravenous, 70–2
Paediatric radiography, patient care,
    52, 53–7, 58–60
  neonate, hazards and their treatment
    in, 53–6
    attached tubes, drips, leads etc.,
      55–6
    dehydration, 54
    hypothermia, 54
    infection, 54
    respiratory difficulties, 54–5
  baby, 56
  6 month–3-year-old infant, 56
  3–6-year-old child, 56–7
    help from mother, 57
  6–14-year-old child, 57
  handicapped child, 57
  importance, 50, 66
    room planning, 50–1
  patient, preparation, 53, 56, 57
Paget's disease (osteitis deformans),
    132–4
  deformities, remodelling, 132
    skull enlargements, 132, 133
    tibia bowing, 132, 133
  radiographic techniques, 134
Paraplegics, male predominance, 115
Pelvic floor muscle, 104
Pelvis radiography, 81
  paediatric see Paediatric radiography
Pericardial mass, 158
Pericardium, tapping and effusions,
    177
Persistent ductus arteriosus, 154, 155
Perthe's disease, 65
Petrous bone, 29–48 (see also Ear)
  arteries adjacent, 32
  development, 29–30
  geometry, 30–2
  nerves adjacent, 32
  pathological effects, 33
  patient care, 48
  tomography see Petrous bone
    tomography
  veins adjacent, 32–3
Petrous bone tomography (general
    features), 33–48
  considerations and problems, 33–48
    eye protection, 34
    layer thickness, 34
    magnification, 34
    movement of subject, 33–4
    patient co-operation, 38
  projections and views see Petrous
    bone tomography, projections
    and views
  sizes of middle/inner ear, 35
Petrous bone tomography, projections
    and views, 29, 34–48
  antero-posterior, 35–7

bilateral, 37
  features demonstrated, 35, 37
  head position, 35
  radiographic views, 36
  single, 35–6
  of both bones simultaneously, 44
  radiographic view, 48
Chausse III, 35, 39–40
  head position and rotation, 39
  radiography anatomy, diagram of,
    39
Guillen's (transorbital), 35, 37–9
  features demonstrated, 39
  head position and rotation, 37–9
  radiographic views, 38
lateral, 35, 42–4
  features demonstrated, 43–4
  head position and rotation, 43
  radiographic views, 45, 46
Poschl's (axial pyramidal), 35, 42,
    43
  head position and rotation, 42
  radiographic views, 43
Stenver's (and Chausse IV), 35,
    40–2
  features demonstrated, 40–2
  head position and rotation, 40
  radiographic views and anatomy,
    40, 41
sub-mento vertical, 35, 44–8
  difficulty, 44
  features demonstrated, 44–8
  head position, 46
  radiographic views and anatomy,
    47–8
Pharyngiogram, lateral, with croup, 59
Philips Medical UPI system see UPI
    system
Pre-vertebral space, widening,
    diagnostic value, 9
Profunda femoris artery, 184, 185
Pulmonary artery
  angiography, 186
    left branch views, difficulty with,
      186
  stenosis, 156
Pulmonary vein, anomalous drainage,
    156
Pylorus, stenosis, 62


Radiation, protection
  of angiographers, 162
  of children, 51, 53, 64–5, 66
  in petrous bone tomography, 34
  in research, 200–1
    keeping records, 201
Radiography (general features) (see also
    specific tissue, e.g. Cervical
    spine)
  in A and E see Accident and
    emergency radiography
  bone measurement, 208–9
  of children see Paediatric
    radiography

neuro-*see* Neuroradiography
skeleton *see* Skeleton radiography
soft tissue *see* Soft tissue radiography
techniques *see* Radiography
    techniques
with urodynamic studies,
    comparisons, 115–17 (*see also*
    Urodynamics)
Radiography techniques, 24–7
  bi-planar, 5
  ciné and video tape, 5–6
    angiography, 143, 144, 145 (*see*
    *also* Angiography)
  discography, 24
  flicker image, 5
  image intensification, 152, 181, 182
  myelography, 24, 25
  in research, examples, 207–9
  soft tissue and xero-, comparisons,
    21
  stereo-, 5
  tomography *see* Tomography
  X-ray TV system in urodynamics,
    107
Radiology, therapeutic, 189–97
Radionuclide imaging
  cervical vertebrae fracture, 26–7
  myeloma detection, 123–4
  radiography complementation, 124
Renal radiography, 70–2
  pre-/post-embolisation, 191–2
Research, 198–209
  on cadavers, 2–5
  findings, authority of, 199
  grant application, 201–7
    considerations, 201–7, 209
    funding, 206
    sample form, 202–5
    types, choice, 201
  hypotheses, 200
  paper/thesis writing, 206–7
    statistics, use, 207
  personal experience, drawbacks in
    relying on, 199
  preparation, 198–9, 200
    references, cataloguing, 198–9
  radiation hazards *see* Radiation
    protection
  radiography, uses, 207–9
  reasoning
    deductive, 199
    inductive, 200
Respiration, child, complications,
    58–60
Rickets *see* Osteomalacia

Sacrum, cord lesion, 115
Sarcoma, osteogenic, 122–3
Scoliosis, 64
Shoulder radiography, 80
Sinuses, paediatric radiography, 67–8
Skeleton
  function, 119, 120
    processes maintaining, 120

maturity, disease affecting, 132–7
radiography *see* Skeleton radiography
tumours *see specific tissue or disease*
  metastases, 121–2
Skeleton radiography, 119–39
  malignant disease, 120–4
    views required, 121
  maturity, assessment, 131–2
    views required, 132
  metabolic disorder, 125–37
    views required, 125
  non accidental injury *see* Paediatric
    radiography
  requirements for, 119
  techniques, 119
  uses, 119
Skull
  A and E radiography, 77, 82–4
    projections required, 83
    standards, maintenance, 82–3
  nerve and blood supply, 32–3
  paediatric radiography *see* Paediatric
    radiography
  traction, 10–11
Soft tissue
  neck radiography, 20–1
  NMR visualization, 97
Sphincter, urethral
  female, 104
  male, 103
  relationship to detrusor, 113
Sphincterotomy, trans-urethral, 116–17
Spinal canal narrowing (stenosis), 92–3
  acquired, 93
  developmental, 93
  risks, 5
Spinal cord (*see also* Cervical spine)
  blood supply, 5–6, 7
    subdivisions, 5–6
    venous drainage, 6
  flat, 94
  injury/trauma *see* Spinal cord injury
  normal and atrophic, CT viewed,
    92, 94
  puncture, 95
  round large, 94
  schematic diagram, 103
Spinal cord injury, 1–27, 85–6 (*see*
    *also* Cervical Spine injury;
    Cervical vertebrae)
  bladder dysfunction, 112, 114–15
  common sites, 103
  effects, 6–7
  damaged tissues, 6
  life threatening conditions, 7
    high level lesion, 7–8
    improved treatment, 7–8
  oedema, 89, 90
  regeneration, 27
  rehabilitation, 1, 27
  repair, future techniques, 27
  severe, urodynamic studies *see*
    Urodynamics
  spondylosis caused *see* Spondylosis
  venous drainage, 6

Spine radiography *see* Cervical spine
    radiography; Spinal cord
    radiography
  computed tomography, high
    resolution, 88–95
    applications, 89–95
    averaging, before and after, 92
    considerations, 96
    contrast media, 95
    coronal re-formation, 93
    dual windowing, 90
    filtering, before and after, 92
    historical development, 87–8
    sagittal re-formation, 93
    sector scanning, 89
  injuries in A and E, 85–6
  paediatric radiography *see* Paediatric
    radiography
Spine (and spinal cord) tumours, 90–2
  extracanalicular, 91–2
  intracanalicular, 90–1
Spinous complex, anterior, 3
Spondylosis (cervical and ankylosing),
    22–4
  pre- and post-injury radiography, 23
Stapes, 30, 31
Statistics, use, 207
Sterispon, embolisation, 190
  injection, 190, 191
Stone retrieval, bile duct, 194–5
Stridor, acute inspiratory, 59
Subluxation, 12
Syringomyelia, 94


Tear-drop fracture, 14,16
Thorax radiography, 81–2
  aortography, 184, 185, 186
  paediatric *see* Paediatric radiography
Tomography
  advantages, 48
  computed, spine *see* Spine
    radiography
    cervical *see* Cervical spine
    radiography
  linear, 19–20
    assymetrical, 20
    multidirectional, 20
  petrous bone *see* Petrous bone
    tomography
  thin section, 34–5
  transverse axial, 24–5
Traction
  care, 24
  skull, 10–11
Transducer stand, 110
Tricuspid valve disease, 160, 161
Trigone, 103
T-tube, 194–5
Tumours
  bone, 122–4
  spinal, 90–2
Tympanic cavity (middle ear), 30, 31,
    34
  areas, size of, 35

Tympanic cavity (middle ear) (contd)
    lateral half, 30, 31
    medial part, 31

U-arm, 145–7, 148, 167, 168, 169, 170, 174, 182
Ultrasound, paediatric, advantages, 72–3
UPI system, 180, 181
Urethra
    catheters see Catheters
    dysfunction see Bladder
    female, 104
    profile of see Urethral profile
    smooth muscle activity, 113
Urethral profile, 101
    measurement, 105
        errors, 105
        requirements, 104–5
Urinary tract, lower, 102
Urination see Micturition
Urodynamics, measurement in severe cord injury, 101, 105–18
    clinical considerations, 113
    information gained by, 113–14
        application, 114–15
        assessment for surgery, 114

methodology, 109–12
    fluid introduction, 109–11
    multiple refills of bladder, 111–12
    peristaltic pumping, physiological simulation, 111
    recording, 112
parameters measured, 109
patient availability, 109
post-surgical problems, 115–17
provocative techniques, 111
requirements, 107–9
    equipment, 107–9
    integrated team, 107
    time, 107
X-ray image/video trace, correlation, 108–9
Urography, paediatric, 70–2
Urology, survey before urodynamic measurement, 105–7

Valve (heart)
    disease, 158–61
    replacement, 177
Vasopressin infusion, 189–90
Venepuncture (paediatric), 71
Ventricles
    aneurysm, 163, 164

left, angiography, 163–4
    projections, 163
    septal defect, 156
Vertebrae
    cervical see Cervical vertebrae
    simulated models, forces on, 2
Vertebral body
    collapse, 121
    fracture, 89
Vestibule, 31
Vitamin C deficiency, 139
Vitamin D metabolism, 125
    deficiency and malabsorption, 128–9
Volvulus, 61–2

Whiplash injury, 19

Xero-radiography, neck, 21, 27
X-ray film see Film
X-ray tube supports, in A and E, 77

Zeumatography see Nuclear magnetic resonance
Zimmer halter, 14–15
Zygomatic arch, 84, 85